A BASIC/INTERMEDIATE COURSE
FOR WATER SYSTEM OPERATORS

Volume 2

INTRODUCTION TO

Water Treatment

PRINCIPLES and PRACTICES of
WATER SUPPLY OPERATIONS

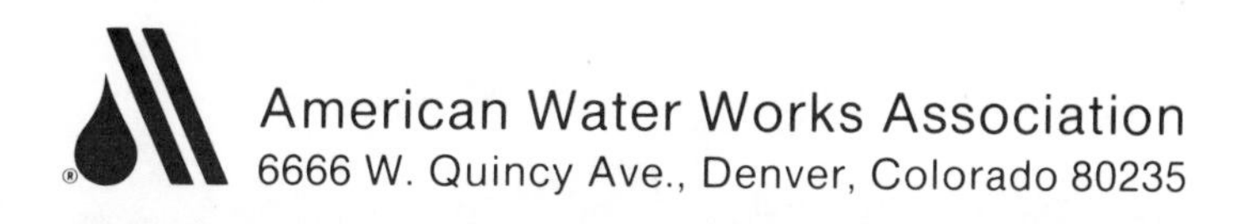
American Water Works Association
6666 W. Quincy Ave., Denver, Colorado 80235

ISBN 0-89867-180-9

Foreword

Introduction to Water Treatment is the second volume in a five-part handbook series designed for use in a comprehensive training program entitled "Principles and Practices of Water Supply Operations." Volume 2 has been written to examine the operations and problems associated with conventional water treatment processes and the importance of maintaining each unit process to ensure that water quality standards are being met.

Other student volumes in the series include:

Volume 1	*Introduction to Water Sources and Transmission*
Volume 3	*Introduction to Water Distribution*
Volume 4	*Introduction to Water Quality Analyses*
Reference Handbook	*Basic Science Concepts and Applications*

Instructor guide and solutions manuals have been prepared with detailed lesson plan outlines, resource materials, and examination questions and answers for volumes 1 through 4.

Course content in the water supply operations series has been developed to meet training requirements of basic to intermediate grades of certification in water treatment and water distribution system operations. The modular format used throughout the series provides the flexibility needed to conduct both short- and long-term vocational training.

The reference handbook is a required companion textbook that is correlated with volumes 1 through 4 through footnote references. The purpose of a separate reference book is to provide the student with supplementary reading in the areas of mathematics, hydraulics, chemistry, and electricity.

The development of the training materials in the water supply operations series was made possible by funding from the US Environmental Protection Agency, Office of Drinking Water, under Grant Agreement No. T900632-01, awarded to the American Water Works Association.

Disclaimer

Several photographs and illustrative drawings that appear in this volume have been furnished through the courtesy of various product distributors and manufacturers. Any mention of trade names, commercial products, or services does not constitute endorsement or recommendation for use by the American Water Works Association or the US Environmental Protection Agency.

Acknowledgments

Publication of this volume was made possible through a grant from the US Environmental Protection Agency, Office of Drinking Water, under Grant Agreement No. T900632-01, as part of a national program strategy for providing a comprehensive curriculum in water treatment plant and water distribution system operations. John B. Mannion, Special Assistant for Communications and Training (currently on an intergovernmental personnel assignment to the staff of the American Water Works Association Research Foundation), represented the Environmental Protection Agency, Office of Drinking Water, as project officer; and Bill D. Haskins, Director of Education, served as project manager for the American Water Works Association.

The American Water Works Association gratefully acknowledges the work contributed by Jack W. Hoffbuhr, Acting Assistant Regional Administrator, Policy and Management, USEPA, Region VIII, whose dedication to rewriting major portions of the original manuscript made the publication of this book possible.

Special recognition is also extended to the following individuals who provided manuscript as identified:

Michael D. Curry, President, Curry & Associates, Nashville, Ill.—Stabilization Module

Ralph W. Leidholdt, Project Manager, Ladd Engineering, Denver, Colo.—Disinfection Module

Nancy E. McTigue, Water Supply Engineer, Drinking Water Branch, USEPA, Region VIII, Denver, Colo.—Fluoridation Module

Thomas E. Braidech, Aquatic Biologist, Drinking Water Branch, USEPA, Region VIII, Denver, Colo.—Preliminary Treatment Module involving aquatic plant control.

A special thanks to Mary Kay Cousin, AWWA Technical Editor, who prevailed throughout the many manuscript revisions, and to Lois M. Sherry, Secretary to the project manager, AWWA, for preparing manuscripts, sourcing illustrations, and deciphering cryptographic handwriting. Credit is also due to Jane Olivier for typesetting and to Robert J. Love for graphic art support, including paste up and preparation of many original illustrations. Finally, appreciation is expressed to all who gave liberally of their time and expertise in providing technical review of manuscripts. In particular, the following are credited for their participation on the review committee or as an independent volunteer reviewer:

Donald B. Anderson, Program Director, Water Supply Operations Training, Environmental Resources Training Center, Southern Illinois University at Edwardsville

Charles R. Beer, Superintendent of Water Treatment, Denver Water Department, Denver, Colo.

James O. Bryant Jr., Director, Environmental Resources Training Center, Southern Illinois University at Edwardsville

Hugh T. Hanson, Chief, Science and Technology Branch, Criteria and Standards Division, Office of Drinking Water, USEPA, Washington, D.C.

James T. Harvey, Superintendent of Production, Water Works, Little Rock, Ark.

William R. Hill, Associate, Camp Dresser & McKee, Inc., Boston, Mass.

Kenneth D. Kerri, Professor of Civil Engineering, School of Engineering, California State University at Sacramento

Russell W. Lane (retired), Head, Chemistry Section, State Water Survey Division, Illinois Department of Energy & Natural Resources, Champaign, Ill.

Jack E. Layne, Director of Engineering, Denver Water Department, Denver, Colo.

O. Thomas Love, Chief, Organics Control—Exploratory Activities, Drinking Water Research Division, Municipal Environmental Research Laboratory, Cincinnati, Ohio.

Andrew J. Piatek Jr., Chief Operator & Superintendent, Borough of Sayreville, N.J.

Donald C. Renner, Superintendent of Utilities, Village of Arlington Heights, Ill.

John F. Rieman, Senior Technical Editor, American Water Works Association, Denver, Colo.

Frank W. Sollo (retired), Principal Chemist, State Water Survey Division, Illinois Department of Energy & Natural Resources, Champaign, Ill.

Alan A. Stevens, Chief, Organics Control—Chemical Studies Section, Drinking Water Research Division, Municipal Environmental Research Laboratory, Cincinnati, Ohio

James M. Symons, Professor of Civil Engineering, Environmental Engineering Program, University of Houston, Texas

Robert L. Wubbena, President, Economic & Engineering Services, Inc., Olympia, Wash.

Introduction for the Student

Good drinking water is so readily available in the United States that people often take it for granted. However, this does not mean that the nearly 250,000 public water systems in the United States are entirely free of deficiencies and are able to consistently produce safe drinking water. On the contrary, most of the waterborne-disease outbreaks result from deficiencies in public water systems, deficiencies such as inadequate treatment and improper operation. In addition, water sources are becoming increasingly contaminated by natural and man-made pollutants, making effective, well-operated treatment processes essential to the production of safe drinking water.

Why Water Needs Treatment

The primary function of water treatment is to provide a continuous supply of safe, palatable drinking water—safe water that is free of contaminants that can cause disease or be toxic to a consumer; and palatable water that is free, or practically free, of unpleasant characteristics such as color, turbidity, taste, and odor.

Few raw-water sources can supply water of this quality without some treatment. Ground water, for example, generally requires at least disinfection for complete public health protection. Some ground water needs additional treatment to reduce hardness or remove iron and other constituents that cause staining, color, tastes, and odors. Surface water should never be used for drinking without complete treatment. In addition to disease-causing organisms, surface water often contains turbidity and aquatic life (such as algae), as well as domestic and industrial wastes. These constituents must be removed through proper treatment processes.

The Operator's Role in Treatment

The unquestioning trust of the public in the quality of its drinking water is a credit to treatment plant operators. Because of this trust, the operator's role in water treatment is an important and challenging responsibility. Operators must maintain constant vigilance to assure that safe water is delivered to the consumer. Even with the simplest types of treatment and in the smallest plant, only qualified, responsible personnel should be in charge.

The duties of water treatment operators are wide ranging. As part of their typical duties, operators may be expected to

- Operate mechanical equipment such as meters, pumps, filters, and feeders.
- Operate electrical and electronic equipment such as motors, controllers, automatic monitors, recorders, and standby power systems.

- Calibrate, maintain, service, repair, and replace various mechanical, electrical, and electronic equipment.
- Determine proper chemical dosages and control chemical applications for various treatment processes.
- Inventory, order, and store chemicals.
- Inventory spare parts and standby equipment.
- Keep accurate and complete records on various aspects of water treatment.
- Collect water samples for testing.
- Perform certain laboratory analyses.
- Maintain a safe working environment.
- Perform regular preventative maintenance on various types of equipment.
- Perform general plant maintenance and housekeeping.
- Keep informed as to what regulations of local, state, and federal government apply to water treatment.

To perform these duties successfully an operator must have good judgment and sufficient education in the fundamentals of mathematics, hydraulics, electricity, bacteriology, biology, and chemistry. The operator also needs training and experience in the operation, maintenance, repair, and replacement of water treatment plant equipment.

Water Treatment Processes

There are various treatment methods available for making water safe and appealing to consumers. The methods used depend primarily on the characteristics of the raw water. Table 1 summarizes the water treatment processes commonly used today and identifies the main purpose of each. Surface-water sources require more extensive treatment than ground-water sources because of greater exposure to contamination. Therefore, with the exceptions of disinfection and flow measurement, most of the following processes are generally used only in the treatment of surface waters. The order in which these processes are often performed is shown in Figure 1.

A water treatment system usually begins with some form of PRELIMINARY TREATMENT.* Screening, presedimentation, and microstraining are preliminary treatment processes intended to remove materials that can damage or clog plant equipment or otherwise foul the major treatment processes. Chemical pretreatment involves the addition of a chemical, such as copper sulfate, to control the growth of algae.

A FLOW MEASUREMENT is taken somewhere near the point where water enters the plant, often just prior to the major treatment processes. This measurement

*Words set in SMALL CAPITAL LETTERS are glossary terms. Definitions for these terms can be found in the glossary at the end of this volume.

Table 1. Complete Water Treatment Processes

Process	*Purpose*
Preliminary Treatment	
Screening	Removes large debris that can foul or damage plant equipment
Chemical pretreatment	Conditions the water for the eventual removal of algae and other aquatic nuisances that cause taste, odor, and color
Presedimentation	Removes gravel, sand, silt, and other gritty material that can foul or damage plant equipment
Microstraining	Removes algae, aquatic plants, and small debris that can clog or foul other processes
Flow Measurement	Measures the amount of water being treated
Main Plant Processes	
Aeration	Removes odors and dissolved gases, adds oxygen to improve taste
Coagulation/flocculation	Converts nonsettleable particles to settleable particles
Sedimentation	Removes settleable particles
Softening	Removes hardness-causing chemicals from water
Filtration	Removes finely divided particles, suspended flocs, and most microorganisms
Adsorption	Removes organics and color
Stabilization	Prevents scaling and corrosion
Fluoridation	Adds flouride in order to harden tooth enamel
Disinfection	Kills disease-causing organisms

provides the operator with valuable information regarding the amount of water being treated.

If AERATION, the mixing of air into the water, is used at a treatment plant, it is often the first major water treatment process. Aeration is designed to remove certain dissolved gases in the water; it can also be used to increase the dissolved oxygen content of the water, which is the first step in the removal of iron and manganese.

COAGULATION/FLOCCULATION is a chemical treatment process designed to convert small, lightweight, nonsettleable particles into larger, heavier particles that will settle. Coagulation/flocculation must occur before sedimentation in order to improve the performance of that process.

The purpose of SEDIMENTATION is to settle out many of the suspended particles in water, thus reducing the load on the filters. In the sedimentation process, the flow rate is slowed as the water passes through a tank (called a sedimentation basin, or clarifier), which allows the particles to settle by gravity.

Some water may be exceptionally "hard," meaning that it has high concentra-

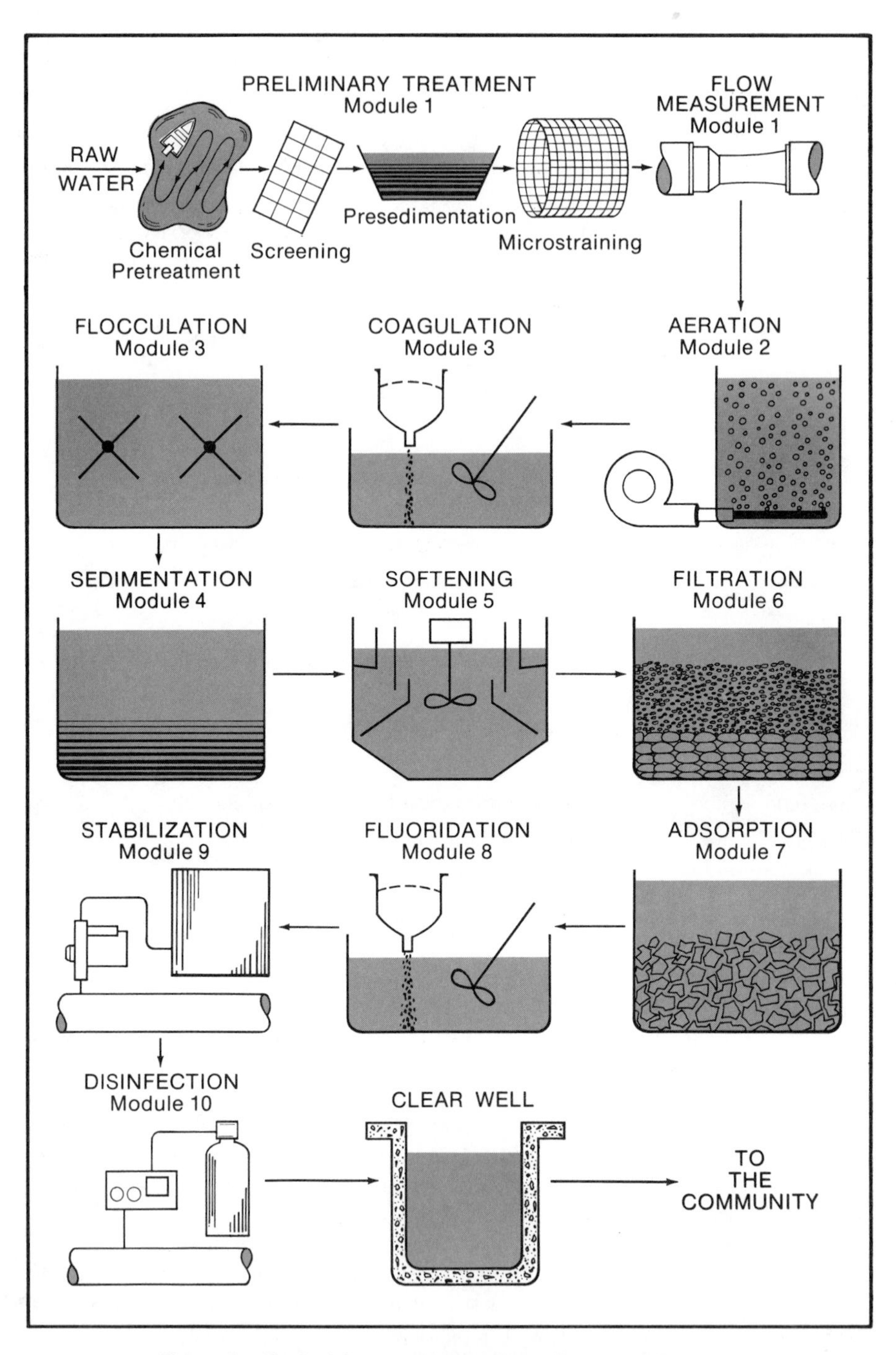

Figure 1. Typical Arrangement of Water Treatment Process

tions of natural chemical compounds containing calcium and magnesium. Although the chemicals causing hardness generally pose no health threat, they can cause scale buildups in plumbing fixtures (especially hot water heaters) and result in the formation of soap scum in sinks and tubs. Hardness can be removed through a process called SOFTENING.

Some extremely fine particles, including harmful microorganisms, may still be in suspension after sedimentation. For this reason, FILTRATION should always follow the sedimentation process in order to "polish" the water by removing the remaining particles.

Although filtration is effective in removing most of the small particles, it cannot remove certain dissolved organic materials that, if present, can affect health and cause taste and odor problems. Where such organics (or the chemicals that result in their formation) cannot be eliminated at the water source, they may be removed by activated carbon through a process called ADSORPTION.

Many waters contain constituents that cause corrosion of metal or buildup of scale in pipelines and boilers. STABILIZATION is the treatment process used for reducing the scaling and corrosion tendency of water, usually through the addition of certain chemicals.

FLUORIDATION is the practice of adding small quantities of fluoride to the water in order to strengthen tooth enamel and help prevent tooth decay.

DISINFECTION is the treatment process used to kill disease-causing (pathogenic) organisms; it is usually accomplished by adding chlorine to the water. Although sedimentation and filtration remove bacteria and viruses, these processes do not remove all of the organisms. For this reason, it is vital that water be disinfected before it leaves the plant for distribution to the consumer.

The modules in this volume discuss each of the treatment processes in detail. However, not every treatment plant will have all of the processes. The types and arrangements of processes in a water treatment plant vary from community to community, depending on the special characteristics of the raw water being treated.

Table of Contents

Water Treatment

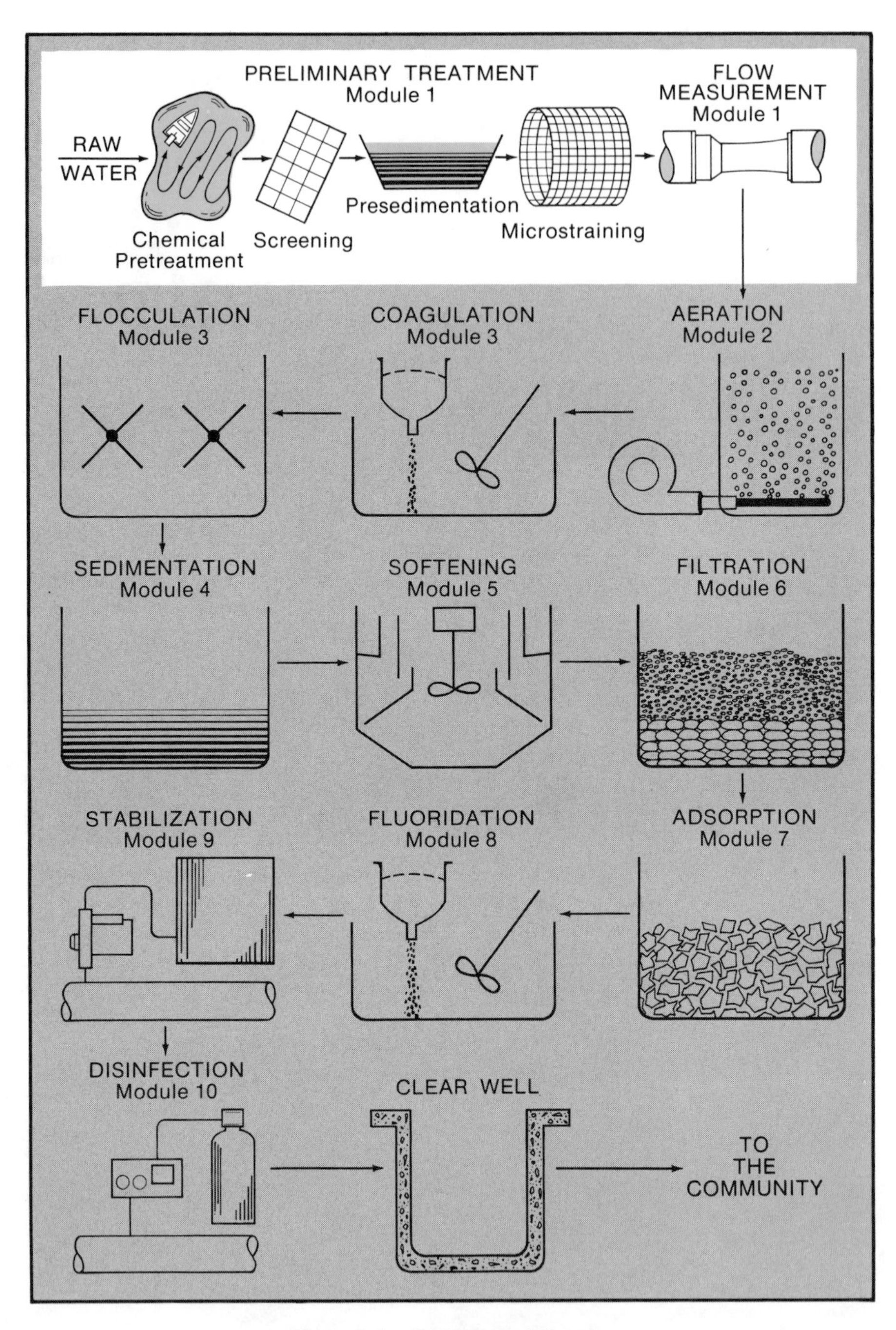

Figure 1-1. Pretreatment Process

Module 1

Preliminary Treatment Processes and Flow Measurement

Water supplies, whether from surface- or ground-water sources, often contain impurities or objectionable characteristics. The main treatment processes, such as coagulation, flocculation, sedimentation, filtration, and disinfection, are used to treat water at the plant. The load on these processes may be reduced by removing some impurities from the water or altering some of the objectionable characteristics of the water, before it reaches the treatment plant. This is the purpose of PRETREATMENT, also known as PRELIMINARY TREATMENT.* Pretreatment is any physical, chemical, or mechanical process used on the water before it is subjected to the main treatment process. As shown in Figure 1-1, pretreatment processes include

- Screening: Debris, such as rocks, sticks, limbs, logs, and other large objects that can clog or damage plant equipment, is trapped on screens and removed.
- Chemical pretreatment: Algae, which can cause tastes and odors, clog equipment, and cause other interferences, are treated with chemicals to control their growth.
- Presedimentation: Silt, sand, and grit, which can damage or overload later processes, are removed by gravity in a basin.
- Microstraining: Nuisance particles of small size (such as algae) are removed with a straining device.

*Words set in SMALL CAPITAL LETTERS are glossary terms. Definitions for these terms can be found in the glossary at the end of this volume.

Two other treatment processes are sometimes used in preliminary treatment: aeration and chlorination. However, since these are more commonly considered to be major processes, they are discussed separately in the Aeration and Disinfection Modules. It should be noted that properly developed ground-water sources seldom require pretreatment of any kind.

A discussion of flow measurement is included in this module because a flow measurement is often made immediately after preliminary treatment of the water, as illustrated in Figure 1-1. Although this may not be the precise location of flow measurement in every system, the measurement is almost always made somewhere near the front end of the plant, as well as at other plant locations.

The first part of this module contains discussions of purpose, required equipment, and monitoring for each of the preliminary treatment processes. In the second part of the module, several types of flow meters common to water treatment are discussed. After completing this module you should be able to

- Identify the various types of preliminary treatment processes.
- Explain what each process accomplishes.
- Describe the facilities or equipment used in each process.
- Explain what tests and controls are appropriate for each process.
- List the various types of flow metering devices commonly used in water treatment operations.
- Describe the basic operating principle of each metering device.

1-1. Screening

SCREENING is often the first pretreatment water receives; screening facilities should always be provided at points where water is withdrawn from the source. Coarse screens located on intake structures prevent clogging and protect the rest of the water system by removing sticks, limbs, logs, and other large debris commonly found in rivers, lakes, and reservoirs. These coarse screens are sometimes called TRASH RACKS or DEBRIS RACKS. It is common for finer screens to be located within the headworks of treatment plants to remove smaller debris.

Types of Screens

There are two types of screens used in pretreatment:

- Bar screens
- Wire-mesh screens.

Each type is available in both manually and automatically cleaned models.

The simplest type of screen, the BAR SCREEN, consists of straight steel bars welded at their ends to two horizontal steel members. The screens can be ranked as fine, medium, or coarse, depending on the spacing between bars: a fine screen has 1/16 to ½ in. (1.5 to 13 mm) between bars; a medium screen has ½ to 1 in. (13 to 25 mm) between bars; and a coarse screen has 1¼ to 4 in. (32 to 100 mm) between bars.

A bar screen assembly is installed in a waterway at an angle of about 60 to 80 degrees from the horizontal (see Figure 1-2). This angle is important, particularly in manually cleaned bar screens, because it makes it convenient to rake debris up the screen and onto the concrete operating platform for draining and eventual disposal. The slope also helps keep the screen from clogging between cleanings. As debris stacks up against the screen, the passing water lifts and pushes it up the slope, leaving the submerged part of the screen open and clear.

Manually cleaned bar screens are used at the raw-water intake structures and at the headworks of small treatment plants that receive only small amounts of debris. Automatically cleaned screens are used at plants receiving large amounts of debris and at any plant or intake structure where it is not practical or convenient to reach the screen for cleaning.

Automatically cleaned bar screens are available in a variety of styles. The bar screen in Figure 1-2 is equipped with an automatic rake, a horizontal piece of metal that moves up the face of the screen. It is pulled by a continuous chain-and-sprocket drive attached at both ends of the rake. Figure 1-3 is a side view of an automatically cleaned bar screen. Note how the several rakes move up past the screen, dump the screenings into the collecting hopper, and then return into the water to repeat the cycle.

Courtesy of Envirex Inc., a Rexnord Company

Figure 1-2. Bar Screen Assembly Installed in a Waterway

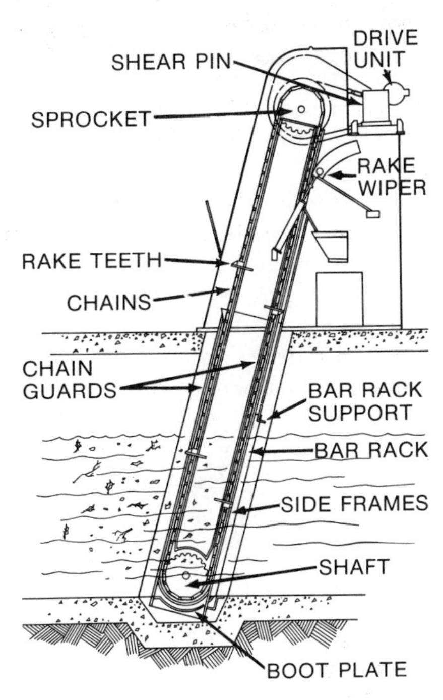

Courtesy of Envirex Inc., a Rexnord Company

Figure 1-3. Side View of an Automatically Cleaned Bar Screen

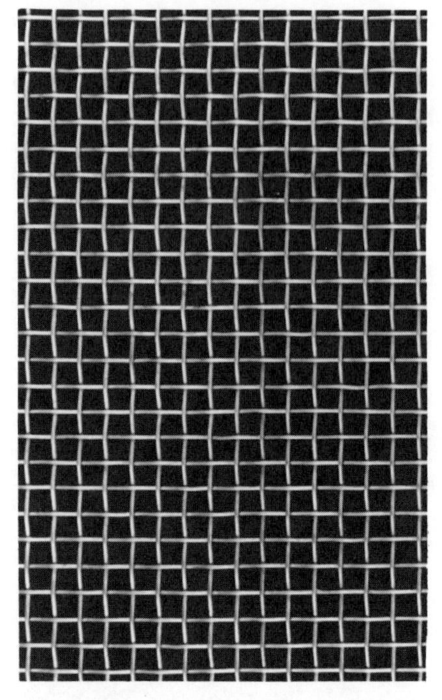

Courtesy of FMC Corporation, MHS Division

Figure 1-4. Wire-Mesh Screen Material

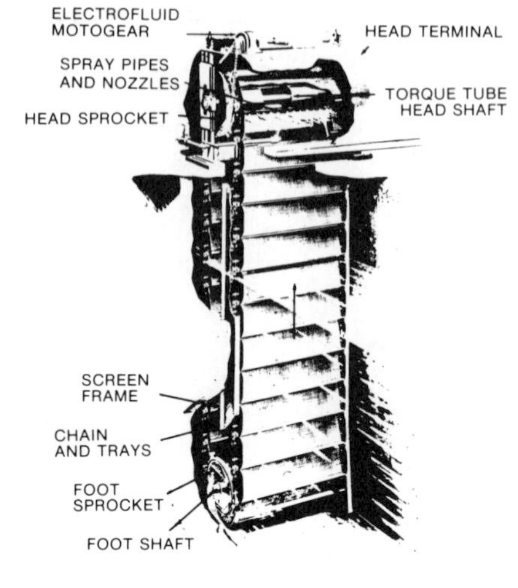

Courtesy of FMC Corporation, MHS Division

Figure 1-5. Automatically Cleaned Wire-Mesh Screen

Even the openings in the fine bar screen are too large to remove some debris found in water. In such cases, WIRE-MESH SCREENS are used. These screens are made of fabric woven from stainless steel or other corrosion-resistant, wire-like material (Figure 1-4). The fabric may have openings as large as 3/8-in. (10-mm) square or as small as 1/60-in. square (60 mesh, or 0.4-mm).

Because debris can accumulate quickly on wire mesh, automatically and continuously cleaned wire-mesh screens are favored over manually cleaned units. Figure 1-5 shows an automatically cleaned wire-mesh screen. It is mounted vertically in the water and moves continuously. As illustrated in the figure, spray nozzles located in the head terminal are used to wash away the screenings. Screenings and wash water fall away from the screen, behind the head terminal, and are conveyed to the disposal area.

Routine Operation

Clogging and corrosion are two problems commonly associated with screening. To prevent these problems, routine cleaning and inspection are required.

Manually cleaned screens should be checked and cleaned as needed. The frequency of inspections should depend on local weather conditions and type of watershed. Surface-water sources will generally produce more debris during the fall of the year, when trees lose their foliage, and during the spring and rainy seasons, when high waters carry debris. Heavily wooded watersheds will usually produce more debris than those that are more open.

Automatic screening devices should not be neglected—routine inspection of these units is essential in keeping them functional. Automatically cleaned screens are usually equipped with mechanical and electrical protection devices designed to shut down the unit in case of emergency. A mechanical protection device is incorporated into the unit shown in Figure 1-3; if a piece of debris becomes jammed between the screen and the automatic rake, a shear pin is designed to break. When the shear pin breaks, the motor drive shaft disconnects from the rake, preventing damage to the equipment. Shutdown due to the operation of a device such as this may not be apparent unless the unit is inspected. Electrical protection devices include circuit breakers, which open and automatically shut down the system when an electrical malfunction threatens to overload the motor circuit. In systems where this does not trigger a failure light or an alarm bell at the main control panel, the failure may not be noticed until visual inspection of the equipment is performed.

The constant wetting and drying of the screen equipment creates ideal conditions for accelerated corrosion. Therefore, in addition to inspection for clogging and related problems, screen equipment should be inspected at least monthly for signs of corrosion. Replacement parts should be available and repairs made routinely at the operator's convenience, instead of under emergency conditions following an equipment failure.

Suggested Records

Records should be kept on the type and quantity of screened material that is removed. Reference to these records will help identify an appropriate schedule of inspection and cleaning frequency, and also may help identify activities in the watershed that are causing excessive amounts of debris and require further investigation. Complete, up-to-date operating records should include:

- Date of inspection
- Amount of material removed from screens (in cubic feet or cubic metres)
- Notations regarding unusual or unexpected types of debris or water conditions.

Maintenance records should also be kept, identifying the type and location of equipment, a list of required spare parts, a checklist of spare parts on hand, and the date and description of maintenance performed.

1-2. Aquatic Plant Control

Although aquatic plants can cause problems for the water treatment plant operator, it is important to remember that these plants are normal inhabitants of the aquatic environment and have a definite role in maintaining the ecological balance in lakes, ponds, and streams. For example, algae help purify the water by adding oxygen during the process of PHOTOSYNTHESIS. Algae and rooted aquatic plants (water weeds), in moderate amounts, are essential in the food chains of fish and waterfowl. Water weeds store large amounts of nutrients, which would otherwise be available to initiate algal blooms. Therefore, overcontrol of aquatic plants can result in long-term problems more serious than the weeds.

The problems encountered at water treatment facilities from aquatic plants usually arise from the overproduction of a few types of plants at certain times of the year. Some species can cause taste and odor problems in the finished water; others can be the source of operational problems at the treatment plant. The aim of a well planned control program is to control aquatic plants only to the extent necessary to prevent water quality and treatment problems.

Algae Problems

All surface waters contain algae and other microscopic organisms that can foul pipelines, cause taste and odor problems, clog filters, and create potentially dangerous, slimy growths on treatment plant structures.

There are about 30,000 species of algae, ranging in size from microscopic one-celled organisms to marine kelp ,which can attain a length of 200 ft (61 m). The four major groups of algae of interest when considering water supplies are blue-green algae, green algae, diatoms, and pigmented flagellates.

Algae cause two major operational problems:

- Taste and odor
- Filter clogging.

Algae can also cause additional problems, including:

- Slime
- Color
- Corrosion
- Interference with other treatment processes
- Toxicity.

Some of these problems may make the final treated water unpalatable and unsafe. Many of the problems become time-consuming and cause unnecessary additional work for the operating staff, thereby greatly increasing operating costs. Some of the algae that may cause problems in water treatment are illustrated in *Basic Science Concepts and Applications*, Appendix D, Algae Color Plates.

Taste and odor. Although the exact mechanism of algae taste and odor production by algae is not completely understood, the problems are probably caused by certain complex organic chemical compounds that are by-products of the life cyle of algae. Whatever the cause, operational and economic considerations suggest that algae are better controlled at their source than in the treatment plant.

The types of tastes caused by algae in the water have been categorized as sweet, bitter, and sour. Algal-caused tongue sensations are categorized as oily or slick, metallic or dry, and harsh or astringent. The human sense of smell is generally much more sensitive than, and closely related to, the sense of taste (odors can usually be detected at the microgram per litre level). Because of this, tastes and odors are primarily classified and measured by the specific type of odor-causing algae together with the type of taste, odor, and tongue sensations they cause.

Filter clogging. Algae can shorten filter runs by forming a mat on the filter's surface, which essentially seals it (known as blinding the filter). The filter then must be backwashed (reversal of flow, passing previously filtered water back through the filters to clean them) to restore their filtering capability. The more backwashing that must be done, the less treated water that can be produced for delivery to the community. DIATOMS are usually the primary group associated with this problem.

Slimes. Slimes come from the layer that surrounds the algal cell. The blue-green algae, as a group, are slime producers and can produce a slimy, slippery layer on concrete or other surfaces. The slime is unsightly, has a bad odor, and can be hazardous to anyone walking on such a surface. Since most algae require sunlight to grow, slime accumulation in dark portions of distribution systems is usually due to bacteria rather than algae. Several diatoms, as well as green algae and a few flagellates, also produce slime, as noted in Table 1-1.

Color. Color can be caused by algal by-products and is usually an indicator that taste and odor problems will also occur. Almost any type of algae can cause color. Table 1-1 shows the types of algae usually associated with color problems.

Consumers will notice and usually complain when their drinking water has a yellowish or other colored tinge to it. Colors ranging from yellow-green to green, blue-green, red, and brown could all be caused by algae. Color in water may also be caused by substances other than algae, such as iron, manganese, organic matter from natural runoff, and industrial wastes; it is important to determine the cause before attempting to solve the problem.

Corrosion. Algal-caused corrosion of concrete and metal structures can become a serious problem. Algae may contribute to corrosion, either directly on surfaces where they grow or indirectly through their modification of the water by physical or chemical changes. Some types of algae that can grow on submerged concrete and cause it to pit and crumble are listed in Table 1-1. Algae are not usually the direct cause of corrosion of iron or steel pipes because most algae are not capable of growth in the absence of light.

Interferences with other treatment processes. The presence of algae in water can also create problems because of the change in water characteristics. As algae grow and die, they cause changes in pH, alkalinity, hardness, dissolved oxygen, and organic matter. These changes can interfere with water treatment processes. Algae also add organic matter that increases the chlorine demand and that may make additional treatment such as carbon adsorption necessary. Changes in pH caused by the chemical processes of algae affect the dosages required for adequate coagulation.

Toxicity. Certain freshwater algae found in the United States are TOXIC (poisonous). Several of these toxic algae are listed in Table 1-1. Skin problems such as dermatitis are believed to be caused by the blue-green algae *Anabaena*, and hay fever allergies have been associated with the presence of the blue-green algae *Anacystis*. Outbreaks of gastrointestinal illness involving thousands of people have been reported in the same areas where extensive algal blooms were present. It is possible that the presence of large numbers of blue-green algae on filters has led to the release of toxic products into the distribution system resulting in gastrointestinal disturbances.

Blooms of blue-green algae have been known to cause fish kills by their toxic by-products or by the dissolved-oxygen depletion, which occurs when the algal cells die, sink to the lake bottom, and consume oxygen as they decompose. There have also been many reports of livestock poisoning due to blooms of blue-green algae in watering ponds.

Table 1-1. Additional Problems Caused by Algae in Water Supplies*

Problem and Algae		*Algal Group*
Slime-producing algae:		
Anacystis (Aphanocapsa, Gloeocapsa)		Blue-green
Batrachospermum		Red
Chaetophora		Green
Cymbella		Diatom
Euglena sanguinea var. furcata		Flagellate
Euglena velata		Flagellate
Gloeotrichia		Blue-green
Gomphonema		Diatom
Oscillatoria		Blue-green
Palmella		Green
Phormidium		Blue-green
Spirogyra		Green
Tetraspora		Green
Algae causing coloration of water:	*Color of water*	
Anacystis	Blue-green	Blue-green
Ceratium	Rusty brown	Flagellate
Chlamydomonas	Green	Flagellate
Chlorella	Green	Green
Cosmarium	Green	Green
Euglena orientalis	Red	Flagellate
Euglena rubra	Red	Flagellate
Euglena sanguinea	Red	Flagellate
Oscillatoria prolifica	Purple	Blue-green
Oscillatoria rubescens	Red	Blue-green

*Many of these algae are pictured in *Basic Science Concepts and Applications*, Appendix D.
Source: *Algae in Water Supplies*. US Public Health Service, Pub. 657 (reprinted 1962).

Table 1-1. Additional Problems Caused by Algae in Water Supplies *(continued)*

Problem and Algae	*Algal Group*
Algae causing corrosion of concrete:	
Anacystis (Chroococcus)	Blue-green
Chaetophora	Green
Diatoma	Diatom
Euglena	Flagellate
Phormidium	Blue-green
Phytoconis (Protococcus)	Green
Algae causing corrosion of steel:	
Oscillatoria	Blue-green
Algae persistent in distribution systems:	
Anacystis	Blue-green
Asterionella	Diatom
Chlorella	Green
Chlorococcum	Green
Closterium	Green
Coelastrum	Green
Cosmarium	Green
Cyclotella	Diatom
Dinobryon	Flagellate
Elaktothrix gelatinosa	Green
Epithemia	Diatom
Euglena	Flagellate
Gomphosphaeria aponina	Blue-green
Scenedesmus	Green
Synedra	Diatom
Algae interfering with coagulation:	
Anabaena	Blue-green
Asterionella	Diatom
Euglena	Flagellate
Gomphosphaeria	Blue-green
Synedra	Diatom
Algae causing natural softening of water:	
Anabaena	Blue-green
Aphanizomenon	Blue-green
Cosmarium	Green
Scenedesmus	Green
Synedra	Diatom
Toxic marine algae:	
Caulerpa serrulata	Green
Egregia laevigata	Brown
Gelidium cartilagineum var. robustum	Red
Gonyaulax catenella	Dinoflagellate
Gonyaulax polyedra	Dinoflagellate
Gonyaulax tamarensis	Dinoflagellate
Gymnodinium brevis	Dinoflagellate

(Table continues on the next page.)

Table 1-1. Additional Problems Caused by Algae in Water Supplies ***(continued)***

Problem and Algae	*Algal Group*
Toxic marine algae (continued):	
Gymmodinium veneficum	Dinoflagellate
Hesperophycus harveyanus	Brown
Hornellia marina	Flagellate
Lyngbya aestuarii	Blue-green
Lyngbya majuscula	Blue-green
Macrocystis pyrifera	Brown
Pelvetia fastigiata	Brown
Prymnesium parvum	Flagellate
Pyrodinium phoneus	Dinoflagellate
Trichodesmium erythraeum	Blue-green
Toxic freshwater algae:	
Anabaena	Blue-green
Anabaena circinalls	Blue-green
Anabaena flos-aquae	Blue-green
Anabaena lemmermanni	Blue-green
Anacystis (Microcystis)	Blue-green
Anacystis cyanea (Microcystis aeruginosa)	Blue-green
Anacystis cyanea (Microcystis flos-aquae)	Blue-green
Anacystis cyanea (Microcystis toxica)	Blue-green
Aphanizomenon flos-aquae	Blue-green
Gloeotrichia echinulata	Blue-green
Gomphosphaeria laeustris (Coelosphaerium kuetzingianum)	
Lyngbya contorta	Blue-green
Nodularia spumigena	Blue-green
Parasitic aquatic algae:	
Oodinium limneticum	Dinoflagellate
Oodinium ocellatum	Dinoflagellate

Algae Control

A program instituted for the control of algal populations should consist of more than routine periodic chemical treatment to reduce algal numbers. Such routine, unmonitored treatment may upset the natural balance in the water body and create conditions conducive to the growth of problem species.

Before any control procedures are attempted, the operator should establish a routine program of collecting raw-water samples at least once a week. The numbers of the different types of algae in each sample should be counted. By performing such testing, the operator can determine which types of problem-causing algae are present in a raw-water source. The best time to initiate algal control procedures can be determined by following the trends in the numbers of algae present. It would take about one year to determine the algal trends for a body of water; during this period, the operator would be able to relate the algal conditions in the raw water to problems in the finished water.

Several biological and chemical methods have been experimented with for control of algae in large water bodies. However, two effective methods commonly used for drinking water supplies are (1) use of the chemical copper sulfate ($CuSO_4$, also known as blue vitriol, or blue stone) and (2) use of powdered activated carbon.

Copper sulfate. The control of algae using copper sulfate has been practiced since 1904. Because not all algae are killed by this chemical, the effectiveness of the treatment depends on the type of algae; it is therefore important that the problem-causing algae be identified accurately. Table 1-2 summarizes the effectiveness of copper sulfate against various types of algae.

The effectiveness of copper sulfate treatment is also dependent on its ability to dissolve in water, which is in turn dependent on pH and alkalinity. Therefore, the dosage required depends upon the chemical characteristics of the water to be treated. The best and most lasting control will result if the water has a total

Table 1-2. Relative Toxicity of Copper Sulfate to Algae

Group	*Very Susceptible*	*Susceptible*	*Resistant*	*Very resitant*
Blue-green	*Anabaena, Anacystis, Gomphosphaeria, Rivularia*	*Cylindrospermum, Oscillatoria, Plectonema*	*Lyngbya, Nostoc, Phormidium*	*Calothrix, Symploca*
Green	*Hydrodictyon, Oedogonium, Rhizoclonium, Ulothrix*	*Botryococcus, Cladophora, Oscillatoria Enteromorpha, Gloeocystis, Microspora, Phytoconis, Tribonema, Zygnema*	*Characium, Clorella, Chlorococcum, Coccomyxa, Crucigenia, Desmidium, Draparnaldia, Golenkinia, Mesotaenium, Oocystis, Palmella, Pediastrum, Staurastrum, Stigeoclonium, Tetraedron*	*Ankistrodesmus, Chara, Coelastrum, Dictyosphaerium, Elakatothrix, Kirchneriella, Nitella, Pithophora, Scenedesmus Testrastrum*
Diatoms	*Asterionella, Cyclotella, Fragilaria, Melosira*	*Gomphonema, Navicula, Nitzschia, Stephanodiscus, Synedra, Tabellaria*	*Achnanthes, Cymbella, Neidium*	
Flagellates	*Dinobryon, Synura, Uroglenopsis, Volvox*	*Ceratium, Cryptomonas, Euglena, Glenodinium, Mallomonas*	*Chlamydomonas, Peridinium, Haematococcus*	*Eudorina, Pandorina*

Source: *Algae and Water Pollution.* US Environmental Protection Agency (Dec. 1977).

alkalinity less than or equal to approximately 50 mg/L as $CaCO_3$ and a pH between 8 and 9.

The following suggested dosages for copper sulfate are general recommendations only and may not be the best dosages for every situation.

- Bodies of water with a total methyl-orange alkalinity equal to or greater than 50 mg/L as $CaCO_3$ are usually treated at a dosage of 1 mg/L, calculated for the volume of water in the upper 2 ft (0.6 m) of the lake, regardless of the depth of the lake. This converts to about 5.4 lb (2.3 kg) of commercial copper sulfate per acre of surface area.[1] The 2-ft (0.6-m) depth has been determined as the effective range of surface application of copper sulfate in such waters. The chemical tends to precipitate after this depth.
- For water bodies having a total methyl-orange alkalinity less than 50 mg/L as $CaCO_3$, a dosage of 0.3 mg/L is recommended. This dosage is based on the total lake volume and converts to about 0.9 lb (0.4 kg) of commercial copper sulfate per acre-foot of volume.

The minimum copper sulfate dosage depends on the alkalinity of the water; the maximum safe dosage depends in part on the toxic effect on fish life. A safe dose for most fish is 0.5 mg/L; however, trout are very sensitive and can be killed by doses greater than 0.14 mg/L. It would be wise to check with the state department of game and fish to determine if any special precautions need to be taken.

The simplest method of copper sulfate application to control algae in small lakes, ponds, or reservoirs is to drag burlap bags of the chemical behind a boat. The boat is guided in a zigzag course for overlapping coverage over the water, as shown in Figure 1-6. Wind, waves, and natural diffusion mix the chemical with the water, giving fairly uniform coverage. This method works well if the chemical is evenly spread over the entire surface. For larger areas, a boat should follow a path such as that shown in Figure 1-7. Very large treatment areas might be handled better by power spray application from shore (Figure 1-8) or from a motor boat.

The effect of the copper sulfate treatment on algal populations can be noticed soon after the chemical has been added. Within a few minutes, the color of the water will change from dark green to grayish-white. At no time are all the algae in the lake entirely eliminated, but the water should be visibly free of cells for two or three days following a complete application.

Proper treatment of a body of water ensures that the major portion of the algae will be eliminated so that a long time must pass before the algae can again create problems. The frequency of treatment depends on local climate and the amount of nutrients in the water. Warm temperatures, plentiful sunlight, and a high nutrient concentration all tend to encourage a rapid regrowth of algae. In general, one to three complete treatment applications per season should be

[1] *Basic Science Concepts and Applications*, Chemistry Section, Chemical Dosage Problems (Milligrams-per-Litre to Pounds-per-Day Problems).

Courtesy of Phelps Dodge Refining Corporation, 300 Park Ave., New York, NY 10022

Figure 1-6. Path Taken for Copper Sulfate Application to Small Water Bodies

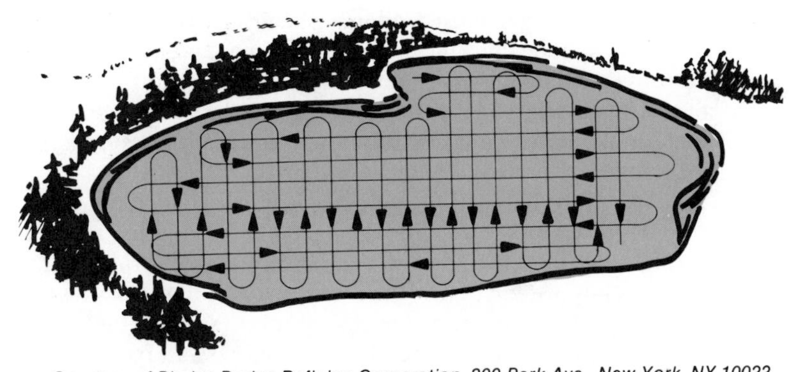

Courtesy of Phelps Dodge Refining Corporation, 300 Park Ave., New York, NY 10022

Figure 1-7. Path Taken for Copper Sulfate Application to Large Water Bodies

Courtesy of Applied Biochemists, Inc., Mequon, Wis.

Figure 1-8. Power Spray Application of Copper Sulfate

sufficient. The actual length of time between applications can best be found by taking periodic algal counts.

Powdered activated carbon. Powdered activated carbon (PAC) also can be used to control algae. This is not a form of chemical treatment because it operates by a physical rather than a chemical process. The activated carbon forms a black blanket, cutting off sunlight penetration, which is vital for algal growth. However, a large amount is needed to effectively block the sun, and powdered carbon is messy and very difficult to handle from a boat. Powdered carbon can also be added manually or with a chemical feeder before the water enters the treatment plant, in order to adsorb algal by-products responsible for taste and odor problems.

Pond covers. A method used to control algal growth in smaller water bodies, such as presedimentation impoundments, is to cover the pond. This greatly reduces the amount of sunlight available to the algae for photosynthesis. The cover in Figure 1-9 is made of a synthetic rubber fabric that floats on the surface of the water. The fabric should be a type that can be in contact with water without causing a taste and odor. Although they effectively control algae, floating covers can interfere with normal dewatering and cleaning operations.

Chemical control of algal-caused odors. Chemicals used in the control of algal-related odors include chlorine, chlorine dioxide, and potassium permanganate. Note that these chemicals are generally not used to control algae, but are used to control algal-related problems. Some are best suited for particular odors. For example, chlorine effectively removes fishy, grassy, haylike, and septic odors, but it makes earthy odors worse.

Special note should be made of the possible problems associated with the use of chlorine for chemical pretreatment. Although prechlorination has long been used to control tastes and odors, recent studies show that chlorine can combine with organic materials present in raw water to form TRIHALOMETHANES (THMs). These complex organic compounds are suspect CARCINOGENS, that is, possible cancer-causing chemicals, and the USEPA drinking water regulations (see

Courtesy of Commissioner of Public Works, Charleston, S.C.

Figure 1-9. Pond Cover to Control Algal Growth

Introduction to Water Quality Analysis, Drinking Water Standards Module) specify a maximum contaminant level (MCL) of 0.10 mg/L for total THMs in finished water. For this reason, serious consideration should be given to the use of a chemical other than chlorine (such as powdered activated carbon) for pretreatment, particularly if taste, odor, and color are continuing problems.

Problems Caused by Rooted Aquatic Plants

Rooted aquatic plants (water weeds) are different from algae in that they are plants with defined leaves, stems, and root systems. They may be classified as

- Emergent weeds
- Floating or surface weeds
- Submerged weeds.

EMERGENT WEEDS, such as those illustrated in Figure 1-10, grow in shallow water on or near the shoreline. They root in the bottom mud and can extend several feet above the water surface. Cattails, water willows, and rushes are familiar examples. Figure 1-11 shows emergent weeds around an entire lake.

FLOATING OR SURFACE WEEDS generally are plants with their leaves floating on the surface of the water. They can be rooted in the bottom mud or free floating. Sometimes they are mixed with the emergent weeds (Figure 1-12). The most common example of the rooted types of these plants are the water lilies; duckweed and similar plants are free floating.

SUBMERGED WEEDS grow entirely under water and are rooted in the bottom mud (Figure 1-13). The depth to which they will grow is limited primarily by the depth of sunlight penetration. The clearer the water, the taller the plants are likely to grow. Coontail and blatterwort are examples of the many types of submerged weeds.

Courtesy of Applied Biochemists, Inc., Mequon, Wis.

Figure 1-10. Emergent Weeds

Courtesy of Applied Biochemists, Inc., Mequon, Wis.

Figure 1-11. Emergent Weeds Around a Lake

Courtesy of Applied Biochemists, Inc., Mequon, Wis.

Figure 1-12. Floating Weeds and Emergent Weeds

Courtesy of Applied Biochemists, Inc., Mequon, Wis.

Figure 1-13. Submerged Weeds

Water weeds can cause the same problems with plant operations and treatment as algae (that is, clogging, color, tastes, and odors). And rooted aquatic vegetation can act as a habitat (breeding area) for disease-causing and nuisance insects, which can create problems for operators and consumers alike. Water weeds serve many beneficial functions as well; they provide shelter and attachment surfaces for small beneficial organisms; they provide spawning and schooling areas for fish; and they produce dissolved oxygen (DO). In addition, they consume and temporarily store nutrients (such as phosphorus) that could otherwise support undesirably large algal populations.

Ecologically, aquatic plants are a necessary part of the living community of any surface-water source. It is best to control water weeds only in the specific areas where they can create problems for plant operations.

Control of Rooted Aquatic Plants

Rooted aquatic plants are a beneficial and necessary part of the aquatic environment. The main benefit derived, from a water treatment aspect, is the ability of these plants to consume large amounts of nutrients that would otherwise be available to stimulate algal growth. Procedures to control these plants should be initiated whenever the plants begin to cause operational control problems at the treatment plant or add color to the raw water.

There are three methods for controlling rooted aquatic plants in lakes, reservoirs, and other surface water:

- Physical
- Biological
- Chemical.

Physical methods for control of aquatic plants include:

- Harvesting
- Dewatering
- Dredging
- Shading
- Lining.

HARVESTING methods depend on the extent and amount of control desired. The technique employed can vary from hand pulling or hand cutting and raking, to using power-driven harvesting machines.

DEWATERING, the complete or partial draining of a body of water to kill the weeds, is effective in controlling the growth of aquatic plants. It is most practical if the weed problem is in a small impoundment, such as a presedimentation basin. It may be impractical if the weeds are in a major lake or reservoir. In dewatering, the water level is lowered and the impoundment is allowed to dry. Then, if the lake bottom is stable enough to support heavy equipment, a scraper or front-end loader can be used to clear away the dried plant material; otherwise the dewatered area can be left exposed for a period of time (several months) so that the root systems will dehydrate. In Figure 1-14, a disc plow is being used to break up and aerate drying vegetation on the bottom of a partially dewatered reservoir.

DREDGING, using a clam shell, drag line, or hydraulic dredge, is useful in the control of aquatic weeds, because dredging the bottom mud removes any plants that are growing there. Also, dredging the near shore areas of a water body increases the water depth, lessening the area of suitable habitat for the plants.

SHADING can be accomplished in two ways. First, the turbidity of the water in the water body can be raised by adding clay to the water to form a colloidal suspension. This allows for less sunlight penetration into the water and thereby

Courtesy of Ragnarok

Figure 1-14. Disc Plow Being Used on a Partially Dewatered Reservoir

reduces the amount of growth. Second, shading can be accomplished by placing sheets of black plastic on the lake bottom. This effectively shades areas of the bottom from the sun and reduces plant growth.

LINING the pond with a synthetic rubber material prevents the growth of rooted aquatic plants and prevents water loss due to seepage through the bottom of the pond (Figure 1-15).

Biological controls using specific species of crayfish, snails, and fish have proven extremely effective for control of rooted aquatic plants. However, before instituting a program using these organisms, the state department of game and fish should be contacted to determine if they allow transplanting of the species in the state.

Chemical controls of aquatic plants using herbicides should be used only when the problem becomes unmanageable by other means. The chemicals of choice in this case are diquat and endothall. Both have been registered for use in drinking water supplies and are relatively safe if the application instructions are followed. The main drawback to using these herbicides is that their safe use requires a waiting period of several days after application before the water can be used for human consumption. Diquat has a recommended ten-day waiting requirement and endothall a seven-day requirement; state regulations should be checked to determine if they differ from the manufacturers' recommendations. The length of the waiting period would make it difficult for a utility with only one raw-water storage reservoir to use the chemical for aquatic plant control.

For treating small areas near shore, hand broadcasting of dry chemical herbicides or spraying the chemicals using a shoulder strap sprayer (Figure 1-16) are the methods of choice. For large areas, the use of a power sprayer from shore or from a motor boat would be the most efficient means for applying a herbicide.

Courtesy of Kimberly-Clark of Canada Ltd.

Figure 1-15. Pond Lining

Courtesy of Applied Biochemists, Inc., Mequon, Wis.

Figure 1-16. Shoulder Strap Sprayer for Broadcasting Chemicals

Suggested Records

Keeping records is an important part of an algal and water weed control program. Daily records should be kept of threshold odor test results and of any complaints of taste and odor registered by consumers. When chemical treatment is performed, records should also include:

- Reason for pretreating (such as taste and odor problems or filter clogging)
- Type of algae or weed treated
- Algal count or estimated weed coverage
- Chemical used, concentration, and dosage
- Date of pretreatment
- Length of time since last treatment
- Weather conditions
- Other water conditions (such as temperature, pH, alkalinity)
- Method of application
- Personnel involved
- Results of pretreatment (such as taste and odor following treatment, and filter conditions).

Records such as these will serve as a guide to solving similar problems if they occur. With good records, the operator may save considerable time that might be wasted experimenting with different chemicals and dosages in an attempt to reproduce a previous result. Residual algal counts or estimates of weed coverage can serve as reliable guides in deciding when the next treatment is needed.

1-3. Presedimentation

PRESEDIMENTATION is a pretreatment process used to remove gravel, sand, silt, and other gritty material from raw water before it enters the main treatment facility. This material is potentially damaging to a treatment plant. The heavy gravel-sized particles can jam equipment; and, through abrasion, sand and silt can quickly destroy submerged moving parts such as pump impellers, drive shafts, and bearings. A properly operating presedimentation system should remove 60 percent or more of the settleable material.

Sediment is a far more prominent problem in surface-water supplies than in ground-water supplies. Gritty material can enter into surface-water supplies through the natural process of erosion, whereas sand and other gritty material can be prevented from entering well water through proper grouting, screening, and well development techniques. (See *Introduction to Water Sources and Transmission*, Developing the Water Supply Module, Ground Water Development.)

Types of Presedimentation Systems

There are three types of presedimentation systems:

- Presedimentation impoundments
- Sand traps
- Mechanical sand-and-grit removal devices.

Presedimentation impoundments. The most common type of sediment removal system is the PRESEDIMENTATION IMPOUNDMENT. It is normally used with river or stream supplies. A presedimentation impoundment can be a simple earthen reservoir providing 24-hr storage capacity or it can be a concrete impoundment large enough to provide a holding capacity of several weeks. Such large presedimentation impoundments serve three functions: (1) sediment removal, or presedimentation, (2) storage, and (3) reduction of the impact of changes in water quality on later treatment processes.

Any raw-water supply exposed to sunlight will probably develop nuisance growths of algae and other aquatic plants. Presedimentation impoundments are no exception; however, since the surface areas are relatively small, chemical pretreatment is not difficult. If dual basins are used, then aquatic plant life can be controlled by dewatering one unit at a time without interrupting the raw-water supply. Growth of rooted plants can also be controlled by lining the pond, and algal growth can be controlled by covering the pond.

Sand traps. A SAND TRAP is a depression in the bottom of a structure, such as the one illustrated at the bottom of the wet well in Figure 1-17. Water from the inlet pipe enters the sand trap. Then, because the wet well is larger than the inlet pipe, the water slows so that the sand and gritty material in suspension settle to the bottom. Water flowing out of the trap is pumped to the treatment facilities.

The inlet baffle illustrated in Figure 1-17 is installed in sand traps to prevent SHORT-CIRCUITING and to act as a barrier, causing the flow to move downward

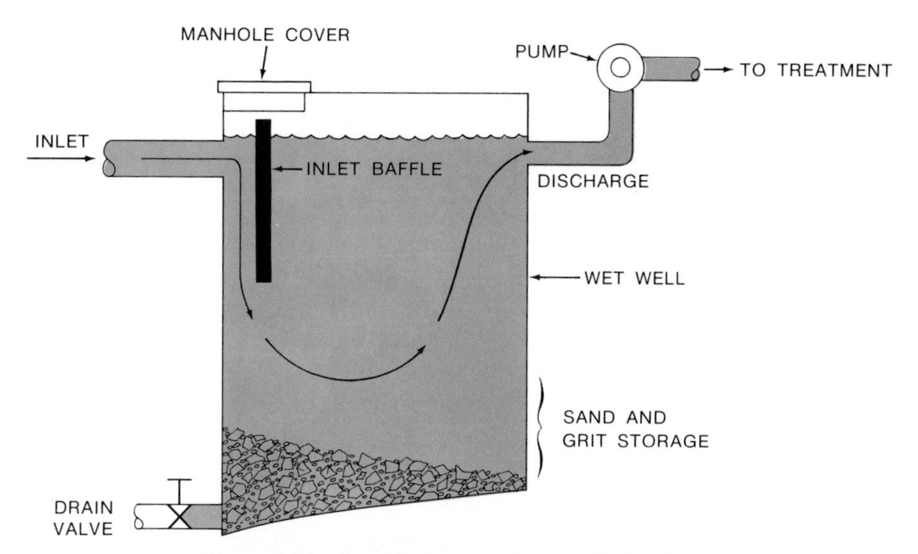

Figure 1-17. Sand Trap at the Bottom of a Wet Well

through the wet well. At the bottom of the wet well is a drain valve that must be opened periodically to flush out the sand; otherwise, the accumulated sand can build up to the bottom of the baffle and be carried into the pump.

Because sand traps must be cleaned manually and have relatively small holding capacity, they are best suited for raw water that contains relatively little sand and grit (less than 100 mg/L).

Mechanical sand-and-grit removal devices. Mechanical sand-and-grit removal devices most often are used when raw water is high in suspended solids. Large amounts of suspended solids occur in surface supplies seasonally, during the spring runoff or during periods of high water flows in a stream or river. High loads also occur in ground-water supplies when well screens fail and sand is drawn into the well.

Typical mechanical sand-and-grit removal devices, sometimes called cyclone degritters, work on the principle of centrifugal force. Figure 1-18 illustrates how a centrifugal sand-and-grit removal device operates. First, sand-laden water is pumped into the feed entry. The water begins to travel in a spiral path inside the cylindrical section toward the vortex finder. As this happens, extremely high centrifugal forces develop (many times the force of gravity), which throw the sand particles toward the wall of the cylinder. The sand particles move in a spiral path toward the discharge and are discharged, along with some water, into the sand-accumulator tank. The clean water leaves the unit through the vortex finder pipe and moves on toward the treatment plant.

Figure 1-19 illustrates the simplicity of a typical cyclone installation. For the operator, one of the best features is that cyclone separators have no moving parts, so they require relatively little maintenance even though they regularly handle gritty material.

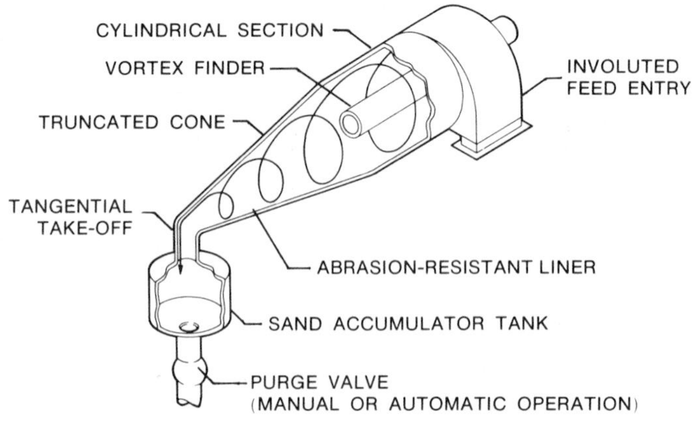

Courtesy of Krebs Engineers

Figure 1-18. Centrifugal Sand-and-Grit Removal Device

Courtesy of Claude Laval Corporation (Lakos Separators)

Figure 1-19. Typical Cyclone Separator Installation

Routine Operation

To ensure successful removal of sand and grit, any presedimentation system must be tested regularly and cleaned routinely. Influent and effluent samples must be collected and tested regularly for settleable solids.

The frequency of sampling and testing varies from plant to plant, depending on the amount of sand and gritty material in the raw water. During peak flow periods, sampling and testing may be required daily because of the rapidly increasing sand and grit loads; at other times, weekly or even monthly testing may be adequate.

Any presedimentation system must be cleaned routinely. If cleaning is not performed often enough, the water flowing through the system can mix the grit and sand back into suspension and carry it into the treatment plant. Deposits can also become ANAEROBIC (lacking free oxygen), resulting in tastes and odors, a common problem of infrequently cleaned impoundments.

A wet-well type of sand trap is cleaned by allowing accumulations to discharge through a drain line. A manhole is provided so that the wet well can be hosed down during draining.

Cyclone separators are cleaned automatically and continuously. The operator must see that the sand-and-grit discharge storage bin or hopper is emptied routinely and that the material is buried or otherwise properly disposed of.

To clean a presedimentation impoundment, it must be completely drained and dried. Then the sand can be removed by scrapers, dozers, or front-end loaders. If the impoundment has a floating cover, the cover must be carefully rolled out of the way of cleaning equipment.

Suggested Records

Detailed record keeping is important. Information about the type and variations in the amount of grit carried by the raw water and removed by presedimentation helps to determine how frequently sampling and testing must be done and during what time of year sand and grit become a problem. Records can be used to monitor the continued efficiency of sand and grit removal. A gradual decrease in removal efficiency may signal the need to clean away accumulated deposits.

Detailed records dealing with presedimentation should cover:

- Date of sampling and testing
- Suspended solids in raw water: mg/L, or mL/L if SETTLEABILITY TEST is used
- Suspended solids in presedimentation effluent: mg/L, or mL/L if settleability test is used
- Cleaning
  - Date
  - Time required
  - Estimated quantity of sand and grit removed (usually in cubic feet).

1-4. Microstraining

A MICROSTRAINER (Figure 1-20) is a very fine screen used primarily to remove algae, other aquatic organisms, and small debris that can clog treatment plant filters. The most commonly used type of microstrainer unit consists of a rotating drum that is lined with finely woven material such as stainless steel wire fabric. One commonly used fabric has about 160,000 openings per square inch. Each opening is about 23 μm in size—smaller than the openings in tightly woven clothing fabric. Other fabrics may have larger or smaller openings.

The microstrainer drum rotates slowly, usually 4 to 7 rpm, as water enters the inside of the drum and flows outward through the fabric. Algae and other aquatic organisms deposited on the inside of the fabric form a tight mat of debris, which adheres to the fabric and rotates up to the backwash hood area.

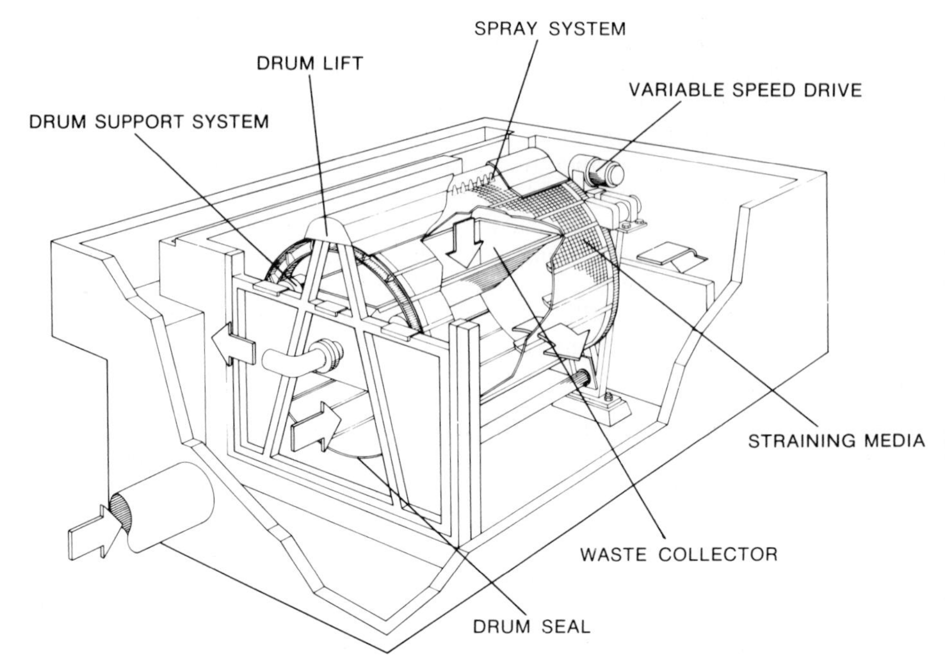

Courtesy of Zurn Industries, Inc.

Figure 1-20. Fine-Screened Microstrainer

Then, high-pressure (25–50 psig, 172–345 kPa) backwash jets spray the back side of the fabric, causing the matted debris to break away. The debris and backwash water fall by gravity into a debris trough inside the drum and flow either directly to a point of disposal or to a pond or tank that separates the debris from the water.

One of the major advantages of microstrainers is the improvement they make in the operation of sand filters. Microstrainers generally remove about 65 percent of the filter-clogging material in the water, with removal efficiency varying from a low of about 50 percent to a high of about 90 percent, depending on the type of algae. By reducing the filter-clogging load on sand filters, filters can be operated longer without backwashing. This saves backwash water and increases the amount of filtered water available to the community.

There are definite advantages to using microstrainers in certain situations, but there are limitations also. Straining only contributes partly to the total water treatment process by removing the coarsest particles. Microstrainers cannot remove all algae, and they do not remove bacteria, viruses, or most suspended matter contributing to turbidity. Even eggs of tiny aquatic animals can pass through the fabric. Also, microstrainers have no effect on the removal of dissolved substances such as inorganic and organic chemicals; these constituents pass through the straining process. Therefore, microstrainers should not be thought of or used as a substitute for coagulation/flocculation and filtration.

Although microstrainers are often made of stainless steel (particularly the fabric) the constant wet/dry duty they serve makes them subject to corrosion. Even under the best of conditions, they require relatively high maintenance, including routine painting, lubrication, replacement of worn parts, and fabric repair and replacement. Because chlorine is particularly corrosive, serious corrosion can result when a microstrainer is used after chlorination.

1-5. Flow Measurement

Because a flow measurement is often made immediately after preliminary treatment of water, a discussion of flow measurement is included here. Although this may not be the practice at every plant, flow measurements are almost always made somewhere near the upper end of a plant. Water treatment plant operators use flow measurements to:

- Control the flow rate to each treatment process
- Adjust chemical feed rates
- Determine pump efficiencies and power requirements
- Calculate detention times
- Monitor the amount of water treated
- Calculate the unit cost of treatment.

For the treated water to be safe to drink and be of consistently high quality, it is absolutely essential that flow measurements be made dependably and accurately. For most purposes, an accurate flow rate measurement is one that registers within 2 percent of the actual flow rate. If the actual flow rate is 100 gpm, a sufficiently accurate measurement can indicate a flow rate between 98 and 102 gpm. Inaccurate readings are misleading and costly. For example, disinfection controlled by a flow meter can be seriously inadequate if the flow meter reads 10 percent low, and can be causing a costly waste of disinfectant if the flow meter reads 10 percent high.

Flow measurements can be made in either pipelines or open channels. This section focuses specifically on four types of measuring devices suitable for pipelines:

- Pressure differential meters
- Velocity meters
- Magnetic flow meters
- Ultrasonic flow meters.

Although there are applications for open-channel flow measuring devices in treatment plants, almost all flow measurements at the treatment plant are performed in pipelines. The open-channel devices (weirs, Parshall flumes, etc.) are discussed in *Introduction to Water Sources and Transmission*, Water Transmission Module.

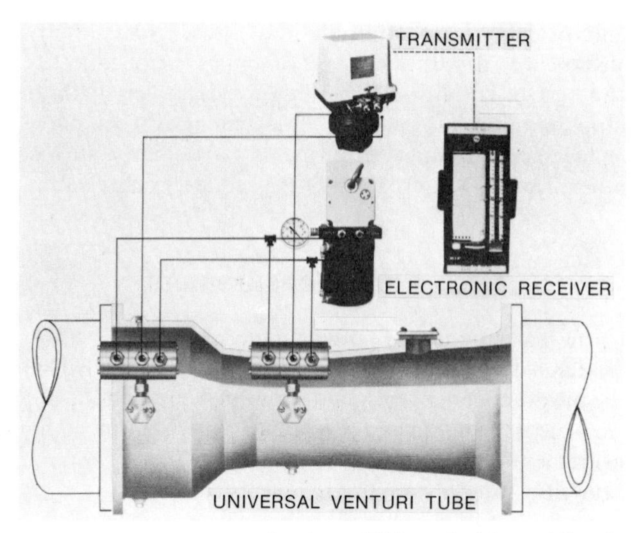

Courtesy of BIF, a unit of General Signal

Figure 1-21. Pressure Differential Meter

Pressure Differential Meters

A PRESSURE DIFFERENTIAL METER is one of the most common types of meters used at a treatment plant. As shown in Figure 1-21, these meters usually consist of three basic components: a primary element, a transmitter, and a receiver. The purpose of the PRIMARY ELEMENT is to create a signal proportionate to the water velocity, a signal that can be sent to the TRANSMITTER. This signal is measured by the transmitter, which then sends another signal to the RECEIVER, where that signal is converted to a flow rate.

Three commonly used primary elements for pressure differential are:

- Venturi tubes
- Flow tubes
- Orifice plates.

Figures 1-22 through 1-24 show examples of the three primary elements. The pressure difference created by each of these primary elements varies with the velocity of the water, and velocity is related to flow rate by the $Q = AV$ equation.[2] Nomographs are often used to determine flow rate given the pressure difference across the metering element.[3]

VENTURI TUBES (Figure 1-22) and FLOW TUBES (Figure 1-23) look and operate similarly; they are both designed to keep friction (HEAD LOSS) to a minimum by

[2] *Basic Science Concepts and Applications*, Hydraulics Section, Flow Rate Problems (Instantaneous Flow Rate Calculations).

[3] *Basic Science Concepts and Applications*, Hydraulics Section, Flow Rate Problems (Flow Measuring Devices).

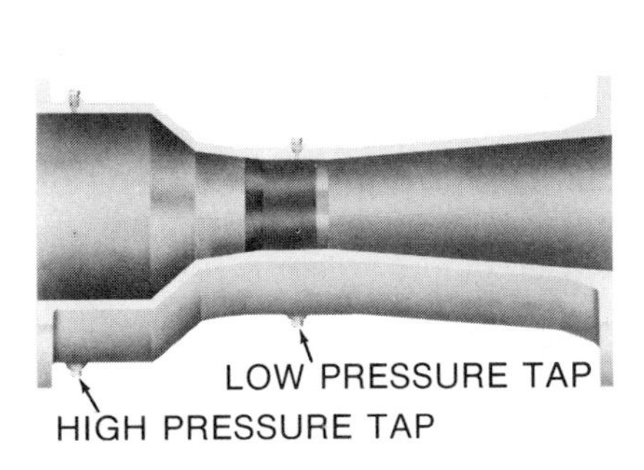

Courtesy of BIF, a unit of General Signal

Figure 1-22. Venturi Tube

Courtesy of The Bethlehem Corporation

Figure 1-23. Flow Tubes

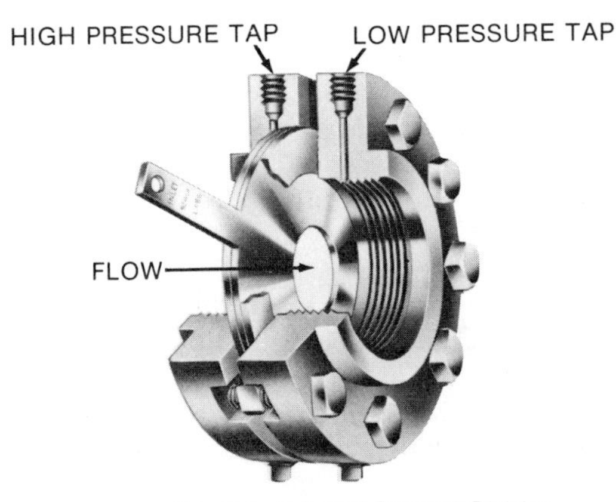

Courtesy of Bristol-Babcock Division,
Acco Industries Inc., Waterbury, Conn.

Figure 1-24. Orifice Plate

having very smooth interiors. The pressure loss caused by either type of primary element is relatively small, usually between 5 and 20 percent of the pressure available. Both types of flow meters are available in a wide range of sizes, from those suitable for very small flow measurements up to those for very large flow measurements. They are widely used as main plant flow meters, as well as flow meters for individual treatment processes. Either of these primary elements can also be used in the rate-of-flow controllers for monitoring filter effluent flow rate and backwash flow rate.

An ORIFICE PLATE primary element consists of a thin plate, usually stainless steel, with a precisely sized hole drilled through it (Figure 1-24). The plate is mounted between two flanged sections of pipe. Water flowing through the plate creates a pressure that is higher on the inlet side than the outlet. The resulting pressure difference is measured, transmitted, received, and converted to a flow measurement. Unlike the venturi and flow tubes, the orifice plate is a high head-loss device. In other words, more energy is used moving the water through the tiny opening in the orifice plate than is used moving it through the smooth venturi and flow tubes. In addition, because of the small opening, the orifice meter can be easily plugged, clogged, or fouled. Orifice meters are used where space is too limited for a flow tube and where greater head loss is not a problem. They are also used to measure gas flows, such as compressed air in the aeration process and carbon dioxide in the stabilization process (see Module 2, Aeration and Module 9, Stabilization).

The transmitter of a pressure differential meter may be electronic or pneumatic, or it may be a simple water column float assembly. The primary elements and receivers of electronic and pneumatic systems can be more conveniently located, and both systems are more widely used than water column float systems. For example, when using electronic systems, the receiver can be located miles from the primary element. This is not so in water column float systems. Water column float systems are used where power is not readily available.

The receivers of pressure differential meters convert the signal from the transmitter and indicate the instantaneous flow rate of the water. Many receivers, such as the one shown in Figure 1-25, also record and keep a running total (with a totalizer) of flow information.

Velocity Meters

A VELOCITY METER uses a rotor, such as a turbine or a propeller, as the main component of its primary element. The speed of the rotor changes with the velocity of the passing water. Each complete turn of the rotor represents a known amount of water. Each rotation is transmitted, either mechanically or electrically, to a receiver, which then indicates, records, and totalizes flow rate.

Courtesy of BIF, a unit of General Signal

Figure 1-25. Receiver of a Pressure Differential Meter

Two basic types of velocity meters are commonly used in water systems:

- Propeller meters
- Turbine meters.

The PROPELLER METER is used to measure flow in large lines (up to 72 in., 1.8 m). It operates at a low head loss, and can be saddle-mounted directly onto a pipeline, as shown in Figure 1-26. The primary element contains a propeller facing upstream, which is rotated by the velocity of passing water. The propeller meter works best as a main-line meter where flow rates do not change abruptly, since the propeller itself cannot change speed abruptly.

The receiver is often a simple totalizer, mechanically linked to the propeller so that no outside power supply is needed. However, propeller meters can be equipped with flow-rate indicators and recorders, and they can be equipped to transmit a flow signal to distant receivers.

The TURBINE METER differs in several respects from the propeller meter. The rotor is a turbine (bladed disc), which is spun by the passing water. The turbine can be located directly in the main flow stream, as shown in Figure 1-27, or it can

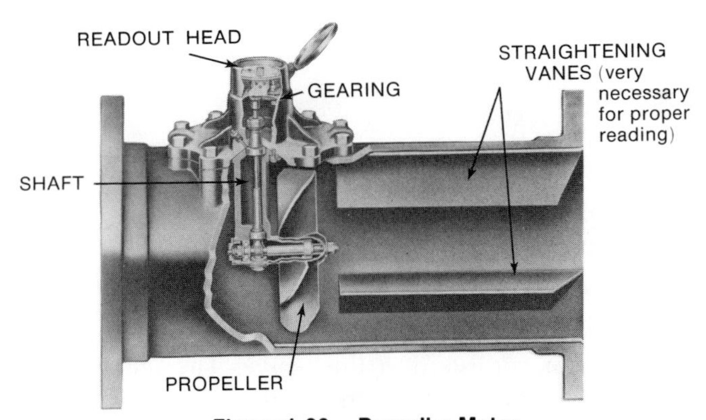

Figure 1-26. Propeller Meter

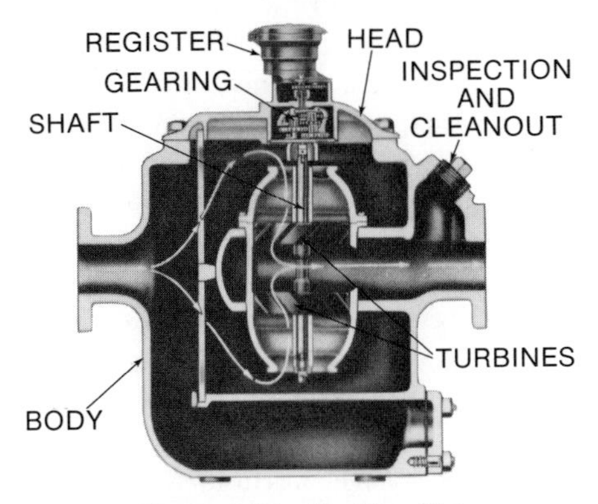

Figure 1-27. Turbine Meter

be located in a bypass chamber, measuring the flow rate of a small stream of water (bypass) directed through the chamber from the main flow (this arrangement is called a PROPORTIONAL METER).

Signals from the spinning turbine are transmitted to a receiver in one of two ways: by direct mechanical linkage, similar to the propeller meter of Figure 1-26 (direct-drive turbine meters); or by a sensor outside the meter body, which detects the passing of a small magnet in one rotor blade each time the rotor makes a complete revolution (magnetic-drive turbine meters).

Turbine meters are available in sizes ranging from small meters that measure flow rates in the 0.001 gpm (0.004 L/min) range to large meters that can measure millions of gallons per day. Since a turbine can clog more easily than a propeller, turbine meters are usually restricted to clean-water applications (for example, filtered water, plant effluent, distribution system).

Magnetic Flow Meters

The MAGNETIC FLOW METER is a relatively new instrument to the water industry, first introduced in 1955. It offers several operational advantages over other meters, in that it:

- Causes no obstruction to flow
- Creates no head loss
- Does not clog
- Handles solids in suspension, such as sludge
- Has no small pipe connections to clean or maintain as do venturis and similar flow tubes
- Can measure flow in either direction.

The basic component parts of magnetic flow meters are shown in Figure 1-28. Structurally, the meter consists of a lined tube. The tube is usually either stainless steel or aluminum and the lining can be neoprene, PTFE (sold under the trademark Teflon), or some other nonmagnetic, nonconducting material. Two electrodes pierce the lining, one on either side of the tube; and two magnetic coils surround the tube.

The electromagnetic coils create a magnetic field across the flow tube.[4] The water flowing through the magnetic field in the tube acts as a conductor, developing (inducing) a voltage between the two electrodes. The voltage, directly related to the velocity of the flowing water, is then transmitted to a recorder that converts the signal and displays the flow rate.

Magnetic flow meters are widely used in plants where it is necessary to measure flow without creating head loss or to measure reversing flows in pipes. Magnetic meters are readily available in sizes from 0.1 to 96 in. The 0.1-in. to 6-in. meters are suited for chemical-feed related, small flow-rate duty. The larger

[4] *Basic Science Concepts and Applications*, Electricity Section; Electricity, Magnetism, and Electrical Measurements.

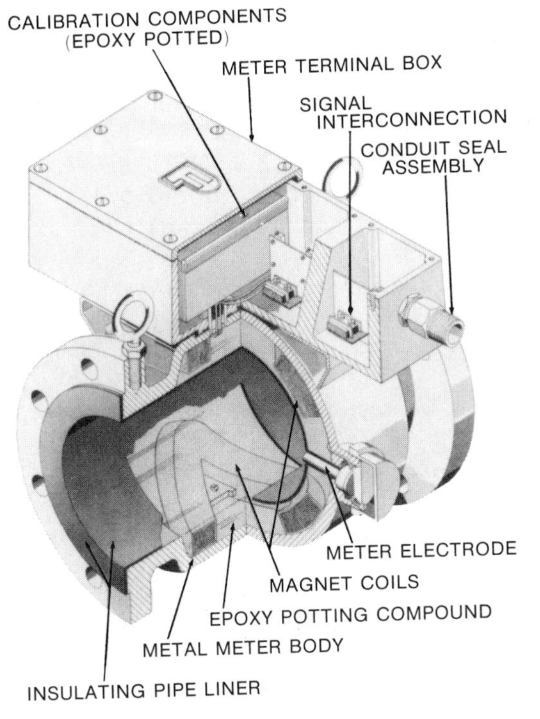

Courtesy of Fischer & Porter Company

Figure 1-28. Magnetic Flow Meter

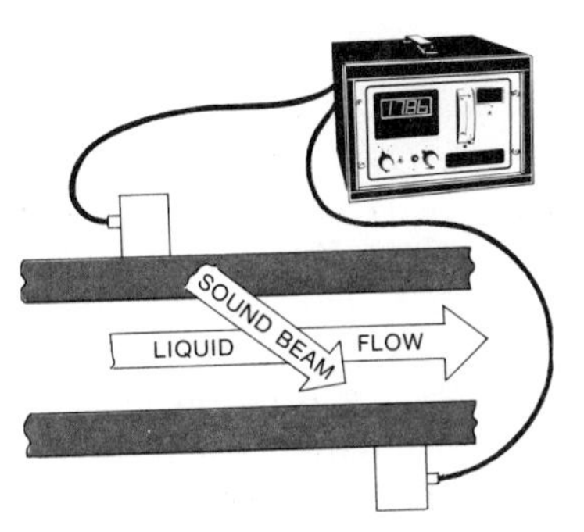

Courtesy of Controlotron Corporation

Figure 1-29. Transducer of an Ultrasonic Flow Meter

meters, 7-in. to 96-in., are mainline flow meters with capacities ranging from 100 gpm (400 L/min) to more than 100,000 gpm (400 kL/min).

Ultrasonic Flow Meters

The ULTRASONIC FLOW METER is one of the latest developments in flow measurement. As illustrated in Figure 1-29, an electronic device (a transducer) sends a beam of ultrasonic sound waves through the water to an opposite transducer. The velocity of the sound beam increases or decreases in relation to the velocity of the water. The difference in the velocity of two sound beams (one directed upstream and one downstream) is measured electronically and converted to a flow rate. One of the advantages of the ultrasonic meter is that there are no obstructions to create head loss or cause maintenance problems. Other advantages include portability and ease of connection.

Ultrasonic flow meters are available in standard diameters from 0.5 to 12 in. and can be specially ordered in diameters up to 196 in. Capacities in the standard sizes are unusually wide ranging, from 15 gpd (60 L/d) for the smallest sizes to 50 mgd (200 ML/d) for the largest. Ultrasonic flow meters can be used for mainline water metering, chemical feed metering, and sludge metering.

Selected Supplementary Readings

Al-Layla, M.A.; Ahmad Shamin; & Middlebrooks, E.J. *Water Supply Engineering Design*. Ann Arbor Science Publishers Inc., Ann Arbor, Mich. (1977). p. 143.

Babbit, H.E.; Doland, J.J.; & Cleasby, J.L. *Water Supply Engineering*. McGraw-Hill Book Company, New York. (6th ed., 1962). pp. 569–570.

Beer, C.R. Fall T&O Control Experience at Marston Lake (Denver). *OpFlow*, 5:12:1 (Dec. 1979).

Hale, F.E. *Use of Copper Sulfate in Control of Microscopic Organisms*. Phelps Dodge Refining Corp., New York. (1950 ed.).

Harvey, J.T. Algae—Its Causes and Its Effects on the Treatment Plant. *OpFlow*, 4:7:1 (July 1978).

Herman, Lyle. Operator Offers Tip on Cleaning of Microstrainers. *OpFlow*, 5:5:5 (May 1979).

How to Identify and Control Water Weeds and Algae. Applied Biochemists, Inc., Mequon, Wis. (2nd ed., 1979).

Introduction to Water Sources and Transmission. AWWA, Denver, Colo. (1979).

Mackenthum, K.M. & Ingram, W.M. *Biological Associated Problems in Freshwater Environments*. DOI, FWPCA, Washington, D.C. (1967). Chap. 4, 6, and 12.

Manual of Water Utility Operations. Texas Water Utilities Assoc., Austin, Texas. (7th ed., 1979). Chap. 7.

Palmer, C.M. *Algae in Water Supplies*. DHEW, PHS, Washington, D.C. Publ. No. 657 (Reprinted 1962).

Pros and Cons: The Value of Aquatic Weeds In Reservoirs. *OpFlow*, 2:4:1 (Apr. 1976).

Reid, G.K. *Pond Life*. Golden Press, New York. (1967). pp. 1-67

Spraying Your Reservoir for Weed Control, *OpFlow*, 3:7:5 (July 1977).

Treating the Reservoir for Algae Control. *OpFlow*, 5:4:1 (Apr. 1979).

Water Meters—Selection, Installation, Testing, and Maintenance. AWWA Manual M6. AWWA, Denver, Colo. (1973). Chap. 1 and 2.

Water Quality and Treatment. AWWA Handbook. McGraw-Hill Book Company, New York. (3rd ed., 1971). pp. 472–473.

Wolfner, J.P. Flow Metering in Water Works. *Jour. AWWA* 63:2:117 (Feb. 1971).

Glossary Terms Introduced in Module 1

(Terms are defined in the Glossary at the back of the book.)

Adsorption
Aeration
Anaerobic
Bar screen
Carcinogen
Coagulation/flocculation

Debris rack
Dewatering
Diatom
Disinfection
Dredging
Emergent weed
Filtration
Floating weed
Flow measurement
Flow tube
Fluoridation
Harvesting
Head loss
Lining
Magnetic flow meter
Microstrainer
Orifice plate
Photosynthesis
Preliminary treatment
Presedimentation
Presedimentation impoundment
Pressure differential meter
Pretreatment
Primary element
Propeller meter
Proportional meter
Receiver
Sand trap
Screening
Sedimentation
Settleability test
Shading
Short circuiting
Softening
Stabilization
Submerged weed
Surface weed
Toxic
Transmitter
Trash rack
Trihalomethane
Turbine meter
Ultrasonic flow meter
Velocity meter
Venturi tube
Wire-mesh screen

Review Questions

(Answers to Review Questions are given at the back of the book.)

1. What treatment processes are normally included in preliminary treatment?
2. How does preliminary treatment reduce the load on treatment processes in the main plant?
3. What is the purpose of screening?
4. Name two basic types of screens.
5. Give two problems associated with the routine operation of screens.
6. Should algae be eliminated entirely from the water source? Why or why not?
7. List at least five problems caused by algae in the water supply.
8. What is the method most widely used to control algae?
9. As a general rule, how much commercial copper sulfate should you use to

control algae in a high alkalinity lake? How much in a low alkalinity lake?

10. List three methods used to apply copper sulfate.

11. List three general types of aquatic weeds.

12. Give at least four problems associated with aquatic weeds.

13. Which of the following techniques are used to control aquatic plants?
    (a) Harvesting
    (b) Burning
    (c) Biological control
    (d) Dewatering
    (e) All of these

14. What type of information should be recorded after treating a surface water source?

15. Explain the purpose of presedimentation.

16. Which of the following are types of presedimentation systems?
    (a) Copper sulfate
    (b) Sand traps
    (c) Comminution
    (d) Mechanical sand and grit removal systems
    (e) Impoundments
    (f) Weed harvesting

17. Which of the following features describe a microstrainer?
    (a) A very fine screening device
    (b) Can be used to substitute for filtration
    (c) Removes algae, aquatic organisms, and debris
    (d) Removes material that would clog sand filters
    (e) Removes bacteria and virus

18. What do treatment plant operators use flow rate information for?

19. Which of the following are pressure-differential-type meters?
    (a) Venturi meters
    (b) V-notch weirs
    (c) Orifice meters
    (d) Parshall flumes
    (e) Flow tubes

20. Identify three basic components of a pressure differential meter.

21. What three functions do many receivers perform?

22. List two types of velocity meters.

23. Meters that measure flow by monitoring the voltage induced across a magnetic field are called ________.

24. Meters that measure flow by monitoring the speed of a high-frequency sound beam are called ________.

25. Use a nomograph to find the flow rate through a Venturi meter with an 8-in. throat diameter producing a pressure difference equivalent to 10 ft. Report the answer in million gallons per day. Nomograph can be found in *Basic Science Concepts and Applications*, Hydraulics Section, Flow Rate Problems—Flow Measuring Devices.

26. How many pounds of copper sulfate will be required to treat a 72 acre lake at a dosage of 1 mg/L? The lake averages 10 ft deep and has a total methyl-orange alkalinity of 60 mg/L.

27. Water enters a presedimentation basin at a flow rate of 3.1 mgd. The basin has a volume of 160,000 gal. What is the detention time in the basin?

Study Problems and Exercises

1. Customers complain about a septic odor in the drinking water. Some customers also complain of cramps, diarrhea, and nausea after drinking the water. Upon inspecting your raw water reservoir you note a fairly heavy growth of rooted aquatic plants extending from near the shoreline to about 60 ft from the shoreline, where the water depth is 10 ft. In addition to the rooted plants, there is a layer of algae covering the reservoir which gives the water a pea-soup green color. Your tests and records show the following data:
   - Surface area of reservoir = 1.6 sq mi
   - Average width = 0.75 mi
   - Average length = 2 mi
   - Miles of shoreline = 15 mi
   - Maximum depth = 53 ft
   - Average depth = 12 ft
   - pH = 7.2
   - Temperature = 23° C
   - Hardness = 110 mg/L as $CaCO_3$
   - Alkalinity = 38 mg/L as $CaCO_3$
   - Surface dissolved oxygen = 8.2 ppm
   - Bottom dissolved oxygen = 0.1 ppm

Determine the following:

(a) What is causing the problem?
(b) What control measure(s) would you use to alleviate the problem?
(c) How much chemical(s) would be used in controlling the problem?
(d) How much would the chemical(s) cost?
(e) What would be the best method for applying the chemical?

2. Prepare a list of pretreatment processes used at a water treatment plant in your area. Prepare a report describing the operating procedures used at each process.

3. To meet increased population growth in your community, a new water source is being developed from a nearby stream with the following conditions:
   - The water quality is highly variable and is subject to frequent changes in alkalinity, color, bacterial quality, and turbidity.
   - Turbidity and color are very high during portions of the year.
   - Bacterial quality is very poor during the summer months.

   You have been asked to recommend types of pretreatment processes that will make the raw water easier to treat. Prepare a report on processes you would recommend and include a diagram showing where the processes would be performed. State why you selected the various processes and discuss all aspects of monitoring that would be necessary to determine the effectiveness of pretreatment. In preparing the report, use the references listed in the supplementary readings at the end of this module; manufacturers' literature may also be helpful.

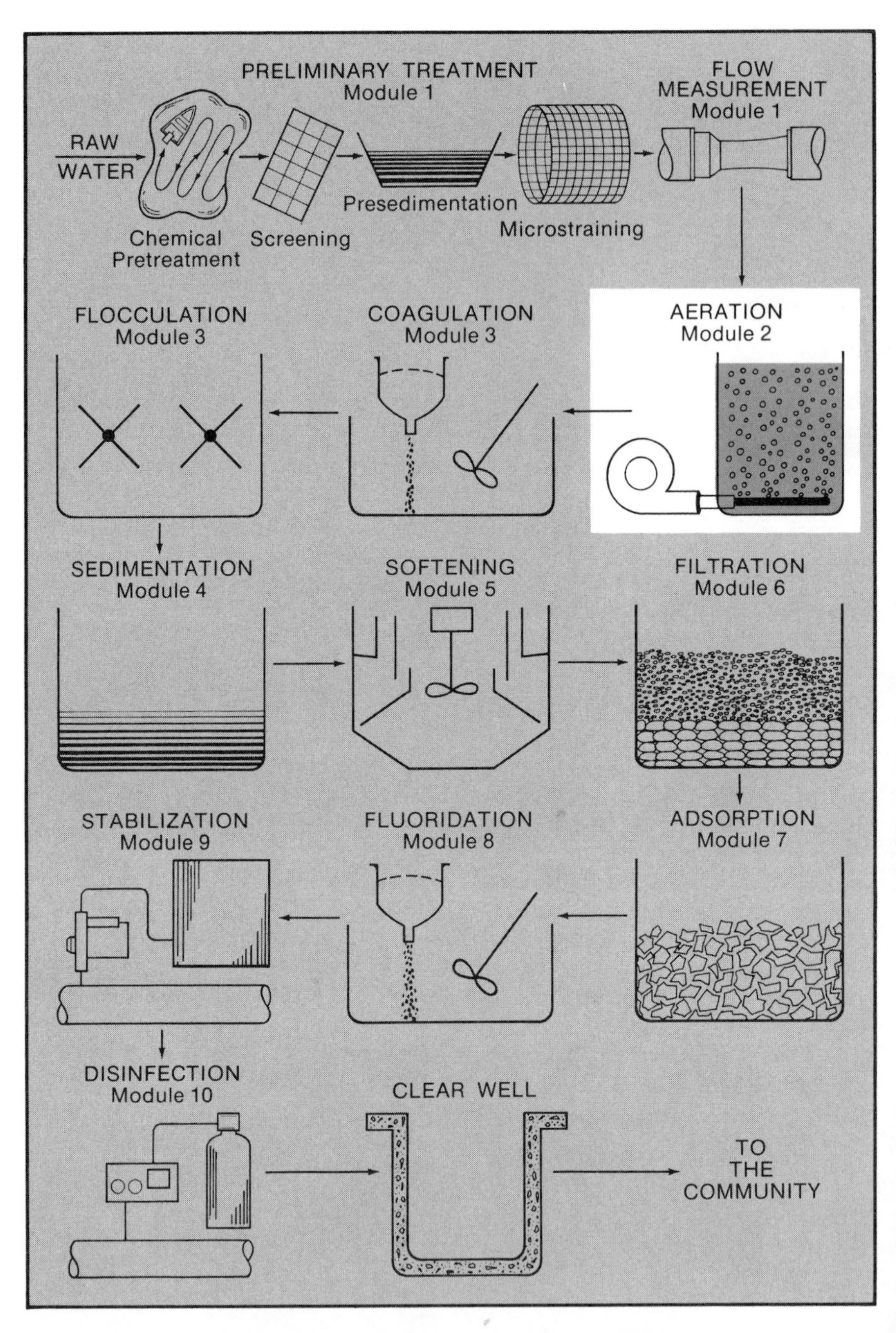

Figure 2-1. Aeration in the Treatment Process

Module 2

Aeration

AERATION is the process of bringing water and air into close contact in order to remove dissolved gases such as carbon dioxide and to oxidize dissolved metals such as iron.

As shown in Figure 2-1, aeration is often the first major process at the treatment plant. During aeration, constituents are removed or modified before they can interfere with other treatment processes.

This module contains discussions of the processes by which aeration removes or changes constituents, what specific constituents aeration removes or changes, and what types of water are aerated. Included in the discussions are illustrations of aeration devices and descriptions of how each device works to remove certain troublesome characteristics of water. Some of the more common operating problems associated with the aeration process are also covered. The module concludes with suggestions for operational control tests and describes some of the important safety considerations in the aeration process.

After completing this module you should be able to

- Describe the aeration process.
- List the troublesome constituents removed from water by aeration.
- Discuss the water quality or operating problems associated with certain constituents.
- Name several types of aerators.
- Identify the operating problems typically associated with aeration and recommend corrective solutions.
- List three basic control tests for aeration process operations.
- Describe hazards and safety precautions involved in the aeration process.

2-1. Description of Aeration

Aeration is the process of introducing air into water. A common example of aeration occurring naturally is a stream tumbling over rocks. The resulting turbulence brings the air and water into close contact and the air dissolves into the water. This process can be used to remove several unwanted constituents from water.

How Aeration Removes or Modifies Constituents

In water treatment, the aeration process brings water and air into close contact by exposing drops or thin sheets of water to the air (water into air), or by introducing small bubbles of air and letting them rise through the water (air into water). For both procedures, the processes by which aeration accomplishes the desired results are the same:

- Sweeping or scrubbing action caused by the turbulence of water and air mixing together
- Oxidizing certain metals and gases.

Undesirable dissolved gases enter water either from the air above the water or as a by-product of some chemical or biological reaction in the water. The scrubbing process caused by the turbulence of aeration physically removes these gases from solution and allows them to escape into the surrounding air.

The second removal process, OXIDATION, can help remove certain dissolved gases and minerals. Oxidation is the chemical combination of oxygen from the air with certain undesirable metals in the water (iron and manganese are the most common). Once oxidized, these substances come out of solution and become very finely divided suspended material in the water. The suspended material can then be removed by filtration.

The efficiency of the aeration process depends almost entirely on the amount of surface contact between air and water. The surface contact is controlled primarily by the size of the water drop or air bubble. For example, the cubic foot of water shown in Figure 2-2A has 6 sq ft of surface area exposed to the air. When the same volume is divided equally into eight pieces, so that each exposed face is

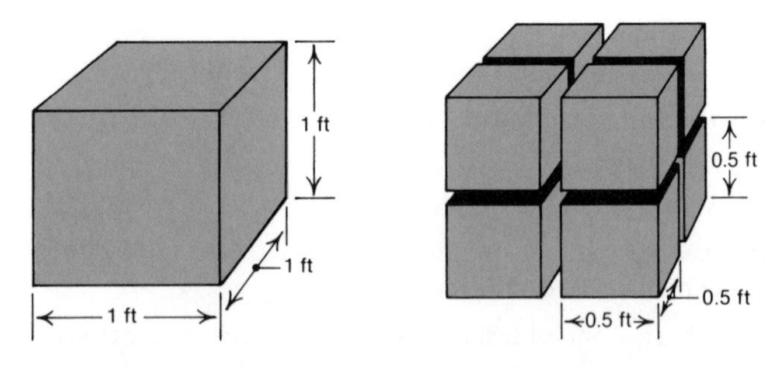

Figure 2-2. Increased Surface Area

1/2 ft by 1/2 ft, as shown in Figure 2-2B, the area exposed to the air increases to 12 sq ft.

If division of the cube continues until each exposed face is 1/100 in. by 1/100 in., the surface area exposed to air will be increased to 7200 sq ft. Because of the increased surface area, the scrubbing action and the oxygen transfer will be far more effective with many small cubes than with one large cube. Similarly, the smaller the water droplets falling through air, or the smaller the air bubbles rising through water, the more efficient the aeration process.

Constituents Affected by Aeration

Aeration of water removes troublesome gases, oxidizes impurities such as iron and manganese so that they can be removed later in treatment, reduces certain types of tastes and odors, and introduces oxygen into the water. The constituents commonly affected by aeration are:

- Carbon dioxide (CO_2)
- Hydrogen sulfide (H_2S)
- Methane (CH_4)
- Iron (Fe)
- Manganese (Mn)
- Various chemicals causing tastes and odors
- Dissolved oxygen (DO).

Carbon dioxide. CARBON DIOXIDE (CO_2) is the gas produced when humans or animals breath (the process of RESPIRATION). It is also produced when fuels are burned, and it is used by plant life in PHOTOSYNTHESIS. Carbon dioxide is very soluble in water, especially in comparison to oxygen. For example, up to 1700 mg/L CO_2 can be dissolved in water at 68 °F (20 °C), whereas only 9 mg/L of O_2 can be dissolved at the same temperature.

Surface waters are usually low in CO_2, in the range of 0 to 5 mg/L. The CO_2 concentration at the water's surface may be very low (in the 0 to 2 mg/L range) due to algae and other plant life using CO_2. However, deep within a lake, the CO_2 concentration may be high (in the 10 to 50 mg/L range) due to the respiration of microscopic animals and lack of abundant plant growth at the lake bottom.

Concentration varies widely, but, as compared to surface waters, ground waters are normally high in CO_2—from one to several hundred milligrams per litre. Deep well waters are usually under 50 mg/L CO_2, whereas shallow well waters may contain 50 to 300 mg/L CO_2. Excessive amounts of CO_2 (above a range of 5 to 15 mg/L) in raw waters can cause three operating problems:

- CO_2 increases the acidity of water, making it corrosive
- CO_2 tends to keep iron in solution, thus making iron removal more difficult
- CO_2 reacts with the lime added to soften water, causing an increase in the cost of softening because of the additional lime that must be added.

Most types of aerators are able to remove CO_2 by the physical scrubbing or sweeping action caused by turbulence. For normal water temperatures and atmospheric composition, aeration can reduce the CO_2 content of the water to as little as 4.5 mg/L. Equilibrium between the CO_2 content of the air and the water prevents further removal.

In addition to removal by aeration, CO_2 can also be removed by extra lime added during chemical softening. Unless the CO_2 concentration is low (below 10 mg/L), it is generally not economical to use lime for removal. For waters with a CO_2 content above 10 mg/L and with an alkalinity less than 100 mg/L, aeration can be used to remove enough CO_2 to significantly reduce the amount of lime needed for further CO_2 reduction. This will result in cost savings in the lime treatment process.

Hydrogen sulfide. HYDROGEN SULFIDE (H_2S) is a poisonous gas and can present dangerous problems in water treatment. Brief exposures to H_2S—less than 30 min—can be fatal if the gas is breathed in concentrations as low as 0.1 percent by volume in air.

Hydrogen sulfide occurs mainly in ground-water supplies. The rotten-egg odor often noticed in well waters is caused by H_2S. Even at concentrations as low as 0.05 mg/L, H_2S in a water supply will disagreeably alter the taste of coffee, tea, ice cubes, or other beverages and foods. The gas (alone or dissolved in water) is corrosive to piping, tanks, water heaters, and any iron, steel, or copper alloy it contacts. Silverware washed in such water may turn black.

Two serious operational problems are caused by H_2S:

- Before disinfecting the water, chlorine reacts with (oxidizes) H_2S. Therefore, to accomplish the desired level of disinfection, more chlorine than would otherwise be necessary is required. This increases the cost of chlorination.
- Since water containing H_2S is corrosive, it will attack (corrode) pipes as well as concrete and metal surfaces in the treatment plant.

The solution to these problems is to remove the H_2S from the water. Since H_2S is very unstable in water, the turbulence created by aeration easily releases it to the atmosphere. However, there must be suitable air movement in the vicinity of the aerator to carry away the released H_2S, otherwise it will accumulate above the water, slowing the removal process and creating a corrosive and hazardous environment. Although aeration removes H_2S primarily by physical scrubbing, a limited amount of H_2S can also be removed by oxidation.

Methane. Commonly called "swamp gas" or natural gas, METHANE (CH_4) can be found in ground water drawn from supplies located near natural gas deposits. Methane is a colorless, odorless, tasteless gas, which is highly flammable and explosive. When mixed with water, methane causes the water to taste like garlic. The gas is only slightly soluble in water and is quite easily removed by aeration.

Iron and manganese. IRON and MANGANESE are two very abundant natural elements in the earth. They are found mostly in ground water, dissolved from various iron and manganese mineral deposits in the earth.

Iron (in the ferrous, or Fe^{+2}, form) and manganese (in the manganous, or Mn^{+2}, form) are objectionable for several reasons.[1] Water containing more than 0.3 mg/L iron stains most things it contacts yellowish- to reddish-brown. This is especially true of plumbing fixtures. At concentrations of 1 mg/L or more, iron gives the water an unpleasant metallic or medicinal taste and can cause turbidity. Iron can deposit on pipe walls, well screens, and valves. Manganese concentrations as low as 0.1 mg/L can produce black stains and cause other problems similar to those caused by iron. Water with both iron and manganese produces stains varying from dark brown to black. Typical consumer complaints are that laundry is stained and that the water is "red" or "dirty."

Iron and manganese problems can usually be controlled by keeping iron concentrations below 0.3 mg/L and manganese concentrations below 0.05 mg/L. Although there are several ways of removing iron and manganese, aeration followed by filtration is a commonly used technique. Aeration provides the dissolved oxygen (DO) needed to convert the iron and manganese from their soluble forms (ferrous, Fe^{+2}; and manganous, Mn^{+2}) to their insoluble forms (ferric, Fe^{+3}; and manganic, Mn^{+3}). It takes about 0.14 mg/L of DO to remove 1 mg of iron and 0.27 mg/L of DO to remove 1 mg of manganese. However, aeration is only the first step. Adequate removal requires a detention time during and after aeration long enough for oxidation to occur completely, followed by filtration (usually pressure filtration) to ensure the removal of the insoluble iron and manganese precipitates.

Tastes and odors. Aeration is effective in removing only those tastes and odors caused by very VOLATILE materials (materials that turn to vapor easily and have low boiling points), or tastes and odors caused by materials that can be readily oxidized. Methane and hydrogen sulfide are two common examples of materials that are removed by aeration. Iron and manganese are examples of easily oxidized minerals.

Many taste and odor problems are thought to be caused by the oils or other by-products that algae produce. Since the oils are much less volatile than the gases previously mentioned, aeration is only partly effective in removing them. On the other hand, odors caused by certain organic industrial wastes can be quite effectively removed by aeration.

Dissolved oxygen (DO). One effect of aeration is to introduce oxygen into the water where it dissolves to become DO. A certain amount of DO is beneficial in that it increases water's palatability by removing the "flat" taste. Too much DO can cause corrosion problems.

The amount of oxygen that can remain dissolved in water depends on the water's temperature—the colder the water the higher the possible concentration of DO. Therefore, depending on the oxygen content and temperature of the water being aerated, oxygen might be either dissolved in or released from the water. For example, water in the lower portions of lakes or reservoirs is often very low in DO. Aerating this water will increase its DO content. Sometimes

[1] *Basic Science Concepts and Applications*, Chemistry Section; Valence, Chemical Formulas, and Chemical Equations (Valence).

water contains very high concentrations of DO, so high that the condition is called SUPERSATURATION. Surface waters containing large algae concentrations are often supersaturated with DO because the algae give off large amounts of oxygen during daylight hours. When supersaturated waters are aerated, the DO levels actually drop since the turbulence created results in the release of the excess DO to the atmosphere. Unless the excess DO is removed intentionally, it can be released later during treatment where it can cause serious operational problems, such as corrosion and air binding of filters. The saturation levels for DO in water at various temperatures are given in Table 2-1.

2-2. Types of Aerators

The general categories of aerators are based on the two main aeration methods: water-into-air and air-into-water. The water-into-air aerator is designed to produce small drops of water that fall through air. The air-into-water aerator creates small bubbles of air that rise through the water being aerated. Both categories of aerators are designed to create a greater amount of contact area between the air and the water than normally occurs.

Within the two main categories, there is a wide variety of aerators. For the removal of a given set of constituents, a certain type of aerator may be more effective than another. The more common types are listed and discussed below, arranged according to whether they are water-into-air or air-into-water aerators. Some aerators operate by a combination of both methods, and these are discussed under the title "combination aerators."

Water-into-air aerators
- Cascade aerator
- Cone aerator
- Slat and coke tray aerator
- Draft aerator
- Spray aerator

Air-into-water aerators
- Diffuser aerator
- Draft-tube aerator

Combination aerators
- Mechanical aerators
- Pressure aerators

Water-Into-Air Aerators

Cascade aerators. A cascade aerator is simply a series of steps that may be designed like a stairway (Figure 2-3) or stacked metal rings (Figure 2-4). The rings in Figure 2-4 are stacked one below the other, surrounding a central vertical feed pipe.

In all cascade aerators, aeration occurs in the splash areas. The aeration action is very similar to that of a natural flowing stream. Splash areas on the inclined

Table 2-1. Oxygen Saturation (or Equilibrium) Levels in Water*

Temperature °C	Saturation Concentration mg/L	Temperature °C	Saturation Concentration mg/L
0	14.60	26	8.09
1	14.19	27	7.95
2	13.81	28	7.81
3	13.44	29	7.67
4	13.09	30	7.54
5	12.75	31	7.41
6	12.43	32	7.28
7	12.12	33	7.16
8	11.83	34	7.05
9	11.55	35	6.93
10	11.27	36	6.82
11	11.01	37	6.71
12	10.76	38	6.61
13	10.52	39	6.51
14	10.29	40	6.41
15	10.07	41	6.31
16	9.85	42	6.22
17	9.65	43	6.13
18	9.45	44	6.04
19	9.26	45	5.95
20	9.07	46	5.86
21	8.90	47	5.78
22	8.72	48	5.70
23	8.56	49	5.62
24	8.40	50	5.54
25	8.24		

*At a total pressure of 760 mm Hg.

Source: *Standard Methods for the Examination of Water and Wastewater.* APHA, AWWA, and WPCF Washington, D.C. (15th ed., 1981).

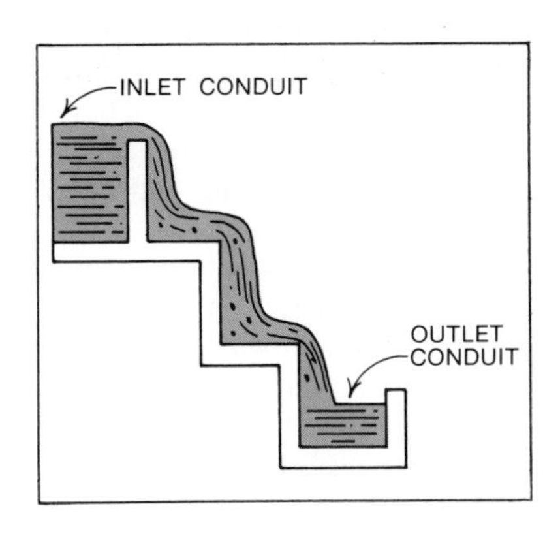

Figure 2-3. Stairway-type Cascade Aerator

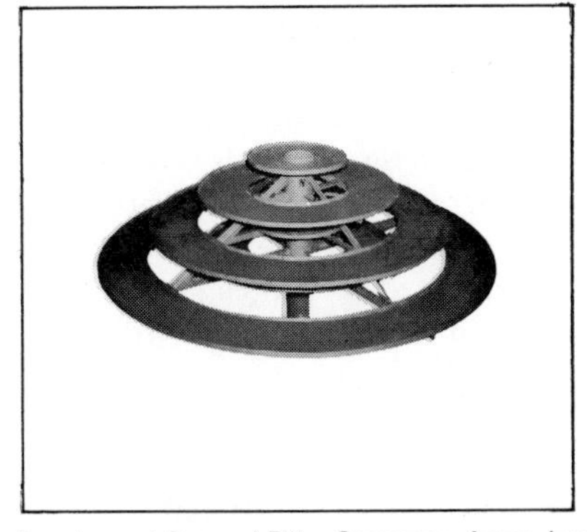

Courtesy of General Filter Company, Ames, Iowa

Figure 2-4. Ring-type Cascade Aerator

cascade are created by placing blocks (riffle plates) across the incline. Cascade aerators can be used to oxidize iron and to partially reduce dissolved gases.

Cone aerators. Cone aerators are similar to the stacked pan cascade type. Air portals draw air into the stacked pans, mixing that air with the falling water. Water enters the top pan through a vertical center feed pipe. The water fills the top pan and begins cascading downward to the lower pans through specially designed cone-shaped nozzles in the bottom of each pan, as shown in Figures 2-5 and 2-6. Cone aerators are primarily used to oxidize iron, although they are also used to partially reduce dissolved gases.

Courtesy of Degremont Inc.
(Water Treatment Handbook, 1979)

Figure 2-5. Cone Aerator

Courtesy of General Filter Company, Ames, Iowa

Figure 2-6. Cone Aerator

Slat and coke tray aerators. The slat and coke tray aerator is very similar in function to the cascade and cone types. As shown in Figure 2-7, they usually consist of three to five stacked trays, which have spaced wooden slats, usually made of redwood or cypress because of their resistance to rot. The trays are filled to about 6 in. (0.15 m) deep with fist-sized pieces of coke. Rock, ceramic balls, limestone, or other materials may be used in place of coke since the primary purpose of these materials is to increase the contact area between air and water and to increase the time of contact.

The aerator is usually constructed with sloping sides, called splash aprons, which are used to protect the splash from wind loss and freezing. Water is introduced into the top, or distributing tray, and moves down through the successive trays, splashing and taking in oxygen each time it hits a layer of coke.

The slat and coke tray aerator is used for oxidation of iron and manganese, and to a limited extent, to lower the concentration of dissolved gases.

Draft aerators. A draft aerator is similar to those already discussed with the added feature of an air flow created by a blower. There are two types of draft aerators: the positive draft type and the induced draft type.

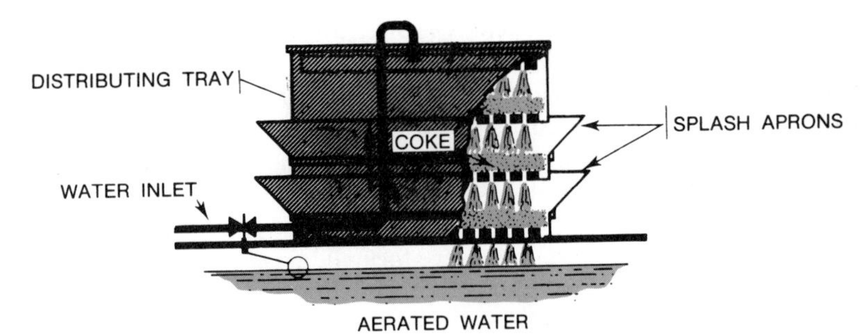

Courtesy of Drew Chemical Corporation.
Copyright © 1977, 1978.

Figure 2-7. Slat and Coke Tray Aerator

As shown in Figure 2-8, a positive draft aerator is a tower of tiered slats (wooden in the illustration) and an external blower, which provides a continuous flow of air. The water is introduced at the top of the tower. As it flows and splashes down through the slats, it is subjected to the high velocity air stream from the blower. The impact caused by the free-falling droplets of water hitting the slats aids in releasing dissolved gases. The high-velocity airflow rushes past the water droplets, carrying away CO_2, methane, or hydrogen sulfide, and constantly renews the supply of dissolved oxygen needed to oxidize iron and manganese.

The induced draft aerator, shown in Figure 2-9, differs from the positive draft in only one major aspect: instead of a bottom-mounted blower forcing air up through the unit, the induced draft aerator has a top-mounted blower, which pulls an upward flow of air from vents located near the bottom of the aeration tower.

Courtesy of General Filter Company, Ames, Iowa

Figure 2-8. Positive Draft Aerator

Courtesy of General Filter Company, Ames, Iowa

Figure 2-9. Induced Draft Aerator

Draft aerators are the most efficient type of aeration for dissolved gas removal and for oxidizing iron and manganese.

Spray aerators. A spray aerator consists of one or more spray nozzles connected to a pipe MANIFOLD. Moving through the manifold under high pressure, the water leaves each nozzle in a fine spray, and falls through the surrounding air, creating a fountain effect. When relatively large high-pressure nozzles are used, the resulting fountain effect can be quite attractive. It is not unusual to find this type of aeration device located in a decorative setting at the entrance to the water treatment plant.

Sometimes spray aeration takes place inside a structure known as a SPRAY TOWER. The spray tower (Figure 2-10) protects the spray from wind-blown losses and also reduces freezing problems.

Spray aerators are sometimes combined with cascade and draft aerators, as shown in Figure 2-11, to capture the best features of each, depending on the application.

In general, spray aeration is successful in oxidizing iron or manganese and is very successful in increasing the DO level of water.

Air-Into-Water Aerators

Diffuser aerator. A typical diffuser aeration system (Figure 2-12) consists of an aeration basin or tank, constructed of steel or concrete. The basin is equipped with compressed air piping, manifolds, and DIFFUSERS. The piping is usually steel or plastic, and the individual diffusers range from plastic or metallic devices, to porous ceramic plates, to simple holes drilled into the manifold pipe.

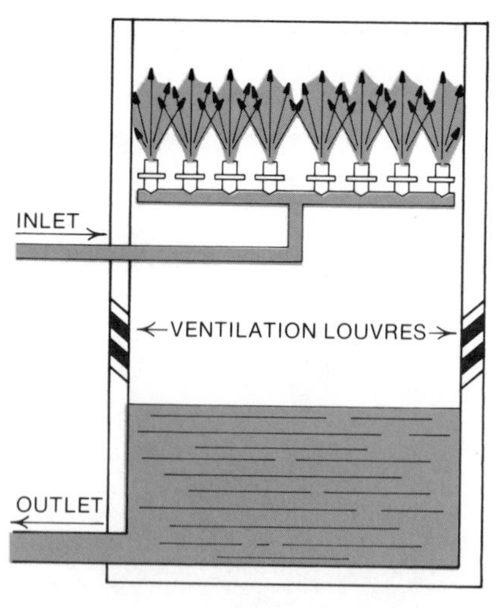

Courtesy of Degremont Inc.
(Water Treatment Handbook, 1979)

Figure 2-10. Spray Tower

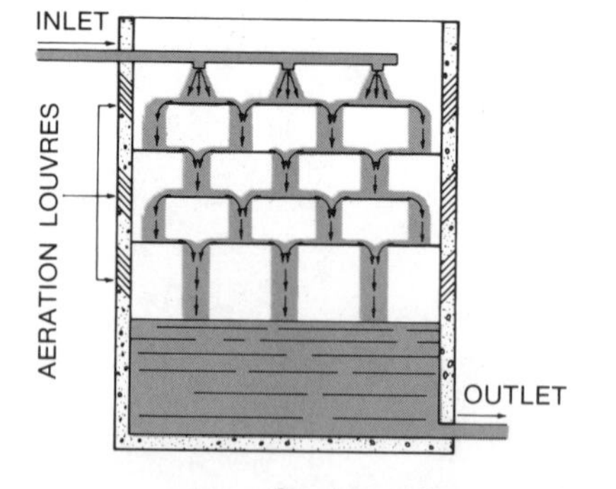

Courtesy of Degremont Inc.

Figure 2-11. Spray Aerator Combined with Cascade and Draft Aerators

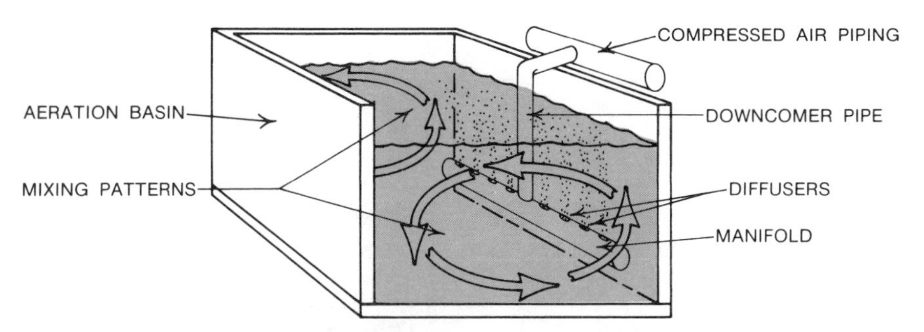

Figure 2-12. Diffuser Aeration System

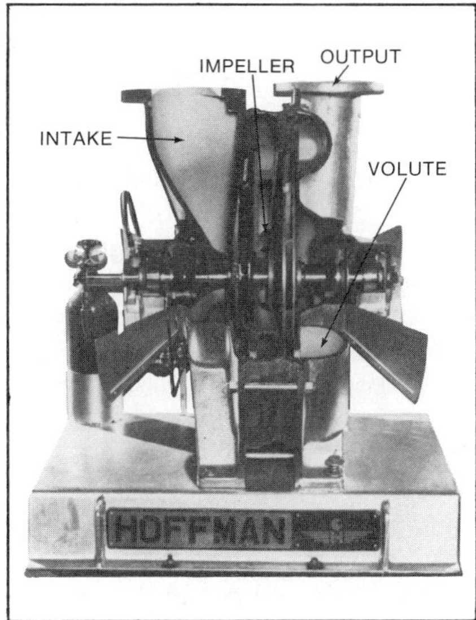

Figures courtesy of Hoffman Air and Filtration Systems, Division of Clarkson Industries, Inc.

Figure 2-13. Centrifugal Blower

Figure 2-14. Cutaway View of Centrifugal Blower

The air diffuser releases tiny bubbles of compressed air into the water, usually near the bottom of the aeration basin. The bubbles rise slowly but TURBULENTLY through the water, setting up a rolling-type mixing pattern. At the same time, each air bubble gives up some oxygen to the surrounding water. This form of aeration is used primarily to increase the DO content in order to prevent tastes and odors from occurring.

The essential piece of equipment in every diffused air system is the air compressor, or blower, of which there are basically two types: (1) centrifugal blowers and (2) positive displacement blowers. A typical unit of the most common type, the centrifugal blower, is shown in Figure 2-13; the cutaway view in Figure 2-14 identifies the major component parts.

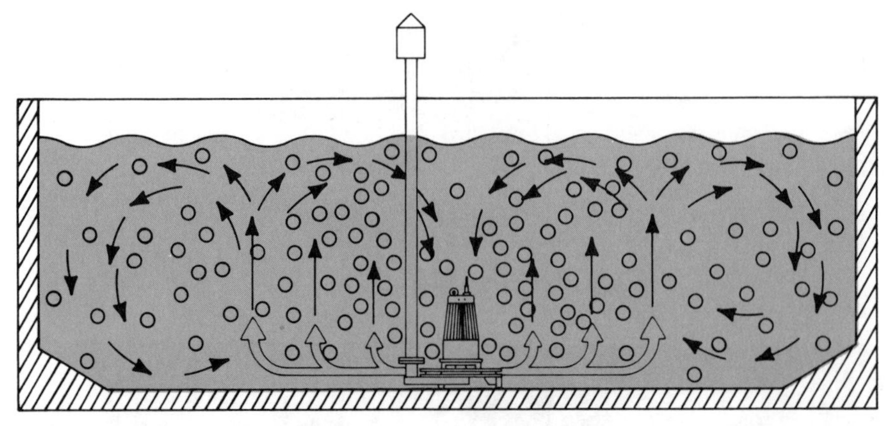

Figure 2-15. Draft-tube Aerator

Draft-tube aerators. The draft-tube aerator is a submersible pump equipped with a draft tube (air intake pipe), as illustrated in Figure 2-15. A partial vacuum is created at the eye of the spinning turbine impeller, causing air to enter through the draft tube and water to enter through the water intake. The air and water are mixed by the turbine impeller and then discharged into the aeration basin. The draft-tube aerator represents a convenient, low-cost method for adding aeration capability to an existing basin or tank.

Combination Aerators

Mechanical aerators. A mechanical aerator consists of a propeller-like mixing blade mounted on the end of a vertical shaft driven by a motor. By rapidly rotating the mixing blade in the water, violent mixing of air and water is created.

Figure 2-16 shows a side-by-side comparison of the four general types of mechanical aerators:

- Surface aerators (water-into-air type)
- Submerged aerators (air-into-water type)
- Combination aerators (combination type)
- Draft-tube surface aerators (water-into-air type).

The purpose of a surface aerator is to draw water into the blade of the aerator and throw that water into the air in tiny droplets, so the water can pick up oxygen. The mixing pattern of a typical surface aerator is shown in Figure 2-17. An operating surface aerator is shown in Figure 2-18. Notice the violent surface white water, necessary for efficient oxygen transfer and release of unwanted gases, tastes, and odors.

A submerged aerator usually consists of two components: (1) a submerged air diffuser (sparger type) and (2) a submerged blade that mixes the air into the

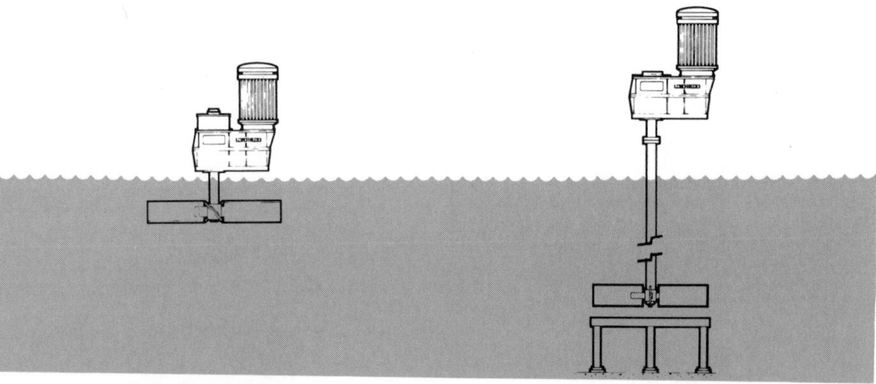

A. Surface Aerator

B. Submerged Aerator

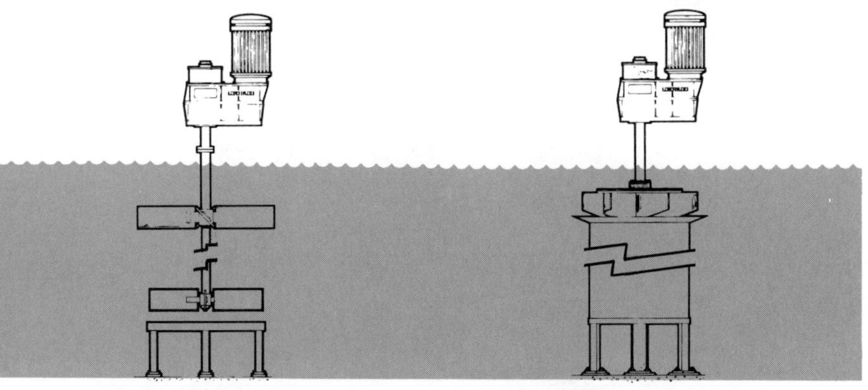

C. Combined Aerator

D. Draft-Tube Surface Aerator

Courtesy of Philadelphia Mixers Division, Philadelphia Gear Corporation

Figure 2-16. Four Types of Mechanical Aerators

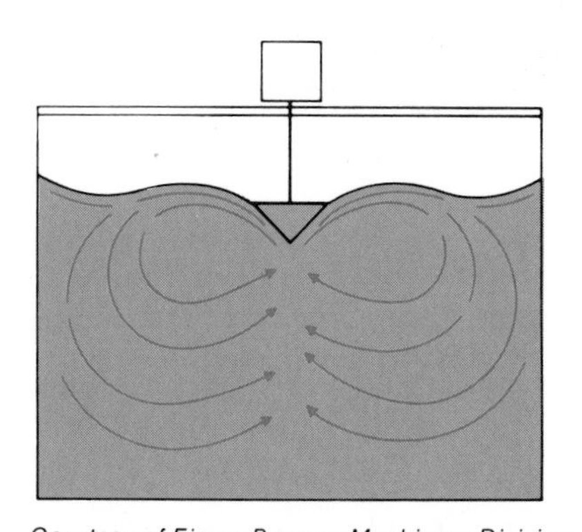

Courtesy of Eimco Process Machinery Division, Envirotech Corporation

Figure 2-17. Mixing Pattern of Surface Aerator

Courtesy of Lightnin Mixers and Aerators, Mixing Equipment Company, Inc.

Figure 2-18. Operating Surface Aerator

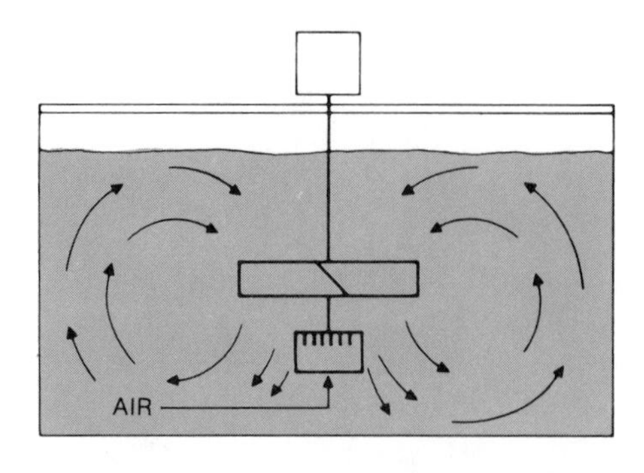

Courtesy of Eimco Process Machinery Division, Envirotech Corporation

Figure 2-19. Mixing Pattern of Submerged Turbine Aerator

Courtesy of Philadelphia Mixer and Aerators, Mixing Equipment Company, Inc.

Figure 2-20. Functioning Submerged Turbine Aerator

Courtesy of Lightnin Mixers and Aerators, Mixing Equipment Company, Inc.

Figure 2-21. Operating Submerged Turbine Aerator

water. Figure 2-19 shows the mixing pattern of a submerged turbine aerator with air entering the water just below the propeller blade. Note that the mixing pattern is opposite to the surface aerator pattern. Figure 2-20 shows more clearly how the air is introduced by the sparger and how the submerged turbine distributes the air.

An operating submerged turbine aerator is shown in Figure 2-21. Notice the relative calm at the water surface as compared with surface aerators. Due to this relative calm, submerged aerators are best used to increase DO levels; this is in contrast to surface aerators whose greater turbulence removes unwanted gases and incidentally acts as a very effective mixer.

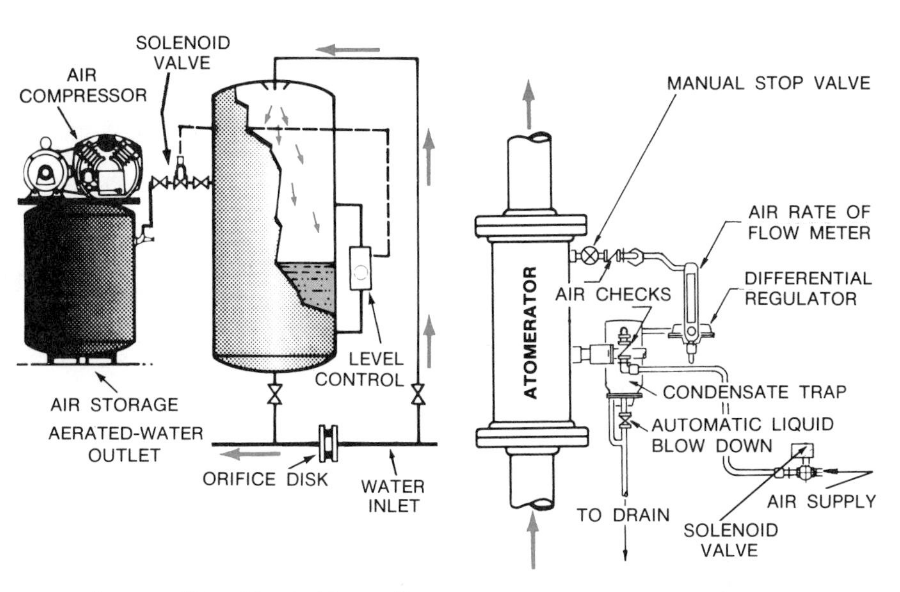

Courtesy of Drew Chemical Corporation, Copyright © 1977, 1978.

Figure 2-22. Pressure Aerator with Pressure Vessel

Courtesy of General Filter Company, Ames, Iowa

Figure 2-23. Pressure Aerator with Air Diffused Directly into Pressure Pipeline

Combination mechanical aerators, like the one diagrammed in Figure 2-16A, offer the features of both the surface and submerged types. They can be used to oxidize iron and manganese and to remove unwanted gases, tastes, and odors.

The draft tube, diagrammed in Figure 2-16D, is used to ensure better mixing when surface aerators are installed in deep aeration basins. The draft tube is open at both ends. When the surface aerator is spinning, water is drawn from the very bottom of the basin into the bottom of the draft tube. The water rises up the tube into the impeller and is thrown out onto the water surface. In this way, the draft tube improves the bottom-to-top mixing and turnover.

Pressure aerators. There are two basic types of pressure aerators. The type diagrammed in Figure 2-22 consists of a closed tank continuously supplied with air under pressure. The water to be treated is sprayed into the high pressure air, allowing the water to pick up DO quickly. Aerated water leaves through the bottom of the tank and moves on to further treatment. Aerators of this type are primarily used to oxidize iron and manganese for later removal by filtration.

The second type of pressure aerator is diagrammed in Figure 2-23. In this type, instead of using a pressure vessel, air is diffused directly into a pressure pipeline. The diffuser inside the special aeration pipe section distributes fine air bubbles into the flowing water. As in any pressure aerator, the higher the pressure, the more oxygen will dissolve in water. The more oxygen there is in solution, the quicker and more complete will be the oxidation of the iron and manganese.

2-3. Common Operating Problems

As discussed throughout this module, aeration easily increases the DO content of water, particularly water that enters the process with a relatively low DO content. In fact, too much DO can be added—this results in SUPERSATURATION, which can cause major operating problems:

- Corrosion of tanks and pipelines
- Floating floc in clarifiers
- False clogging of filters (air binding) as the water warms and releases DO.

In addition to supersaturation, other operating problems in the aeration process are:

- Removal of H_2S
- Algae
- Clogged diffusers
- Wasted energy.

Corrosion

A certain amount of DO is a necessary condition in raw and treated waters. However, DO is also a cause of corrosion. Corrosion can occur whenever water and oxygen come into contact with metallic surfaces. Generally, the higher the DO concentration, the more rapid the corrosion.

The solution to the problem of excessive DO concentration is simply to not over-aerate; unfortunately, there is no definite rule as to what constitutes over-aeration. The amount of aeration needed will vary from plant to plant, and may vary seasonally as water quality changes. The correct amount must be determined from experience. However, it is usually not advisable to supersaturate the water with DO. The saturation levels for DO in water at various temperatures are given in Table 2-l.

There are two things that can be done to prevent or retard corrosion. First, apply and maintain adequate protective coatings on all exposed metallic surfaces. The application and reapplication of paint and other protective coatings is a vital part of any preventive maintenance program. Second, operate the aeration process so that it provides an adequate but not excessive level of DO. Avoid supersaturating the water with oxygen; that is, keep the DO below the oxygen saturation values in Table 2-1. As a general rule, DO concentrations of 2 to 4 mg/L should be acceptable, depending on the situation.

Floating Floc in Clarifiers

Excess aeration, particularly at supersaturated levels, can cause serious sedimentation problems. Small bubbles of excess air will come out of solution and attach to the particles of floc in the clarifier, causing the particles to float. Instead of settling, the floc particles will float to the surface and flow out of the clarifier with the treated effluent. This destroys the efficiency of the sedimentation process and creates a tremendous load on the filters.

False Clogging of Filters (Air Binding)

In the same way that bubbles of excess air can attach to floc particles, they can also attach to the media in a filter. This will occur if the water warms as it passes through the filter, releasing the air that was in solution. The process can continue until the spaces between the media particles begin to fill with air bubbles. This is called AIR BINDING—it causes the filter to behave as though it were clogged and in need of backwashing. Serious disruption of the filter media can also occur as the air bubbles burst within the bed.

Hydrogen Sulfide Removal

HYDROGEN SULFIDE is most efficiently removed by the physical scrubbing action of aeration, not by oxidation. Which of the two removal processes actually takes place depends on the pH of the water. At a pH of 6 or less, hydrogen sulfide exists primarily as the gas H_2S. In this form, it is easily scrubbed away by aeration. However, at pH values of 8 or above, the H_2S gas IONIZES and the ionized form (HS^- and HS^{-2}) cannot be removed by aeration.[2] In fact, the oxygen provided by aeration reacts with the ionized H_2S to release the element sulfur, which occurs as a fine colloidal particle and gives water a milky-blue turbidity.

One of the better procedures for the elimination of hydrogen sulfide is to lower pH to 6 or less before beginning aeration. This will favor H_2S removal by scrubbing. The pH can be lowered by bubbling CO_2 gas into the water. After the H_2S is removed, additional aeration will remove any remaining CO_2.

Algae

In warmer climates, the slat and coke tray and positive draft aerators may be located in the open air and exposed to direct sunlight. The wetted surfaces of these aerators create excellent environments for growth of algae and slime. Although it is not practical to completely eliminate these problems, growth rates can be greatly reduced by providing a roof or canopy to shade the aerator from direct sunlight. The growths can also be controlled by periodic treatment with copper sulfate or other suitable chemicals.

Clogged Diffusers

Air diffusers can become partly clogged, either from dust in the air or from oil, debris, or chemical deposits that can collect around the diffuser opening. Clogging can be minimized by

- Maintaining clean air filters
- Not over-lubricating blowers
- Preventing backflow of water into diffusers.

[2] *Basic Science Concepts and Applications*, Chemistry Section, The Structure of Matter (Ions).

Techniques for cleaning diffusers vary with the type of diffuser. Fine-bubble, porous ceramic types can be placed in a furnace to burn away trapped particulates. Both fine- and large-bubble types can be cleaned with a brush and detergent. Manufacturers of ceramic diffusers should be consulted to determine the best cleaning procedure to follow.

Energy Consumption

Regardless of the method used, the aeration process will require some supply of energy: power supply may be needed to pump the water before or after cascade aeration; diffused aeration requires power to drive the blowers; and mechanical aeration requires power to drive the aerator and power to drive the blowers.

Not over-aerating, keeping motors and blowers properly maintained, and keeping air diffusers clean and free of debris may realize savings of 10 to 100 hp, depending on system size. At 3 to 6 cents per kilowatt hour, a reduced load of 50 hp could result in a savings of over $1000 per month.[3] With energy costs escalating, every effort should be made to achieve energy conservation, efficiency, and savings.

2-4. Operational Control Tests

There are three basic control tests involved in operating the aeration process:

- DO
- pH
- Temperature.

A description of each test can be found in *Introduction to Water Quality Analyses,* Physical/Chemical Tests Module.

The DO test is used to monitor the amount of oxygen being dissolved in the water. Knowing the DO concentration needed and monitoring the amount of DO present will avoid over- or under-aeration.

The pH of water can be used as an indicator of CO_2 removal (pH increases as CO_2 is removed), and it can also be used to monitor the effective pH range for H_2S, iron, and manganese removal. The best pH range for H_2S scrubbing is 6 or less, whereas iron and manganese are best treated in a pH range of 8 to 9.

The amount of oxygen which will dissolve in water (the saturation concentration) varies depending on water temperature, increasing as the water temperature drops. As water temperature drops, the operator must adjust the aeration process to maintain the correct level of DO.

[3] *Basic Science Concepts and Applications*, Hydraulics Section, Pumping Problems (Pumping Costs).

Additional control tests for the following are used to measure how efficient or effective the aeration process is in removing troublesome constituents:

- Iron
- CO_2
- Manganese
- Taste and odor.

Test frequency depends both on the variability of raw-water quality and on the characteristic measured. For example, well water tends to be very constant in quality and very slow to change, so daily testing may not be necessary.

Surface-water quality is subject to more frequent change. Temperature, DO, pH, CO_2, and tastes and odors can vary daily. Therefore, sampling and testing frequencies should be set up so that all significant changes can be detected and monitored.

In addition to these specific control tests, it is advisable to make frequent visual inspections of the process. Clogged diffusers, for example, are common problems that are most easily identified visually.

2-5. Safety and Aeration

Hydrogen sulfide is a poisonous gas, which is colorless, flammable, and explosive. The gas kills by paralyzing the respiratory center; a few minutes of exposure to a concentration of H_2S as low as 0.1 percent by volume in air is fatal. Higher concentrations, above 4.3 percent, are explosive. A very dangerous characteristic of H_2S is that its foul, rotten-egg odor is not evident at the dangerous concentrations, because such concentrations paralyze the sense of smell (the olfactory nerves).

METHANE, or natural gas, is a colorless, odorless, and tasteless gas, which is flammable and highly explosive. If the scrubbed gases containing methane are allowed to accumulate in a confined space, an explosion or fire can result. Methane can also kill by simple asphyxiation (suffocation); and since the gas has no color, taste, or odor, the victim can be completely unaware of its presence. Because methane is odorless, the natural gas piped to homes, industry, and commerce is treated with a strong odor-causing chemical, allowing leaks and accidental discharges to be easily detected.

The safest way to deal with hydrogen sulfide and methane gases is to make sure that the aeration process is well ventilated and that there are no locations within the process where gases could accumulate. Methane is lighter than air and will accumulate at the top of an enclosure. Hydrogen sulfide is heavier than air and will collect at the bottom of confined areas. Ideally, aerators should be located in an open area with good ventilation. Where aerators must be enclosed due to climatic conditions, the enclosure must be well ventilated at all times.

Selected Supplementary Readings

Hardenbergh, W.A. & Rodie, E.R. *Water Supply and Waste Disposal.* International Textbook Company, Scranton, Penn. (1961).

Hartenstein, Alan. Fiberglass Aerators in Florida. *OpFlow*, 3:5:1 (May 1977).

Introduction to Water Quality Analyses. AWWA, Denver, Colo. (1982).

Simplified Procedures for Water Examination. AWWA Manual M12. AWWA, Denver, Colo. (1977).

Treating for Tastes and Odors in Drinking Water—Destruction and Removal. *OpFlow*, 4:8:1 (Aug. 1978).

Water Quality and Treatment. AWWA Handbook. McGraw-Hill Book Company, New York. (3rd ed., 1971). Chap. 2.

Water Treatment Plant Design. AWWA, Denver, Colo. (1969). Chap. 4

Glossary Terms Introduced in Module 2

(Terms are defined in the Glossary at the back of the book.)

Aeration
Air binding
Carbon dioxide
Diffuser
Hydrogen sulfide
Ionize
Iron
Manganese
Manifold
Methane
Oxidation
Photosynthesis
Respiration
Spray tower
Supersaturation
Turbulently
Volatile

Review Questions

(Answers to Review Questions are given at the back of the book.)

1. Aeration is used to remove certain constituents in the water. List the two removal processes.

2. Identify at least four constituents removed by aeration.

3. What are some of the operational problems associated with each of the constituents listed in Question 2?

4. Name at least three types of aerators.

5. List at least five common operating problems associated with the aeration process and briefly describe the reason for each of these problems.

6. For each of the operating problems listed in Question 5, what corrective procedures should be taken?

7. List three control tests associated with the aeration process.

8. Why is good ventilation extremely important in the aeration process?

9. Is hydrogen sulfide heavier or lighter than air, and why is this important?

Study Problems and Exercises

1. Select one type of aerator discussed in this module. Investigate the unit and prepare a report on its routine operation and maintenance requirements. Your investigation should include the review of manufacturer's technical bulletins and information.

2. Your town has just drilled a new well to meet increased demand for water. The water from the new source has an iron content of 3 mg/L. You have been asked for a recommendation on the type of treatment to be used to remove the iron. Prepare a report presenting your recommendation for the selection of an aerator and state what factors you took into consideration insofar as the operation and maintenance of the unit is concerned.

3. In connection with Question 2 above, list and discuss other processes that could be used to reduce the iron content of the water.

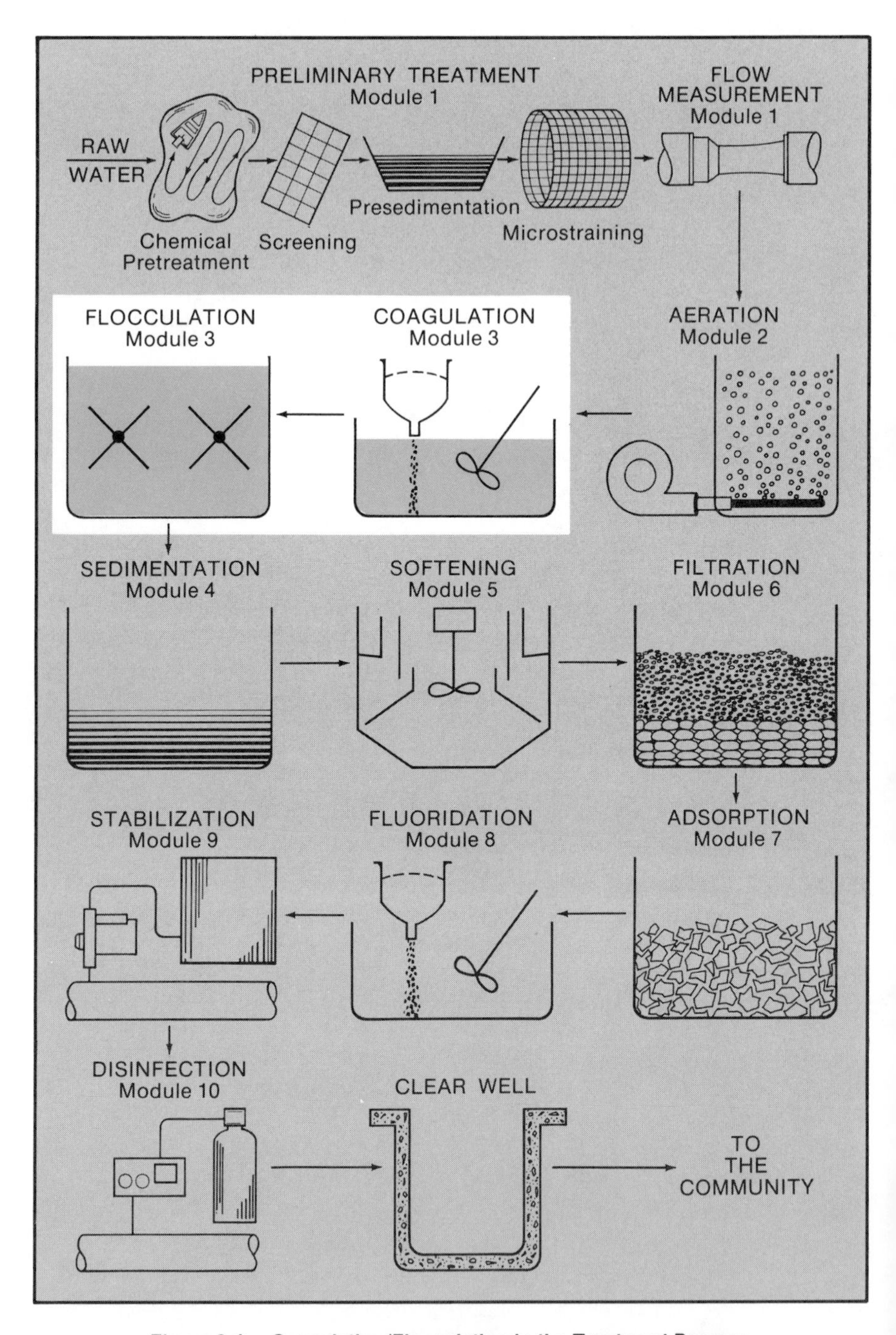

Figure 3-1. Coagulation/Flocculation in the Treatment Process

Module 3

Coagulation/Flocculation

The coagulation/flocculation process (Figure 3-1) helps remove nonsettleable solids. COAGULATION involves the feeding and rapid mixing of one or more chemical COAGULANTS into the water, thereby beginning the formation of particles called FLOC. FLOCCULATION, which partly overlaps the coagulation process, requires the gentle mixing of the water and coagulants for a period of time in order to form larger, heavier, more settleable floc particles. These floc particles are a combination of the coagulant and the particles in the water; they are larger and heavier than the original particles and considerably easier to remove in later treatment processes such as sedimentation and filtration.

This module describes the principles involved in coagulation and flocculation, the common coagulants and coagulant aids in use, the kinds of facilities and equipment used in the coagulation/flocculation process, the operating procedures, and the common operating problems associated with the process. The module concludes with a discussion of the safe handling and use of coagulant chemicals, particularly alum.

After you have completed this module, you should be able to

- Describe the principles involved in coagulation and flocculation.
- List the common chemical coagulants, their advantages and disadvantages.
- Define the following:
    - Turbidity
    - Coagulation
    - Coagulant aid
    - Flash mixing
    - Flocculation
    - Floc.

- Explain why good mixing is important.
- Identify two common operating problems associated with the coagulation/flocculation process, and describe how these problems might be resolved.
- List reasons for using coagulant aids.
- List and briefly describe the purpose of each test used to monitor the effectiveness of the coagulation/flocculation process.
- Describe safe handling practices for coagulant chemicals.

3-1. Description of Coagulation/Flocculation

The coagulation/flocculation process is necessary in water treatment primarily because of NONSETTLEABLE SOLIDS, particles too small to be effectively removed by later treatment processes such as sedimentation and filtration. These nonsettleable solids can be converted into larger and heavier settleable solids by physical–chemical changes brought about by adding and mixing chemical coagulants into the raw water. The settleable solids can then be removed by the sedimentation and filtration processes.

Basic Principles

In order to understand how the coagulation/flocculation process works, it is important to know that nonsettleable particles resist settling for two basic reasons:

- Particle size
- Natural forces between particles.

Particle Size. Untreated, natural water contains three types of nonsettleable solids. From largest to smallest, as shown in Table 3-1, these particles are

- Suspended
- Colloidal
- Dissolved solids.

SUSPENDED SOLIDS are particles carried along (held in suspension) by the natural action of flowing water. The smallest suspended solids (less than 0.01 mm) do not settle quickly and for purposes of water treatment are called nonsettleable. Larger suspended solids (greater than 0.01 mm) are referred to as settleable solids since they will settle unaided to the bottom of a container or sedimentation basin within four hours.

Examples of COLLOIDAL SOLIDS include fine silts, bacteria, color-causing particles, and virus. These colloids do not settle in a reasonable period of time (Table 3-2). Although individual colloidal solids cannot be seen with the naked eye, their combined effect is often seen as color or turbidity in water. These particles are small enough to pass through later treatment processes if not properly coagulated and flocculated.

Table 3-1. Size Range of Solids

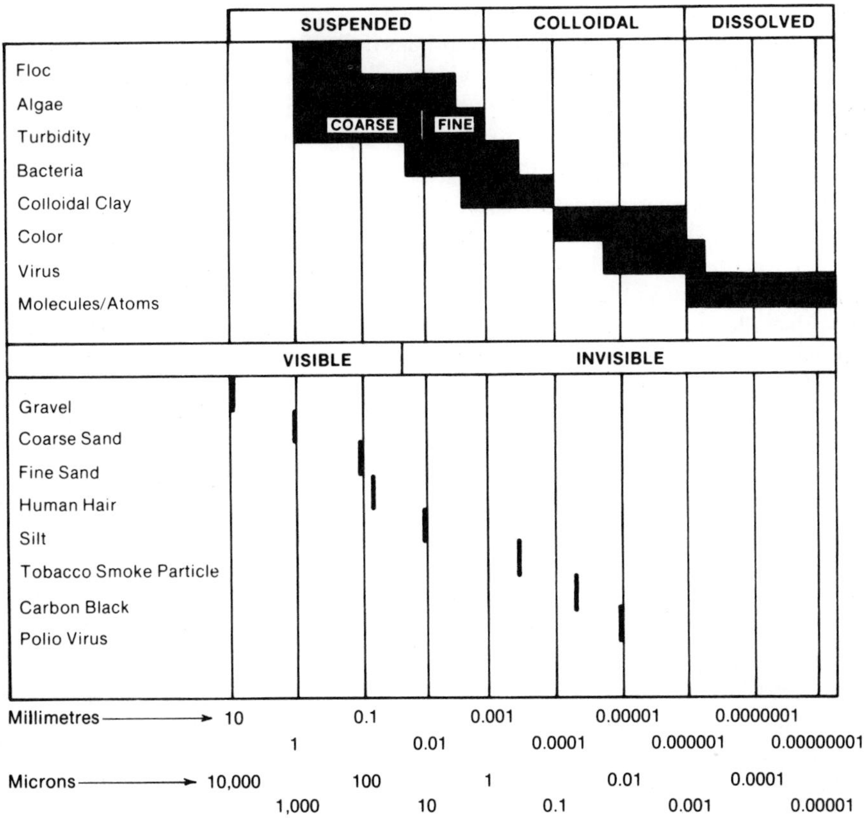

SUSPENDED | COLLOIDAL | DISSOLVED

COARSE | FINE

Floc
Algae
Turbidity
Bacteria
Colloidal Clay
Color
Virus
Molecules/Atoms

VISIBLE | INVISIBLE

Gravel
Coarse Sand
Fine Sand
Human Hair
Silt
Tobacco Smoke Particle
Carbon Black
Polio Virus

Millimetres → 10, 1, 0.1, 0.01, 0.001, 0.0001, 0.00001, 0.000001, 0.0000001, 0.00000001

Microns → 10,000, 1,000, 100, 10, 1, 0.1, 0.01, 0.001, 0.0001, 0.00001

Table 3-2. Natural Settling Rates for Small Particles

Particle Diameter mm	*Representative Particle*	*Time Required to Settle in 1 ft (0.3 m) Depth*
		Settleable
10	Gravel	0.3 sec
1	Coarse sand	3 sec
0.1	Fine sand	38 sec
0.01	Silt	33 min
		Considered nonsettleable
0.001	Bacteria	55 hour
0.0001	Color	230 day
0.00001	Colloidal particles	6.3 year
0.000001	Colloidal particles	63-year minimum

Source: *Water Quality and Treatment.* AWWA, Denver, Colo. (3rd ed., 1971)

DISSOLVED SOLIDS are any organic or inorganic matter—such as salts, chemicals of plant or animal origin, or gases—that is dissolved in water. A dissolved solid is the size of a molecule and is invisible to the naked eye. Most of the trace metals and organic chemicals found in water are dissolved. They are nonsettleable and can cause public health and other problems such as taste, color, or odor. Unless converted to a precipitate by chemical or physical means, they cannot be removed from the water.

One of the main reasons that nonsettleable particles resist settling is their small size. Consider this example: A particle of coarse sand that is the shape of a cube, 1 mm on a side, could be expected to settle quickly, dropping about 1 ft (0.3 m) every 3 sec (see Table 3-2). Now suppose the particle of sand is ground into many smaller particles, each a cube 0.000001 mm on a side. (This grinding simulates the natural forces of erosion.) The weight of all the ground particles together is the same as the original grain of coarse sand, but the exposed surface area has increased from 6 mm^2, the area of the head of a large pin, to 6 m^2, about the area of two pool tables. (See Aeration Module, Figure 2-2, showing another example of subdividing a particle.) The increase in surface area causes a tremendous increase in the forces (often called drag forces) that resist settling. According to Table 3-2, instead of settling 1 ft (0.3 m) in 3 sec, it would now take about 60 years for these tiny particles to settle the same 1-ft (0.3-m) depth—the increase in settling time is due entirely to decrease in the particle size. To promote settling, the particles must be brought together to form larger particles that settle easily and quickly.

Natural forces. Particles in water usually carry a negative electrical charge. Just as like poles of a magnet repel each other, there is a repelling force between any two particles of like charge. In water treatment, this natural repelling electrical force is called ZETA POTENTIAL. The force is strong enough to hold the very small, colloidal particles apart and keep them in suspension.

The VAN DER WAALS FORCE exists between all particles in nature and tends to pull any two particles together. This attracting force acts opposite to the zeta potential. As long as the zeta potential is stronger than the van der Waals force, the particles will stay in suspension.

Effect of coagulation/flocculation. The coagulation/flocculation process neutralizes or reduces the zeta potential of nonsettleable solids so that the van der Waals force of attraction can begin pulling particles together. The nonsettleable particles are then able to gather into small groups of MICROFLOC, as shown in Figure 3-2. These particles, though larger than the original colloids, are held together weakly; individual microfloc particles are invisible to the naked eye, and are still nonsettleable. The gentle stirring action created by flocculation brings the floc particles together to form large and relatively heavy floc particles, which can easily be settled or filtered. The jellylike floc particles are usually visible, and will look like small tufts of cotton or wool.

Coagulants and Coagulant Aids

Coagulants. Since most of the troublesome particles to be removed from water are negatively charged, the coagulants used in water treatment normally

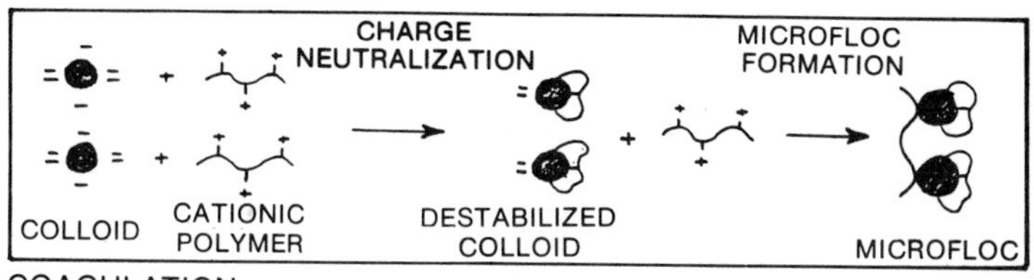

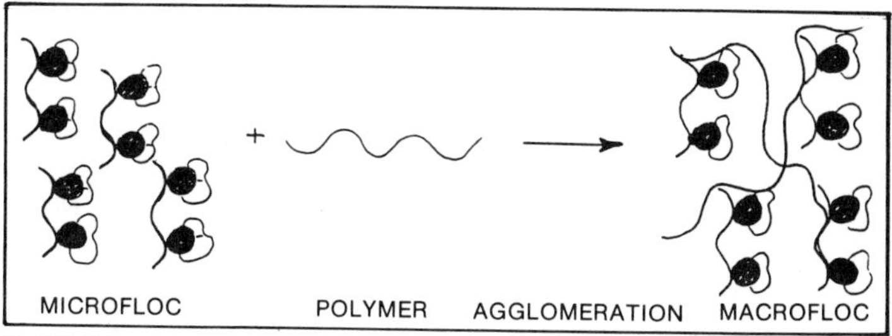

Figure 3-2. Microfloc

consist of positively charged ions. The positive charges neutralize the negative charges and promote coagulation.

Some coagulants contain ions with more positive charges than others. Those consisting of TRIVALENT IONS (ions having three positive charges, such as aluminum, Al^{+3}, and Iron, Fe^{+3}) are 700 to 1000 times more effective as coagulants than MONOVALENT IONS (ions having just one positive charge, such as sodium, Na^{+}) and 50 to 60 times more effective than BIVALENT IONS (those having two positive charges, such as calcium, Ca^{+2}).

The most commonly used coagulants in water treatment are ALUMINUM SULFATE (usually called ALUM), $Al_2(SO_4)_3$; and FERRIC SULFATE, $Fe_2(SO_4)_3$. When either of these two coagulants dissolve and ionize in water, they form the trivalent ions of aluminum or iron (Al^{+3} or Fe^{+3}). Five of the more common coagulants are shown in Table 3-3, with typical dosage ranges. Table 3-4 summarizes some of the common combinations of coagulants.

Since alum is used the most, it is important to understand how it promotes settling to help remove bacteria and other particles that cause turbidity, taste, odor, and color. Alum promotes settling through the coagulation/flocculation process as follows:

- Alum added to the raw water reacts with the alkalinity naturally present, or the alkalinity added (lime, soda ash, etc.) to form jellylike floc particles of aluminum hydroxide, $Al(OH)_3$. (See *Introduction to Water Quality Analyses,* Physical/Chemical Tests Module—Alkalinity.) A certain level of alkalinity is necessary for the reaction to occur.[1]

[1] *Basic Science Concepts and Applications*, Chemistry Section, Chemistry of Treatment Processes (Coagulation).

Table 3-3. Common Coagulants and Doses

Coagulant	*Chemical Formula*	*Typical Dose Range* mg/L
Aluminum sulfate	$Al_2(SO_4)_3$	15–100
Copper sulfate	$CuSO_4$	5–20
Ferric sulfate	$Fe_2(SO_4)_3$	10–50
Ferrous sulfate	$FeSO_4$	5–25
Sodium aluminate	$NaAlO_2$	5–50

Table 3-4. Coagulants in Combination

Coagulants	*Typical Dose* ratio of the first to the second
Aluminum sulfate + caustic soda	3:1
Aluminum sulfate + hydrated lime	3:1
Aluminum sulfate + sodium aluminate	4:3
Aluminum sulfate + sodium carbonate	1:1 to 2:1
Copper sulfate + hydrated lime	3:1
Ferric sulfate + hydrated lime	5:2
Ferrous sulfate + hydrated lime	4:1
Ferrous sulfate + chlorine	8:1
Sodium aluminate + ferric chloride	1:1

- The positively charged trivalent aluminum ion neutralizes the negatively charged particles of color or turbidity. This occurs within one or two seconds after the chemical is added to the water, which is why rapid, thorough mixing is critical to good coagulation. This neutralization of electrical charges marks the beginning of coagulation.

- At this time, the particles begin to come together to form larger particles.

- The floc that is first formed is made up of small particles (microfloc) which still have a positive charge from the coagulant added. They continue to neutralize negatively charged particles, and in this way become neutral particles.

- Finally, the microfloc particles begin to collide and stick together (agglomerate) to form larger settleable floc particles. This process is called flocculation.

There are many physical and chemical factors that affect the success of a particular coagulant, including mixing conditions and the pH, alkalinity, turbidity, and temperature of the water. For example, alum works best in a pH range of about 4.5 to 8.5. If alum is used outside this range, the floc either does not form completely, or it forms and then is dissolved due to the high or low pH.

Ferric sulfate can operate effectively over the wider pH range of 3.5 to 9.0. However, ferric sulfate is corrosive and requires special facilities for storage and handling.

Both alum and ferric sulfate are affected by the alkalinity of the raw water. The floc particles that are formed—$Al(OH)_3$ for alum and $Fe(OH)_3$ for ferric sulfate—need the OH, or hydroxyl, portion of the chemical compounds that make up alkalinity in the water. Therefore, if the alkalinity in the water is not high enough, an effective floc will not form.

Increases in turbidity, temperature, and mixing energy can also improve coagulation. There are many other controlling factors that are not known or entirely understood. As a result, the chemical coagulant and coagulant dose selected should be based on tests such as the jar test (see *Introduction to Water Quality Analyses,* Physical/Chemical Tests Module—Jar Test). This test evaluates the actual performance of coagulants at various concentrations in a particular water. The test is essential since waters from different sources or samples from the same source at different times of the year seldom respond to the same coagulant dosages, no matter how close their chemical makeup seems.

A general range of dosages found effective for various coagulants is shown in Table 3-3. Using this type of general information and the jar test procedure, it is the operator's responsibility to select the best coagulant and the most effective dosage.

Coagulant Aids. A COAGULANT AID is a chemical added during coagulation to improve coagulation; to build stronger, more settleable floc; to overcome temperature drops that slow coagulation; to reduce the amount of coagulant needed; and to reduce the amount of sludge produced. A key reason for using aids is to decrease the amount of alum used. This in turn decreases the amount of alum sludge produced. Since alum sludge is very difficult to dewater or dispose of, coagulant aids can significantly reduce sludge handling and disposal problems. There are three general types of coagulant aids:

- Activated silica
- Weighting agents and adsorbents
- Polyelectrolytes.

ACTIVATED SILICA has been used as a coagulant aid with alum since the late 1930s and remains in wide use today. Used properly in dosages from 7 to 11 percent of the coagulant dose, activated silica will increase the rate of coagulation, reduce the coagulant dose needed, and widen the pH range for effective coagulation. Activated silica is a chemical that is prepared by the operator at the plant. The chemical actually delivered to the plant is sodium silicate, Na_2SiO_3. The operator "activates" the sodium silicate by adding an acid, typically hypochlorous acid, to reduce the alkalinity.

The chief advantage of using activated silica is that it strengthens the floc, which makes the floc less likely to break apart during sedimentation or filtration. In addition, the resulting floc is larger, denser, and faster settling. Improved color removal and better floc formation at low temperatures can also result.

Activated silica is usually added after the coagulant, but adding it before can also be successful, especially with low-turbidity waters. It should never be added directly with the alum, since they react with each other.

A major disadvantage of using activated silica is the precise control required during the "activation" step to produce a solution that will not gel. Too much silica will actually slow the formation of floc, and can clog the filters.

A WEIGHTING AGENT is a material that, when added to water, forms additional particles that enhance floc formation. Weighting agents are used to treat water high in color, low in turbidity, and low in mineral content. This type of water would otherwise produce small, slowly settling floc.

Bentonite clay is a very common weighting agent. Dosages in the range of 10 to 50 mg/L usually produce rapidly settling floc. Clay added to water increases the turbidity. In water with naturally low turbidity, this speeds floc formation by increasing the number of chance collisions between particles. Powdered limestone and powdered silica can also be used as weighting agents.

POLYELECTROLYTES, both natural and synthetic, are a recent and popular entry into the field of coagulant aids. Polyelectrolytes (POLYMERS) are extremely large molecules that, when dissolved in water, produce highly charged ions.

There are three polyelectrolyte classifications:

- Cationic polyelectrolytes
- Anionic polyelectrolytes
- Nonionic polyelectrolytes.

CATIONIC POLYELECTROLYTES are polymers that, when dissolved, produce positively charged ions. They are widely used because the suspended and colloidal solids commonly found in water are generally negatively charged. Cationic polyelectrolytes can be used as the primary coagulant or as an aid to coagulants such as alum or ferric sulfate. For the most effective turbidity removal, the polymer is generally used in combination with a coagulant. There are several advantages to using this coagulant aid: the amount of coagulant can be reduced, the floc settles better, there is less sensitivity to pH, and the flocculation of living organisms such as bacteria and algae is improved.

ANIONIC POLYELECTROLYTES are polymers that dissolve to form negatively charged ions, and are used to remove positively charged solids. Anionic polyelectrolytes are used primarily as coagulant aids with aluminum or iron coagulants. The anionics increase floc size, improve settling, and generally produce a stronger floc. They are not materially affected by pH, alkalinity, hardness, or turbidity.

NONIONIC POLYELECTROLYTES are polymers having a balanced or neutral charge, but upon dissolving, release both positively and negatively charged ions. Nonionic polyelectrolytes may be used as coagulants or as coagulant aids. Although they must be added in larger doses than the other types, they are less expensive.

Compared with other coagulant aids, the required dosages of polyelectrolytes are extremely small. The normal dosage range of cationic and anionic polymers is 0.1 to 1.0 mg/L. For nonionic polymers the dosage range is 1 to 10 mg/L. The dosage of coagulants and coagulant aids must be monitored, to ensure that effective coagulation is occurring, and to ensure that the chemicals added are safe for potable water. The USEPA evaluates the coagulants and coagulant aids available for use in water treatment, and publishes a list of approved products along with the maximum allowable dosage that may be used.

3-2. Coagulation/Flocculation Facilities

The coagulation/flocculation process requires facilities for:

- Chemical storage
- Solution preparation
- Dosing
- Rapid mixing (also called flash mixing)
- Flocculation.

Chemical Storage

Coagulants and coagulant aids are available both in solid (such as powder and granular) and in liquid form. Often the liquids are easier to mix and use. The chemicals can be purchased in various bulk sizes, ranging from 50-lb (23-kg) bags and 5-gal (19-L) carboys to tank cars.

Chemicals should always be stored in a dry area at a moderate and fairly uniform temperature. Because moisture can cause dry chemicals to harden or cake, these chemicals should be stored on pallets to allow for air circulation beneath the bags. It is also desirable to store dry chemicals on a floor over the solution preparation area so they can be gravity-fed to the area through hoppers. However, this may not be practical in small-to-intermediate-sized plants because the volume of chemicals used does not warrant a major investment in chemical-handling equipment. Equipment used to handle chemicals includes hand trucks, overhead monorails and hoists, lifts, mechanical conveyors, and pneumatic conveyors. Storage and handling requirements will vary depending on the type and amount of chemical used.

In smaller plants, chemicals can be stored near the solution preparation area. Figure 3-3 shows some day-storage tanks at an intermediate-sized treatment plant. In large plants, the chemicals are often stored dry in a separate protected hopper storage area. Instructions for proper storage of coagulants and coagulant aids are available from the chemical supplier, and are usually printed on the shipping container.

Solution Preparation

Coagulants and coagulant aids usually must be diluted at the treatment plant to suitable strength solutions before they are added to the untreated water. At small plants, this can be done by manually dumping bags or siphoning liquid chemicals from carboys into a small solution tank, 50 to 300 gal (190 to 1135 L)

Courtesy of Hercules Inc.

Figure 3-3. Day-Storage Tanks

Courtesy of Lightnin Mixers and Aerators, Mixing Equipment Co., Inc.

Figure 3-4. Solution Mixer

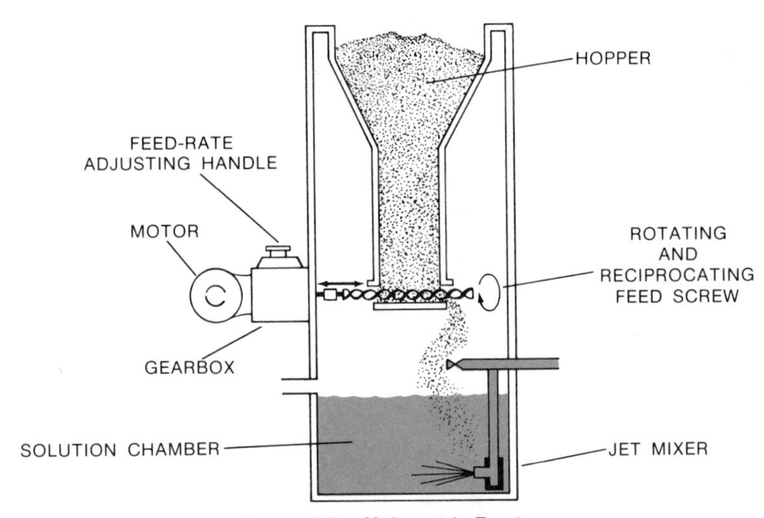

Figure 3-5. Volumetric Feeder

in size, where water is added to prepare the solution. A tank equipped with a solution mixer is shown in Figure 3-4.

In preparing feed solutions from dry polymer, it is particularly important that the operator be careful to mix the dry material evenly into the water. The individual polymer particles must be thoroughly wetted in order to go into solution quickly. If dry polymer is carelessly dumped into the solution water, undissolved clumps of polymer (called "fish eyes") will form. The clumps will not function effectively as a coagulant aid and may clog the chemical feed piping. To eliminate this problem, an ASPIRATOR FEEDER can be used to wet the polymer uniformly. Liquid polymers are often much easier to handle and dilute.

In large plants, an entire building may be devoted to the automatic preparation of coagulant solutions. Such a building might include large hoppers, weighing scales, conveyors, hoists, piping, chemical metering pumps, and holding tanks.

Dosing

Once dissolved or diluted, the coagulants must be accurately metered and fed into the raw water. The dosing system can be a simple one, such as pumping directly from a 55-gal (208-L) solution tank; or it can be a complex, electronically controlled pump metering system that automatically adjusts the dosing rate based on water flow rates, turbidity, or even zeta potential.

Coagulants can be fed to water using either dry or wet feeders. There are two types of dry feeders—volumetric and gravimetric. VOLUMETRIC FEEDERS, like the one shown in Figure 3-5, feed the dry chemical by volume; GRAVIMETRIC FEEDERS (Figure 3-6) feed the dry chemical by weight. Dry feeders usually have their own small hopper, fed from the main storage hopper. The volumetric feeder is less costly but less accurate than the gravimetric type. Once measured, the dry chemical is conveyed to the coagulant flash mixing chamber.

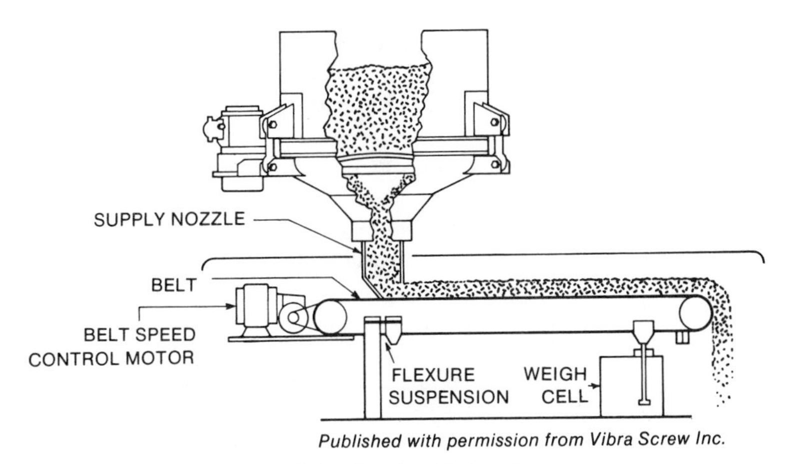

Published with permission from Vibra Screw Inc.

Figure 3-6. Gravimetric Feeder

Wet feeders or solution-feed devices are used to feed liquid coagulants—either directly from storage or indirectly by first diluting the solution. By using a dilute solution, the operator usually has more precise control of dosages.

The most common type of solution feed system is the METERING PUMP, a type of POSITIVE DISPLACEMENT PUMP. It is called a metering pump because with each stroke or rotation it pumps a precisely metered volume of solution. One of the most commonly used types is the DIAPHRAGM-TYPE METERING PUMP shown in Figure 3-7. Metering pumps usually have variable speed motors and can be set manually or automatically to deliver the amount of chemical needed for the water flow rate.

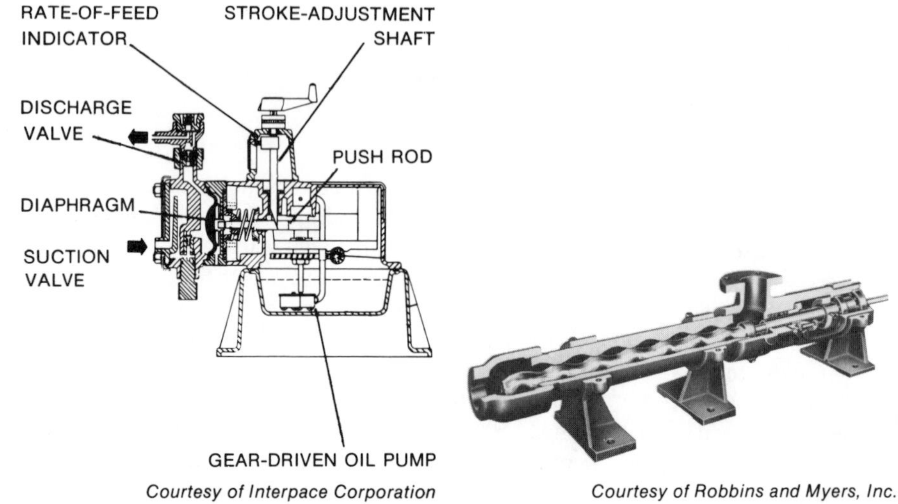

Courtesy of Interpace Corporation

Figure 3-7. Diaphragm-Type Metering Pump

Courtesy of Robbins and Myers, Inc.

Figure 3-8. Progressive Cavity Pump

Another type of positive displacement pump, called the PROGRESSIVE CAVITY PUMP, is also used in pumping chemicals at measured rates. A progressive cavity pump is shown in Figure 3-8.

The complete coagulant feed facility may include solution tanks for the mixing or dilution of the chemicals as delivered, transfer pumps, dilution water lines, solution pipelines, and various weighing scales and mixing equipment. These facilities and equipment are usually provided when the plant is constructed or expanded. Where corrosion is a problem, facilities used in the pumping, storage, and transport of chemicals are made of special materials to prevent corrosion and to create a safe working environment.

Rapid Mixing

One of the most important steps in coagulation is RAPID or FLASH MIXING. Rapid agitation is essential to distribute the coagulant evenly throughout the water. This is particularly true when alum or ferric salts are being used.

The first contact of the coagulant with the water is the most critical period of time in the whole coagulation process. The coagulation reaction occurs quickly—in a fraction of a second—so it is vital that the coagulant and the colloidal particles come into contact immediately. After the coagulant has been added, the water should be agitated violently for several seconds to encourage the greatest number of collisions possible with suspended particles.

There are different types of facilities that can be used to provide rapid mixing, including:

- Mechanical mixers
- Pumps and conduits
- Baffled chambers.

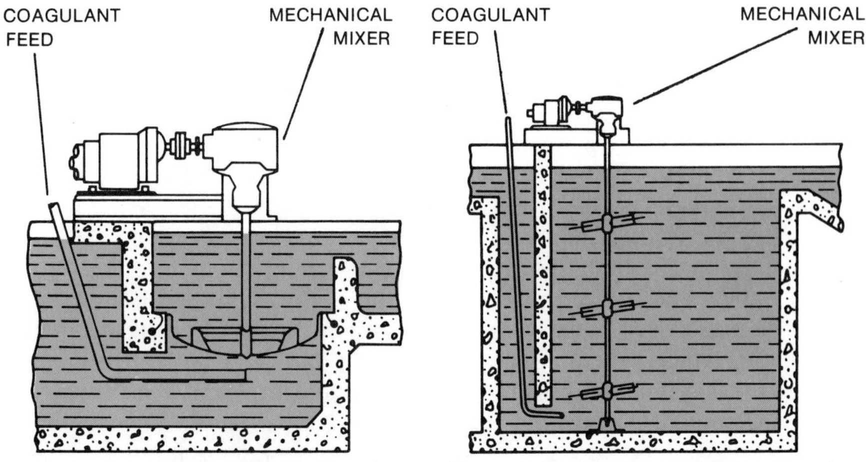

Figure 3-9. Mechanical Mixing Chamber

Figure 3-10. Mechancial Mixing Chamber

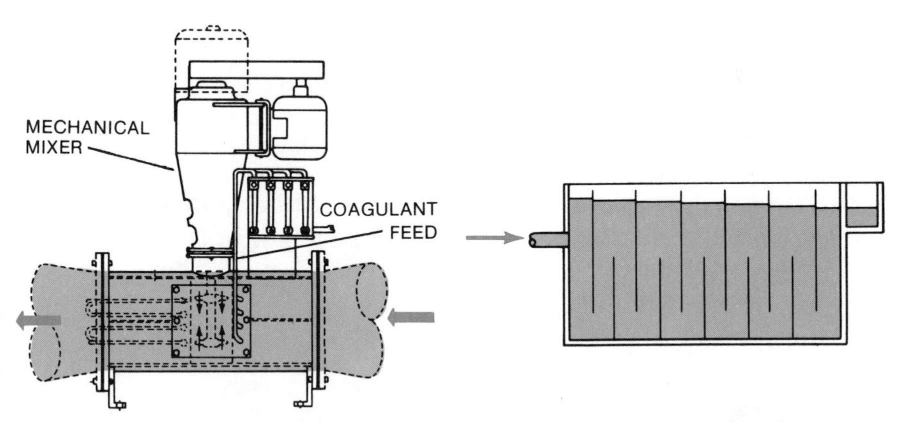

Figure 3-11. In-Line Mixer

Figure 3-12. Baffled Chamber

Because of their positive control features, propeller-, impeller-, and turbine-type mechanical flash mixers are widely used. These mechanical mixers are often placed in small chambers or tanks in which the turbulence (mixing energy) is positively controlled by varying the mixer speed. The detention times[2] in these chambers are very short, usually less than 1 min and often less than 20 sec. Figures 3-9 and 3-10 show two different types of mechanical mixers in chambers.

As shown in Figure 3-11, mechanical mixers can also be mounted directly into a pipeline; they are then referred to as in-line mixers. Unlike pump and conduit type mixers, the in-line mechanical mixer can be adjusted to provide the correct amount of mixing energy. Since the in-line unit requires no separate tank or

[2] *Basic Science Concepts and Applications*, Mathematics Section, Detention Time.

chamber, it is a low cost mixing device that can be added to an existing pipeline. In-line mixers (used commonly in direct filtration) need to be located so that flocculation and settling will not occur within the pipeline.

In the pump and conduit type of flash mixer, coagulant is added to the water in a pipeline before it passes through a pump. The spinning pump impeller supplies turbulent mixing. While this is an efficient use of existing facilities, it is not necessarily efficient flash mixing. The amount of mixing is determined by the speed of the pump and this may not correspond with the actual turbulence required for proper flash mixing. In addition, the coagulating particles can begin to flocculate and settle in the pipeline downstream of the pump. Overall, the pump and conduit type of mixing leaves the plant operator with little or no opportunity to adjust the operation to suit the treatment needs. Instead, mixing, flocculation, and even settling are determined by the rate the water is processed.

The baffled chamber shown in Figure 3-12, although an improvement over the pump and conduit type, also lacks positive operator control. The amount of mixing energy is determined by the turbulence of the water as it races over and under each baffle. The turbulence is determined by the rate of flow, and generally can not be controlled.

Flocculation

The flocculation facility consists of a basin or tank and a means of slow, gentle mixing. Several examples are shown in Figures 3-13 through 3-18.

Since flocculation is a much slower process than coagulation, the flocculation basin must be larger. Floc is quite fragile, so the mixing must be gentle and the velocity of flow through the tank must be slow enough so that the floc particles are not broken or sheared. The flocculation basin must be large enough to provide an adequate detention time of 20 to 60 min. Figure 3-13 illustrates how baffles are used to slow down the water.

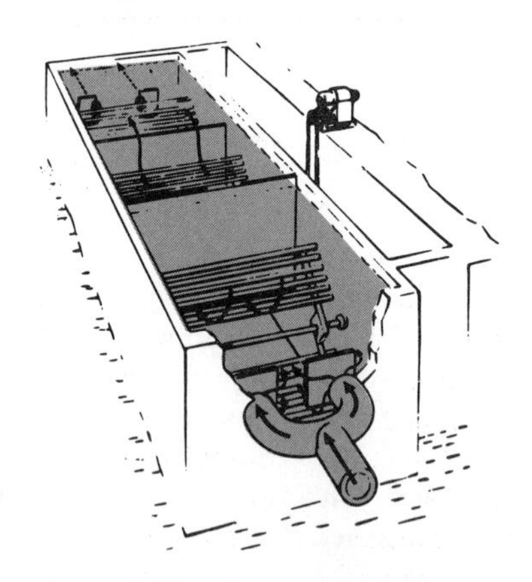

Courtesy of Envirex, Inc., a Rexnord Company

Figure 3-13. Baffles and Horizontal Paddle-Wheel Flocculator

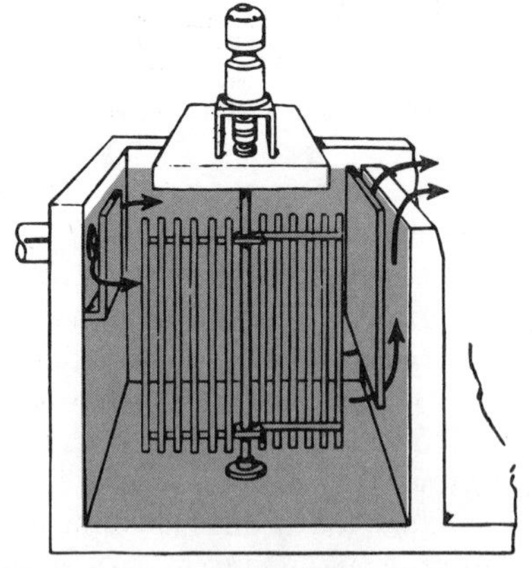

Courtesy of Envirex, Inc., a Rexnord Company

Figure 3-14. Paddle-Wheel Flocculator, Vertical Type

Mixing for flocculation can be done mechanically, using rotating paddles, or hydraulically, resulting from the motion of the water currents. Examples of mechanical flocculators include paddle-wheel flocculators (Figures 3-13 and 3-14), propeller flocculators (Figure 3-15), turbine flocculators (Figure 3-16), and walking beam flocculators (Figure 3-17). Mechanical flocculators are often provided with variable speed drives to give an operator maximum flocculation control.

An example of a hydraulic flocculator is shown in Figure 3-18. The unit combines the coagulation, flocculation, and sedimentation processes in one treatment unit. This equipment is commonly used in the lime-soda ash process for softening water. (See Module 5, Softening.)

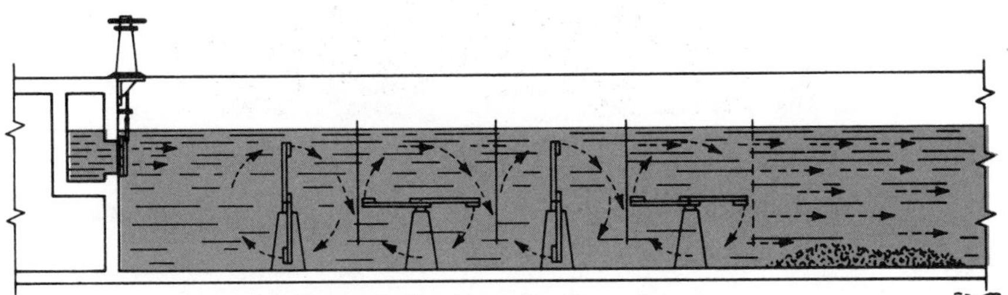

Figure 3-15. Propeller Flocculator

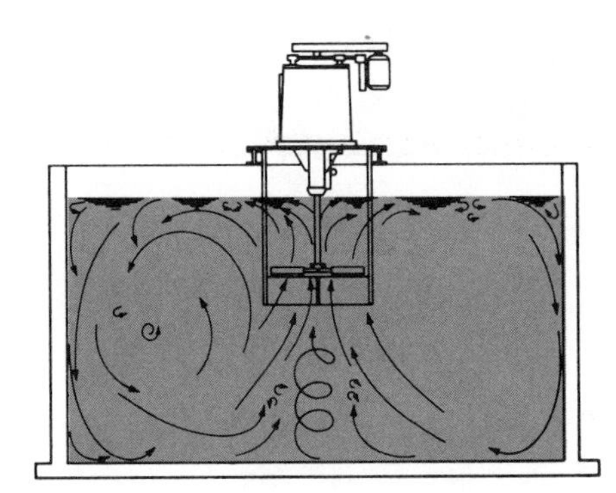

Figure 3-16. Turbine Flocculator

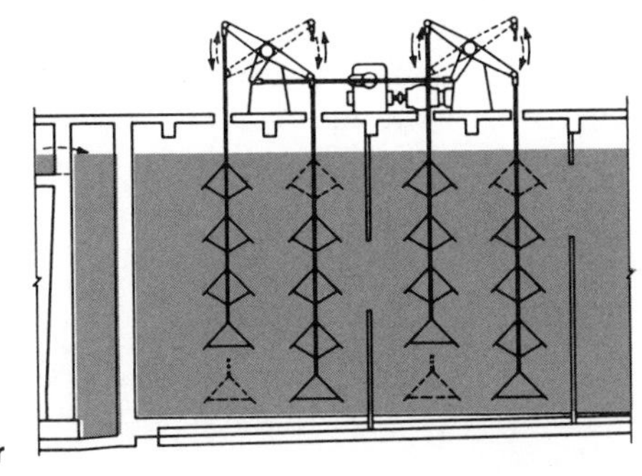

Figure 3-17. Walking Beam Flocculator

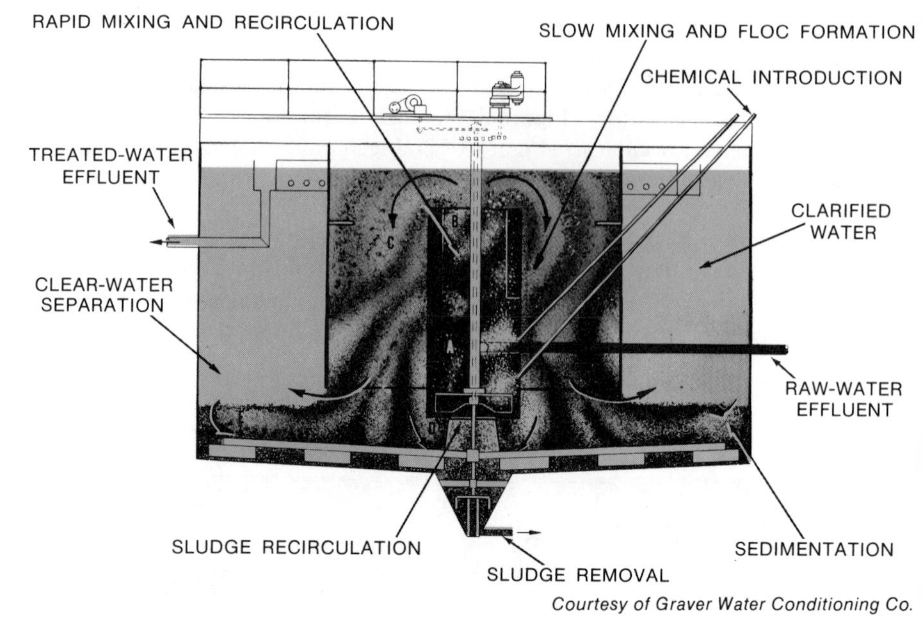

Courtesy of Graver Water Conditioning Co.

Figure 3-18. Hydraulic Flocculator

3-3. Operation of the Coagulation/Flocculation Process

There are three fundamental steps in operating the coagulation/flocculation process:

- Selecting the chemicals
- Applying the chemicals
- Monitoring process effectiveness.

Selecting the Chemicals

The selection of chemical coagulants and aids is a continuing program of trial and evaluation, normally using the jar test. Usually, an operator tests for chemical selection in the treatment plant laboratory. To do a thorough job of chemical selection, the following characteristics of the raw water to be treated should be measured:

- Temperature
- pH
- Alkalinity
- Turbidity
- Color.

The effect of each characteristic on coagulation and flocculation is as follows:

Temperature. Usually, lower temperatures cause poorer coagulation/flocculation and can require that more chemical be used to maintain acceptable results.

pH. Extreme values, high or low, can interfere with coagulation/flocculation. The optimum pH varies depending on the coagulant used.

Alkalinity. Alum and ferric sulfate interact with the chemicals that cause alkalinity in the water, forming complex aluminum or iron hydroxides that begin the coagulation process. Low alkalinity limits this reaction and results in poor coagulation; in this case it may be necessary to add alkalinity to the water.

Turbidity. The lower the turbidity, the more difficult it is to form a proper floc. Fewer particles mean fewer random collisions and hence fewer chances for floc to accumulate. The operator may have to add a weighting agent such as clay (bentonite) to low-turbidity water.

Color. Color is an indication of organic compounds. The organics can react with the chemical coagulants, making coagulation more difficult. Pretreatment with oxidants or adsorbents may be necessary to reduce the concentration of organics.

The effectiveness of a coagulant or flocculant will change as raw-water characteristics change. The effectiveness of coagulant and coagulant-aid chemicals may also change for no apparent reason, suggesting that there are other factors, not yet understood, that affect coagulation and flocculation. For this reason, an operator should begin chemical selection by using the jar test with various chemicals, singly and in combination. The jar test is discussed in *Introduction to Water Quality Analyses,* Physical/Chemical Tests Module—Jar Test.

The jar test, although still the most widely used coagulation control test, is subjective, and depends solely on the human eye for evaluation and interpretation. To gain further information, the operator should also run the pH and turbidity tests and perhaps filtrability and zeta potential tests. All of the control tests are discussed in more detail under Operational Control Tests, later in this module.

Applying the Chemicals

Jar test results help determine the type of chemical or chemicals to use and the best dosage. Jar test results are in milligrams per litre, which must be converted to the equivalent full-scale dosage in pounds per day or gallons per day.[3] The operator then applies the chemical to the water by setting the manual or automatic metering of the chemical feed system to the desired dosage rate.

Monitoring Process Effectiveness

Although the tests listed above (and discussed under Operational Control Tests) provide a good indication of the results to expect, full-scale plant

---

[3] *Basic Science Concepts and Applications*, Chemistry Section, Chemical Dosage Problems (Milligrams-per-Litre to Pounds-per-Day Conversions).

operation may not always match these results. Actual plant performance must be monitored for:

- Adequate flash mixing
- Gentle flocculation
- Adequate flocculation time
- Settled- and filtered-water quality.

Successful coagulation is based on rapid and complete mixing. Though coagulation occurs in less than a second, the chamber may provide up to 60 sec of detention time. Mixing should be turbulent enough so that the coagulant is dispersed throughout the coagulant chamber.

Some experts maintain that during the first one-tenth of a second the coagulant must be thoroughly mixed with every drop of water to start an efficient floc or, from that point, the efficiency of the entire process declines. Most improper coagulant usage shows its effect at this moment, and the problem can usually be corrected by altering either the point of application or method of mixing. If polymers are to be used as prime coagulants, rapid mixing is less critical, but thorough mixing remains very important in encouraging as many particle collisions as possible. Adequate flash mixing is not something that can be observed directly, but the following are indicators of inadequate mixing:

- Very small floc (called pin-point floc)
- Fish eyes
- High turbidity in settled water
- Too frequent filter backwashing.

If any of these conditions appear, the operator should begin to investigate the adequacy of the flash mixing operation, the gentleness of flocculation, and the length of flocculation time.

Proper flocculation requires long, gentle mixing. Mixing energy must be high enough to bring coagulated particles constantly into contact with each other, but not so high as to break those particles already flocculated. For this reason, flocculation basins and equipment are usually designed to provide higher mixing speeds immediately following coagulation, and slower speeds as the water progresses down the basin. Properly coagulated/flocculated particles will appear as minute snowflakes or tufts of wool suspended in very clear water. The water should definitely not look cloudy or foggy—if it does, or if it displays any of the four signs of inadequate mixing listed above, then the speed of the flocculators should be checked.

It takes time to develop large, heavy floc particles. The flocculation basin should provide 20 to 60 min of detention time. Since short circuiting can be a major problem, a minimum of three flocculation basins in series is recommended.

Effective coagulation/flocculation will result in a settled water of less than 10 NTU turbidity. This will result in more efficient use of the filters since they will not have to handle a high suspended solids loading. The product water should be monitored to verify its quality.

Excessive carry-over of floc particles can seal ("blind") a filter, increasing the frequency required for backwashing. Visible floc carry-over from the sedimentation process to the filters is a clear sign that blinding may occur. Frequent backwashing to clean the filter takes the filter out of service for up to 30 min (resulting in loss of filtered-water production), and wastes filtered water in the backwash process. Therefore, frequent backwashing is costly. If blinding occurs, usually a change in chemical dosage is required. If the problem persists, a change in the coagulant or coagulant aids may be required.

Record Keeping

An operator should maintain records of past raw-water quality and of the coagulants and dosages that worked best for that water, as well as recording general observations relating to the operation of the coagulation/flocculation process. This is particularly important for surface-water supplies, because surface-water quality varies seasonally, necessitating periodic changes in chemical dosages and sometimes in the chemicals themselves. If past experiences have been recorded, that information will be readily available to act as a guide when similar situations arise. Figure 3-19 illustrates a suggested record-keeping form for the coagulation/flocculation process.

TYPE OF COAGULANT ______________________ **DATE STARTED** ______________

ITEM	RESULTS (BY DATE)								
	DATE	DATE	DATE	DATE	DATE	DATE	DATE	DATE	DATE
COAGULANT DOSAGE									
RAW WATER									
TEMPERATURE (°F)									
pH									
ALKALINITY (mg/L AS $CaCO_3$)									
TURBIDITY (NTU)									
TASTE AND ODOR									
COLOR (CU)									
SUSPENDED SOLIDS (mg/L)									
ALGAE CONTENT									
COAGULATED WATER									
FILTRABILITY (VOLUME/TIME)									
ZETA POTENTIAL (mV)									
SETTLED-WATER TURBIDITY									
FILTERED WATER									
TURBIDITY (NTU)									
COLOR (CU)									
TASTE AND ODOR									
ALGAE CONTENT									
RESIDUAL COAGULANT (mg/L)									

Figure 3-19. Record-Keeping Form

3-4. Common Operating Problems

There are primarily three operating problems that can interfere with successful coagulation/flocculation:

- Low water temperature
- Weak floc
- Slow floc formation.

Low water temperature. Low water temperatures, particularly temperatures approaching freezing, interfere with the coagulation/flocculation process. As water temperature decreases, the VISCOSITY of the water increases. This slows the rate of floc settling. (Viscosity is the resistance of a fluid to flowing or pouring—syrup has a high viscosity, alcohol has a low viscosity.)

The colder temperature also slows chemical reaction rates, although this effect on coagulation may be too slight to be significant. In addition, the best pH for the chemical reaction drops as temperature drops.

Cold water floc has a tendency to penetrate through the filters, indicating that floc strength has decreased. These problems can be overcome using any of these three techniques:

- Operate the coagulation process as nearly as possible to the best pH value for that temperature water.
- Increase coagulant dosage. This increases the number of particles available to collide and also reduces the effect of changes in pH resulting from the drop in temperature.
- Add weighting agents, such as clays, to increase floc particle density, and add other coagulant aids, such as activated silica or polyelectrolytes, to increase floc strength and encourage rapid settling.

Jar testing will determine which approach will be the most effective.

Weak floc. The strength of floc particles is established during the coagulation/flocculation process, but the effects are not noticeable until filtration. A weak floc does not adhere strongly to the filter media. Instead, it is easily broken up and is carried deeper and deeper into the filter, eventually breaking through the filter and causing poor-quality effluent. Often this condition is the result of inadequate mixing in the rapid mix or flocculation basins. This can be checked by varying the mixing speeds and taking samples from various points to see if the floc settles any better. Jar testing should also be used to determine if other combinations of coagulants and coagulant aids produce a better floc. The effect of weak floc on filter operation and the steps needed to correct the problem are discussed in Module 6, Filtration, Operating Problems (Floc Strength).

Slow floc formation. Low-turbidity waters have very few particles and this condition contributes to slow and inadequate floc formation (pinpoint floc). With fewer particles, there are fewer of the random collisions necessary for floc particles to form and grow.

One way to correct this problem is to recycle previously settled sludge from the sedimentation process. This is the same principle used in the solids contact sedimentation process (discussed in Module 4, Sedimentation). Another way to correct the problem is to artificially increase turbidity by adding weighting agents. This increases the number of chance collisions between particles.

If alum or ferric sulfate is the coagulant used, the slow floc formation can be a result of inadequate alkalinity. Both of these coagulants react with alkalinity constituents during coagulation to form complex aluminum or iron hydroxides. This hydroxide formation marks the beginning of coagulation. Low alkalinities slow the reaction, resulting in poor coagulation and slow floc formation. The solution to this problem is to add chemical alkalinity in the form of lime or soda ash.[4] Improper mixing also reduces the speed of floc formation.

3-5. Operational Control Tests

There are five tests that can help in controlling the coagulation/flocculation process:

- Jar test
- pH
- Turbidity
- Filtrability
- Zeta potential.

Jar Test

The jar test helps determine the chemical or chemical combination and the dosage that will produce the best floc. Testing should not be limited to the chemical or chemicals that the plant has always used. A wide variety of coagulants and coagulant aids should be tried. Suppliers will be able to furnish test samples. Periodically, new chemicals that have come on the market should be tested. This testing can lead to the use of new coagulants that are as effective as currently used chemicals, yet are less costly or produce less sludge.

pH

Coagulants and coagulant aids operate best within a particular range of pH. Knowing the pH of the water helps an operator select the proper coagulant or correctly adjust pH prior to coagulation (see *Introduction to Water Quality Analyses,* Physical/Chemical Tests Module—pH).

Turbidity

In the past, the clarity of the settled jar test samples was often evaluated by eye and described with terms such as "poor," "good," and "very good." This

[4] *Basic Science Concepts and Applications*, Chemistry Section, Chemistry of Treatment Processes (Coagulation).

subjectivity, and the errors that it can cause, can be avoided by measuring the turbidity of the settled jar test samples using the standard nephelometric technique (see *Introduction to Water Quality Analyses,* Physical/Chemical Tests Module—Turbidity). To determine how much turbidity is actually being removed compared to the process efficiency predicted by jar test results, turbidity tests should be run on the untreated water and on samples taken after water passes through the sedimentation basin ("settled water"). Process efficiency should be fairly consistent—if the turbidity of the settled water increases suddenly from one test to the next, the coagulation/flocculation process should be investigated for possible problems.

Filtrability

Filtrability is a method of testing how effectively a coagulated water can be filtered. The test measures the amount of water filtered in a given time. It is usually made by applying a small portion of the actual coagulated and flocculated plant flow (before sedimentation) to a 6-in. (15-cm) diameter filter (a pilot filter) containing the same type and depth of fine media as used in the plant filters. The pilot filter is usually equipped with a recording turbidimeter on the effluent, and continuous turbidity measurements are made of the filtered water. The value of this control technique is twofold:

- The test uses the actual coagulated water from the plant, not a laboratory simulation.
- The water at any point takes about 10 min to go through the miniature filter. Since the full-scale sedimentation tank and filter detention time is about 2 to 3 hours, changes in coagulant doses can be made based on the miniature filter performance before serious deterioration of the plant effluent quality occurs.

Because there is no standard for the filtrability test, the test procedure varies widely. Sometimes the pilot filter is replaced by filter paper, gauze, or a membrane filter. Even the filtrability test name varies; it may be called filtrability index, filtrability number, inverted gauze filter, membrane refiltration, pilot column filtration, silting index, or surface area concentration.

Zeta Potential

The ZETA POTENTIAL (ZP) test has been used successfully to help determine the best pH and dosage for cationic polymers and cationic coagulants such as Al^{+3} and Fe^{+3}. The control procedure requires monitoring the ZP of the coagulated water and changing the chemical dosage when the ZP varies outside a range known to produce the lowest turbidity. This range, though variable from plant to plant, is often −6 to −10 millivolts (mV). Table 3-5 shows the degree of coagulation that occurs within various ZP ranges.

Table 3-5. Degree of Coagulation Within Different Zeta Potential Ranges

Average Zeta Potential	*Degree of Coagulation*
+3 to 0	Maximum
−1 to −4	Excellent
−5 to −10	Fair
−11 to −20	Poor
−21 to −30	Virtually none

3-6. Safe Handling of Coagulant Chemicals

Extreme caution is important in handling all chemicals. In working with a concentrated chemical of any kind, always assume you are dealing with a potential hazard. In their concentrated forms, many chemicals are corrosive or explosive. Handling of dry chemicals can create a dust, which is dangerous if inhaled or if it comes in contact with skin or eyes—this is particularly a problem with lime, soda ash, and alum. Other chemicals, including liquid polymers and polymer solutions, can create dangerously slick areas where spills occur.

Aluminum sulfate (alum) is the most common coagulant chemical. It is an acid salt, and must be handled with care. The acidic action of alum, in dry or liquid form, will irritate eyes, skin, and mucous membranes. Liquid alum is corrosive. When handling the chemical, an operator should wear protective clothing, goggles, and a respirator (for dry alum). Where dry alum feeders are used, dust control equipment should be installed and maintained in proper operating condition.

Mixing dry alum with quicklime (CaO) creates tremendous heat. If the temperatures of these stored mixtures reach 1100° F (593° C) or higher, a highly explosive hydrogen gas is given off. This kind of accidental mixture could occur if the operator used the same storage bin or belt conveyor for both chemicals without first carefully cleaning the bin or conveyor.

Table 3-6 summarizes the physical features of some of the chemicals used in coagulation and briefly summarizes types of shipping containers and suitable handling materials. Table 3-7 gives some suggestions for the safe handling of several chemicals.

Table 3-6. Chemicals Used in Coagulation Process

Chemical Name and Formula	*Common or Trade Name*	*Shipping Containers*	*Suitable Handling Materials*
Aluminum sulfate $Al_2(SO_4)_3 \cdot 14\ H_2O$	alum, filter alum, sulfate of alumina	100–200-lb bags, 300–400-lb bbls., bulk (carloads), tank trucks, tank cars	dry—iron, steel; solution—lead-lined rubber, silicon, asphalt, 316 stainless steel
Ammonium aluminum sulfate $Al_2(SO_4)_3$ $(NH_4)_2SO_4 \cdot 24\ H_2O$	ammonia alum, crystal alum	bags, bbls., bulk	duriron, lead, rubber, silicon, iron, stoneware
Bentonite	colloidal clay, volclay, wilkinite	100-lb bags, bulk	iron, steel
Ferric chloride $FeCl_3$ (35–45 percent solution)	"ferrichlor," chloride of iron	5-, 13-gal carboys, trucks, tank cars	glass, rubber, stoneware, synthetic resins
$FeCl_3 \cdot 6\ H_2O$	crystal ferric chloride	300-lb bbls.	—
$FeCl_3$	anhydrous ferric chloride	500-lb casks; 100-, 300-, 400-lb kegs	—
Ferric sulfate $Fe_2(SO_4)_3 \cdot 9\ H_2O$	"ferrifloc," ferrisul	100-, 175-lb bags, 400-, 425-lb drums	ceramics, lead, plastic, rubber, 18-8 stainless steel
Ferrous sulfate $FeSO_4 \cdot 7\ H_2O$	copperos, green vitriol	bags, bbls., bulk	asphalt, concrete, lead, tin, wood
Potassium aluminum sulfate $K_2SO_4 \cdot Al_2(SO_4)_3 \cdot 24\ H_2O$	potash alum	bags, lead-lined; bulk (carloads)	lead, lead-lined rubber, stoneware
Sodium aluminate $Na_2O\ Al_2O_3$	soda alum	100-, 500-lb bags, 250-, 440-lb drums, solution	iron, plastics, rubber, steel
Sodium silicate $Na_2O\ SiO_2$	water glass	drums, bulk (tank trucks, tank cars)	cast iron, rubber, steel

Source: *Water Treatment Plant Design*. AWWA, Denver, Colo. (1971).

Available Forms	*Weight* lb/cu ft	*Solubility* lb/gal	*Commercial Strength* percent	Characteristics
vory-colored: powder, granule, lump, liquid	 38–45 60–63 62–67 10(lb/g)	4.2 (60° F)	15–22(Al_2O_3) 8(Al_2O_3)	pH of 1 percent solution 3.4
lump, nut, pea, powdered	64–68 62 65 60	0.3 (32° F) 8.3 (212° F)	11(Al_2O_3)	pH of 1 percent solution 3.5
powder, pellet, mixed sizes	60	insoluble (colloidal sol used)	—	—
dark brown syrupy liquid	—	complete	37–47($FeCl_3$) 20–21(Fe)	
yellow-brown lump	—	—	59–61($FeCl_3$) 20–21(Fe)	hygroscopic (store lumps and powder in tight container) no dry feed; optimum pH, 4.0–11.0
green-black powder	—	—	98($FeCl_3$) 34(Fe)	
red-brown powder, 70 or granule 72	—	soluble in 2–4 parts cold water	90–94($Fe_2(SO_4)_3$) 25–26(Fe)	mildly hygroscopic coagulant at pH 3.5–11.0
green crystal granule, lump	63–66	—	55($FeSO_4$) 20(Fe)	hygroscopic; cakes in storage; optimum pH 8.5–11.0
lump, granule, powder	62–67 60–65 60	0.5 (32° F) 1.0 (68° F) 1.4 (86° F)	10–11 (Al_2O_3)	low, even solubility; pH of 1 percent solution, 3.5
brown powder, liquid (27° Be)	50–60	3.0 (68° F) 3.3 (86° F)	70–80 ($Na_2Al_2O_4$) min. 32 ($Na_2Al_2O_4$)	hopper agitation required for dry feed
opaque, viscous liquid	—	complete	38–42° Be	variable ratio of Na_2O to SiO_2; pH of 1 percent solution, 12.3

Table 3-7. Suggestions for Handling Chemicals*

Chemical	*Procedure*
1. Activated carbon: 1, 2, 3, 4, 5, 6, 7, 11	1. Wear masks: 1, 2, 4, 7
	2. Store in clean, dry place: 1, 2, 4
	3. Do not store in large piles: 1, 2
	4. Use fireproof storage area: 1
2. Aluminum sulphate and ferrous sulphate: 1, 2, 3, 8, 9, 10, 11, 12	5. Store away from flammable material: 1, 6
	6. No smoking: 1
	7. Keep sparks and open electric equipment away: 1
	8. Wear protective clothing: 2, 4, 7
	9. Do not use compressed air to clean dry-feed machines: 2
	10. May explode if mixed with quicklime: 2, 4
3. Ammonia, anhydrous: 13, 14, 15, 16, 17	11. Minimize dust: 1, 2, 7
	12. Put antisplatter shields on solution pumps: 2
	13. Handle cylinders carefully and hoist only when necessary: 3, 6
4. Ammonium sulphate: 1, 2, 8, 10	14. Do not store near chlorine: 3
	15. Store cylinders on side in cool dry place; 3, 6
	16. Ventilate storage room: 3,6
5. Carbon dioxide: 18, 19, 20	17. Have masks available: 3,6
	18. Locate generating equipment in ventilated room near application: 5
6. Chlorine: 5, 13, 15, 16, 17, 21	19. Take precaution against carbon monoxide: 5
	20. Prevent flow to low places; it is heavier than air: 5
	21. Store in cool, dry, ventilated, separate rooms: 6
7. Fluorides: 1, 8, 11	

*The numeral following a chemical or a procedure refers to the corresponding numeral preceding a procedure or a chemical. For example, in handling fluorides (chemical 7), follow procedures 1, 8, and 11. If the storage room is to be ventilated (procedure 16), it should be done for chemicals 3 and 6.
Source: Babbitt, Doland, & Cleasby. *Water Supply Engineering*. McGraw-Hill Book Co. (6th ed., 1962).

Selected Supplementary Readings

Babbitt, H.E. & Doland, J.J.; & Cleasby, J.L. *Water Supply Engineering*. McGraw-Hill Book Company, New York. (6th ed., 1962). Chap. 22.

CHEMTREC: 800-424-9300. *OpFlow*, 6:4:1 (Apr. 1980).

Cox, Charles R. *Operation and Control of Water Treatment Processes*. World Health Organization, Geneva, Switzerland (1969). Chap. 5

Finney, J.W. Jr. Polymers Prove Useful in Water Treatment. *OpFlow*, 1:10:1 (Oct. 1975).

Hannah, S.A.; Cohen, J.M.; & Robeck, G.G. Control Techniques for Coagulation-Filtration. *Jour. AWWA*, 59:9:1149 (Sept. 1967).

How to Safely Handle Alum in Water Treatment. *OpFlow*, 4:8:5 (Aug. 1978).

Hudson, H.E., Jr. & Wagner, E.G. *Jar Testing Techniques and Their Use*. Water Technology Conf. Proc., Paper No. 2A-3. AWWA, Denver, Colo. (Dec. 1979).

Industrial Hygiene Prevents "Chemi-Kill" Injuries. *OpFlow*, 8:2:1 (Feb. 1982).

Introduction to Water Quality Analyses. AWWA, Denver, Colo. (1982). Module 5.

Manual of Instruction for Water Treatment Plant Operators. New York State Dept. of Health, Albany, N.Y. (1975). Chap. 7.

Moffett, J.W. The Chemistry of High-Rate Water Treatment. *Jour. AWWA*, 60:11:1255 (Nov. 1968).

Riddick, T.M. Zeta Potential and Its Application to Difficult Waters. *Jour. AWWA*, 53:8:1007 (Aug. 1961).

Safe Handling of Chemicals: Alum. *OpFlow*, 5:9:3 (Sept. 1979).

Safety Practice for Water Utilities. AWWA Manual M3, AWWA, Denver, Colo. (1983).

Sawyer, Clair & McCarty, P.L. *Chemistry for Environmental Engineering*. McGraw-Hill Book Company, New York. (3rd ed., 1978).

Simplified Procedures for Water Examination. AWWA Manual M12, AWWA, Denver, Colo. (1975).

Spink, C.T. & Monscvitz, J.T. Design and Operation of a 200-mgd Direct-Filtration Facility. *Jour. AWWA*, 66:2:127 (Feb. 1974).

Stearns, D.M. Selection and Application of Dry-Chemical Feeders. *OpFlow*, 3:6:1 (June 1977).

Streicher, Lee. Can Organic Polymers Help Reduce the Total Volume of Sludge for Disposal? *OpFlow*, 5:2:1 (Feb. 1979).

Tesar, Mike. Calibration Procedures for Chemical Feed Systems. *OpFlow*, 9:2:1 (Feb. 1983).

Water Works Operator Manual. Alabama Dept. of Public Health, Montgomery, Ala. (3rd ed., 1972). Chap. XIII and XIV.

Yen, C.H.H. & Ghosh, M.M. *Selection and Use of Polymers in Direct Filtration*. Annual Conf. Proc., Paper No. 6-1, p. 155. AWWA, Denver, Colo. (June 1980).

Glossary Terms Introduced in Module 3

(Terms are defined in the Glossary at the back of the book.)

Activated silica
Alum
Aluminum sulfate
Anionic polyelectrolyte
Aspirator feeder
Bivalent ion
Cationic polyelectrolyte
Coagulant
Coagulant aid
Coagulation
Colloidal solid
Diaphragm-type metering pump
Dissolved solid
Ferric sulfate
Flash mixing
Floc
Flocculation
Gravimetric feeder
Metering pump
Microfloc
Monovalent ion
Nonionic polyelectrolyte

Nonsettleable solid
Polyelectrolyte
Polymer
Positive displacement pump
Progressive cavity pump
Rapid mixing
Suspended solid
Trivalent ion
van der Waals force
Viscocity
Volumetric feeder
Weighting agent
Zeta potential

Review Questions

(Answers to Review Questions are given at the back of the book.)

1. List the three types of nonsettleable solids found in water.
2. What is zeta potential?
3. What is the van der Waals force?
4. Define coagulation.
5. What is flocculation?
6. What are floc particles?
7. List the two most common coagulants used in water treatment.
8. List the physical and chemical factors that affect coagulation.
9. What is the test used to select chemical coagulants and determine the best coagulant dose?
10. What is a coagulant aid?
11. List three general types of coagulant aids.
12. List the three categories of polyelectrolytes.
13. In what forms are coagulants and coagulant aids available?
14. List five basic facilities required in the coagulation process.
15. Why are coagulants and coagulant aids sometimes dissolved or diluted before being added to the raw water?
16. List two types of dry chemical feeders. What are the differences between them?
17. What is the most common type of coagulant solution feeder?

18. What are the three fundamental steps that are involved in operating the coagulation/flocculation process?

19. List and describe at least three common operating problems associated with the coagulation/flocculation process. What can be done to solve these problems?

20. List the five operational control tests that the operator can use to help operate the coagulation/flocculation process.

21. Discuss some of the hazards associated with the handling of coagulant chemicals.

22. What is the detention time in a flocculation tank 26 ft wide, 90 ft long, and with 8 ft water depth, if the water being treated is flowing through at a rate of 3.5 mgd? How does this detention time compare with the usual range?

23. Jar test results indicate that the best alum dosage for a particular water is 75 mg/L. Plant flow is currently running at 0.86 mgd. How many pounds of dry alum must be fed per hour in order to dose at 75 mg/L? How many gallons per hour of liquid alum must be added? Liquid alum contains 5.36 lb of dry alum per gallon.

Study Problems and Exercises

1. Select a coagulant from Table 3-3 and prepare a brief report on its use in water treatment. The report should cover:

   - Physical characteristics of the chemical
     - Form (liquid, powder, lump, etc.)
     - Weight (per gallon or per cubic feet)
     - Color
   - Chemical composition
     - Percent of active chemical
     - Percent of impurities
   - Operating characteristics including typical dosage, optimum pH range, and mixing and dosing equipment
   - Handling and storage requirements, including safety considerations
   - Advantages and disadvantages
   - Manufacturer and local source of supply
   - Quantities available (bags, barrels, bulk rail cars, etc.)
   - Current costs.

2. Based on given basin volumes and an average daily flow rate, calculate the detention time provided by the coagulation and flocculation basins at a local water treatment plant. In your opinion, are these detention times appropriate? Why? Or would you recommend different detention times? Why?

3. Describe the coagulation/flocculation process used by a local water treatment plant, including type of equipment in operation. Describe the operational characteristics and any operational problems experienced. Discuss how the problems could be corrected to improve operational performance.

4. You have been employed as a water treatment operator at a treatment plant that has been in operation for approximately two years. The plant is experiencing serious problems with pinpoint floc.

   (a) Describe the procedure you would use to determine what is causing this problem.

   (b) Assuming the problem is caused by an improper choice or use of coagulant, relate how you would determine the proper coagulant or coagulant/polymer combination to correct the problem. Make certain that your selection of chemicals is cost-effective.

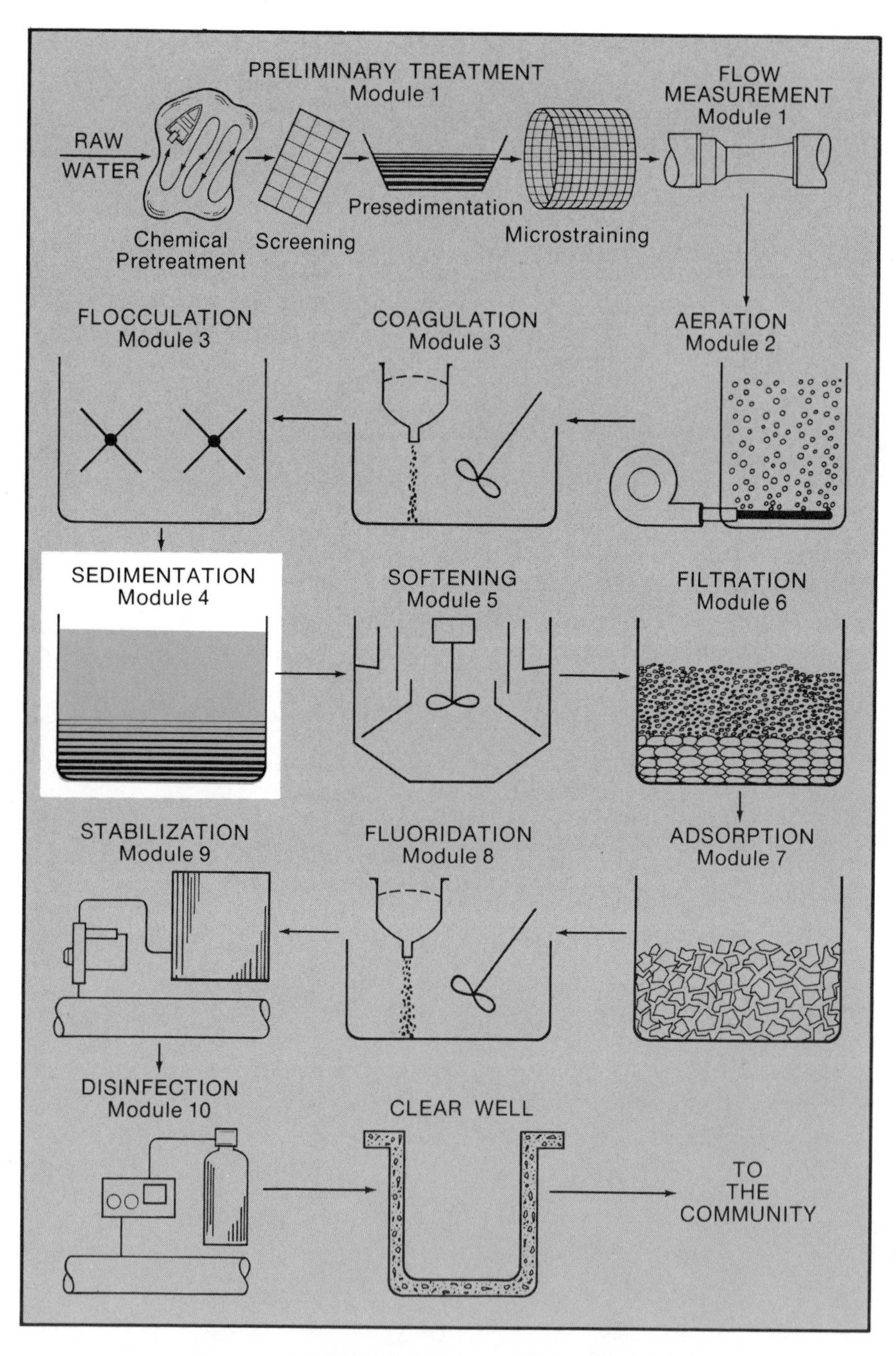

Figure 4-1. Sedimentation in the Treatment Process

Module 4

Sedimentation

Sand, grit, chemical precipitates, pollutants, floc, and other settleable solids are kept suspended in water by the velocity and TURBULENCE (agitation and irregular flow) of the flowing water. Sedimentation removes these solids in order to decrease loading on filtration and other treatment processes. Where sedimentation occurs in the treatment process is shown in Figure 4-1. Basins designed for efficient sedimentation allow the water to flow very slowly. Because velocity is low and turbulence is decreased, the suspended solids settle to the bottom of the basin. SLUDGE, a residue of solids and water, accumulates at the bottom of the basin and is pumped out of the basin for DEWATERING (removing water from sludge) and eventual disposal or reuse.

After completing this module you should be able to

- Describe the role of sedimentation in water treatment.
- Describe the major sedimentation processes.
- List the key parts of settling basins.
- Identify and describe the function of the four zones of sedimentation.
- List important factors in the proper operation of settling basins.
- Identify the records that should be maintained on the sedimentation process.
- Describe the common operating problems associated with settling basins.
- State which tests should be run to properly monitor sedimentation performance.
- Describe the safety considerations associated with the sedimentation process.

4-1. Description of Sedimentation

Sedimentation, sometimes called CLARIFICATION, is the removal by gravity of settleable solids in water. The solids removed include sand, silt, grit, chemical precipitates, pollutants, floc, and other settleable solids. The sedimentation process takes place in a SEDIMENTATION BASIN or SETTLING BASIN. As shown by Figure 4-1, sedimentation occurs after coagulation/flocculation and before filtration in a conventional water treatment plant. Figure 4-2 shows a typical installation.

To help produce drinking water with the lowest possible turbidity, sedimentation reduces the amount of suspended material the filters have to remove. This also makes longer filter runs possible and reduces filter maintenance. Sedimentation is also used to remove the large amounts of chemical precipitates formed during the lime-soda ash softening of water.

Forms of Sedimentation

The two commonly used forms of sedimentation are plain sedimentation and sedimentation following chemical addition.

Plain sedimentation. PLAIN SEDIMENTATION, also called PRESEDIMENTATION, is commonly used as a preliminary treatment process (see Module 1, Preliminary Treatment Processes and Flow Measurement). The process reduces heavy sediment loads by gravity settling without chemical addition or other alterations. This prevents the sediment from damaging pumps and creating maintenance problems in later treatment processes. Plain sedimentation occurs in lakes, reservoirs, or man-made basins designed for this purpose.

Sedimentation following chemical addition. Sedimentation following chemical addition removes by gravity solids that have become larger, heavier, and more settleable due to chemical treatment or conditioning.

Figure 4-2. Typical Installation

Modifications. There are two modifications to the conventional sedimentation process: SHALLOW-DEPTH SEDIMENTATION and the SOLIDS-CONTACT PROCESS. Shallow-depth sedimentation, also called TUBE SETTLING, generally uses inclined tubes in a conventional basin. The solids-contact process combines coagulation, flocculation, and sedimentation in one treatment unit, in which the flow of water is vertical. Both processes are described more fully later in this module.

4-2. Sedimentation Facilities

Types of Sedimentation Basins

Sedimentation takes place in specially designed basins. These basins are known as SETTLING TANKS, SETTLING BASINS, SEDIMENTATION TANKS, SEDIMENTATION BASINS, or CLARIFIERS. They can be rectangular, square, or circular. The most common types of basins are the rectangular tank and the circular basin with center feed.

In rectangular basins (Figure 4-3A), the flow is essentially in one direction and is parallel to the basin's length. This is called RECTILINEAR FLOW. In center-feed, circular basins (Figure 4-3B), the water flows radially from the center to the

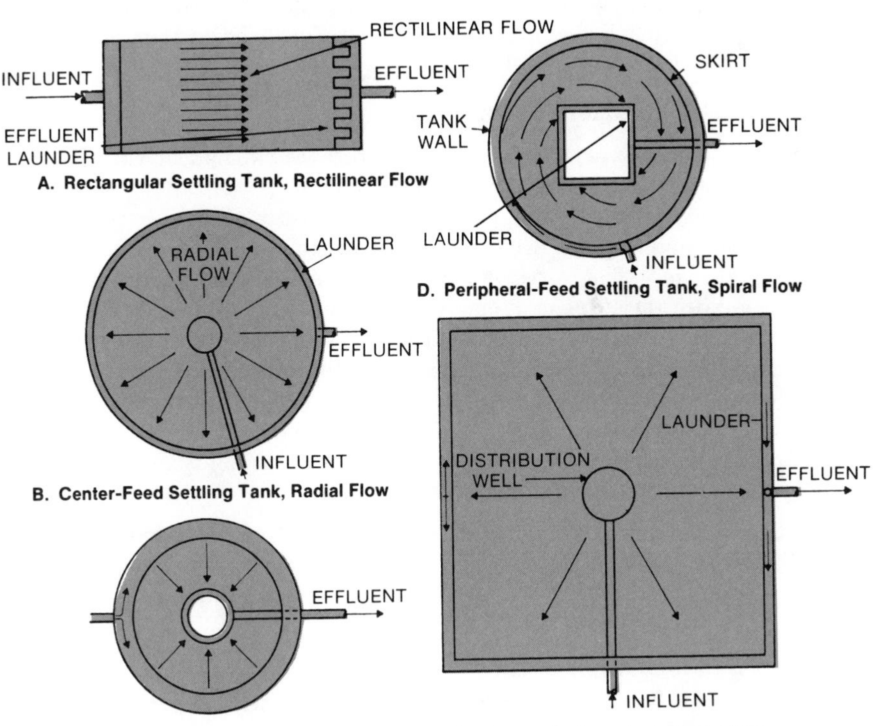

Figure 4-3. Flow Patterns in Sedimentation Basins

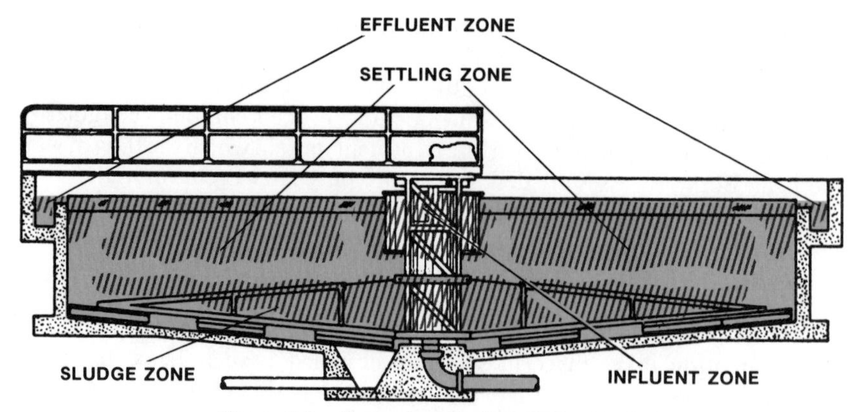

Figure 4-4. Zones in a Sedimentation Basin

outside edges. This is called RADIAL FLOW. Both basins are designed to keep the velocity and flow distribution as uniform as possible to prevent currents and eddies from forming, which would keep the suspended material from settling. Other flow patterns are shown in Figure 4-3C, D, and E.

The basins are usually made of steel or reinforced concrete. The bottom slopes slightly to make sludge removal easier. In rectangular tanks, the bottom slopes toward the inlet end, whereas in circular or square tanks the bottoms are conical and slope toward the center of the basin.

Zones in a Sedimentation Basin

Regardless of the type of basin used, four zones exist, each with its own function. These zones, shown in Figure 4-4, are:

1. *Inlet zone.* The INLET ZONE decreases the velocity of the incoming water and distributes the flow evenly across the basin.
2. *Settling zone.* The SETTLING ZONE provides the calm (quiescent) area necessary for the suspended material to settle.
3. *Outlet zone.* The OUTLET ZONE provides a smooth transition from the settling zone to the effluent flow area. It is important that currents or eddies that could stir up any settled solids and carry them into the effluent do not develop in this zone.
4. *Sludge zone.* The SLUDGE ZONE receives the settled solids and keeps them separate from other particles in the settling zone.

These zones are not as well defined as in the illustration. In actual operation, the settling zone is directly affected by the other three zones, causing less efficient settling. The greater the effect on the settling zone, whether due to poor design or operating practice, the worse the effluent quality.

Parts of a Sedimentation Basin

Equipment used in conventional settling basins varies depending on the design and manufacturer. Figures 4-5 and 4-6 show the parts of a rectangular and circular basin, respectively.

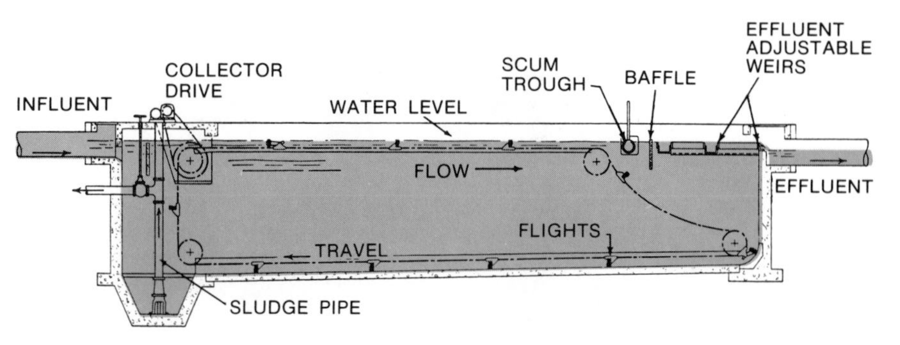

Courtesy of the FMC Corp., Material Handling Systems Division

Figure 4-5. Parts of a Rectangular Basin

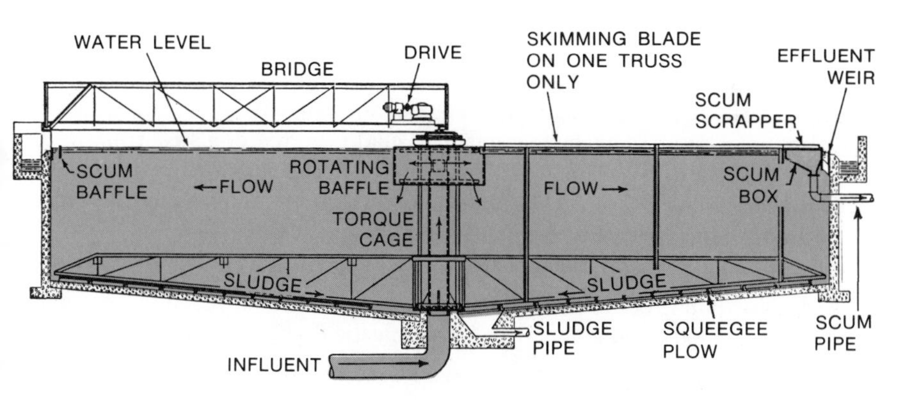

Courtesy of the FMC Corp., Material Handling Systems Division

Figure 4-6. Parts of a Circular Basin

The inlet distributes the influent evenly across (or around) the basin so that the water will flow uniformly. A BAFFLE, installed downstream of the inlet, reduces the velocity of the incoming water and helps produce calm, nonturbulent flow conditions for the settling zone. The water flows underneath the baffle and into the main part of the basin.

The EFFLUENT LAUNDER, or trough, collects the settled water as it leaves the basin and channels it to the effluent pipeline, which carries the water to the next treatment process. Launders can be made of fiberglass or steel, or they can be made of reinforced concrete as an integral part of a concrete clarifier or sedimentation tank.

The launder is equipped with an OVERFLOW WEIR—a steel or fiberglass plate designed to evenly distribute the overflow to all points of the launder. One of the most common types of weirs, the V-notch, is shown in Figure 4-7. The launder

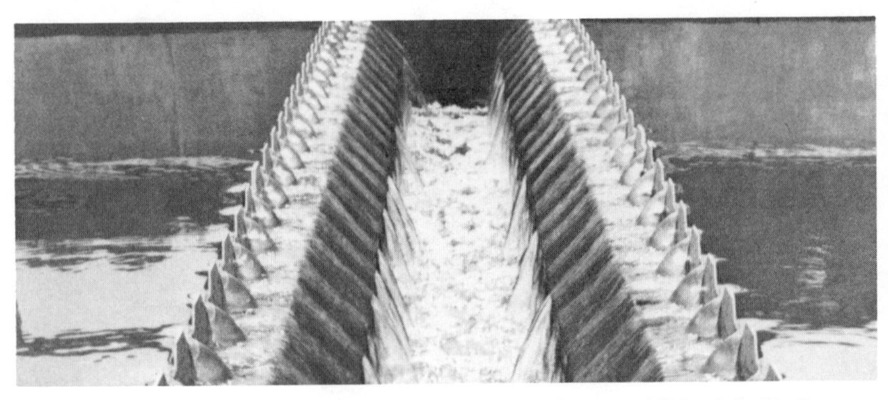

Courtesy of Fisher Scientific Company

Figure 4-7. V-Notch Weir

illustrated in Figure 4-7, commonly used in rectangular sedimentation tanks, allows water to enter from both sides.

Sludge Removal Equipment

As solids settle to the bottom of the basin, a sludge layer develops. This layer must be removed because solids can become resuspended or tastes and odors can develop. The sludge can be removed manually by periodically draining the basins and flushing the sludge to a hopper and drawoff pipe. This is recommended only for small installations or where not much sludge is formed. Continuous removal equipment is installed in most cases.

For rectangular basins, the removal equipment is usually one of the following:

- A *chain and flight collector* (Figure 4-5) consists of a steel or plastic chain and redwood or fiberglass reinforced plastic flights (scrapers).
- A *traveling-bridge collector* (Figure 4-8) consists of a moving bridge, which spans one or more basins. The mechanism has wheels that travel along rails mounted on the basin's edge. In one direction, the scraper blade moves the sludge to the hopper. In the other direction, the scraper retracts and the mechanism skims any scum from the water's surface.
- A *floating-bridge siphon collector* (Figure 4-9) uses suction pipes to withdraw the sludge from the basin. The pipes are supported by foam plastic floats, and the entire unit is drawn up and down the basin by a motor-driven cable system.

Circular and square basins usually are equipped with scrapers or plows (labeled "squeegee plow" in Figure 4-6) that slant toward the center of the basin and sweep sludge to the effluent hopper or pipe. The bridge can be fixed as illustrated or it can rotate with the truss.

Regardless of the collection method, the sludge is washed or scraped into a sludge hopper. It is then discharged (usually by pumping) to the sludge treatment facilities.

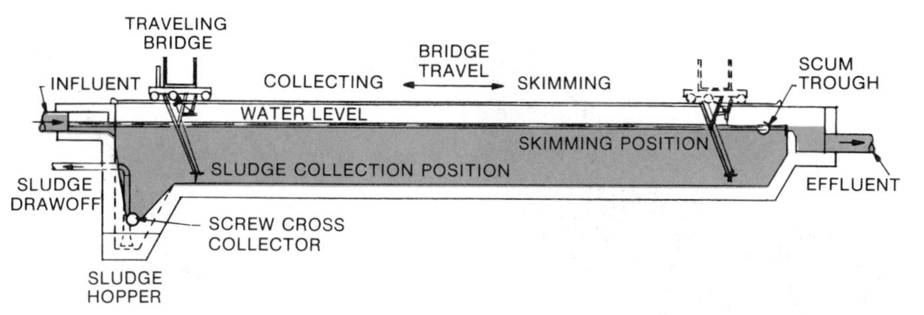

Figure 4-8. Travelling-Bridge Collector

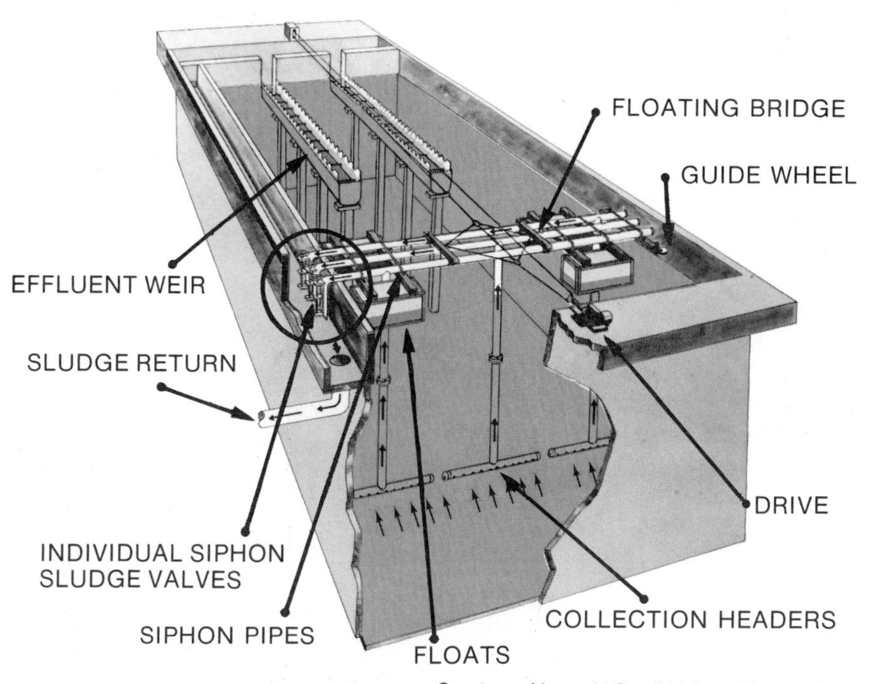

Courtesy of Leopold Co., Division of Sybron Corp.

Figure 4-9. Floating-Bridge Siphon Collector

Equipment for Shallow-Depth Sedimentation

If sedimentation basins were shallower, the settleable solids would be removed more rapidly. This, in turn, would allow shorter detention times and smaller basins. TUBE SETTLERS were developed with these facts in mind. The tubes are made of plastic and are either 2 in. (50 mm) square or chevron-shaped. The tubes are about 2 ft (0.6 m) long and are produced in modules of about 750 tubes, as shown in Figure 4-10.

The tubes are inclined about 60 degrees from horizontal to establish the countercurrent flow shown in Figure 4-11. As the water flows upward, the

Courtesy of Neptune Microflox, Inc.

Figure 4-10. Tube Settlers

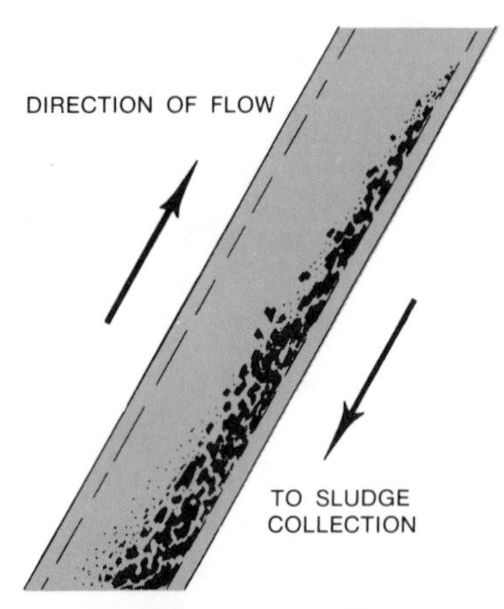

Courtesy of Neptune Microflox, Inc.

Figure 4-11. Counter-Current Flow in Tubes

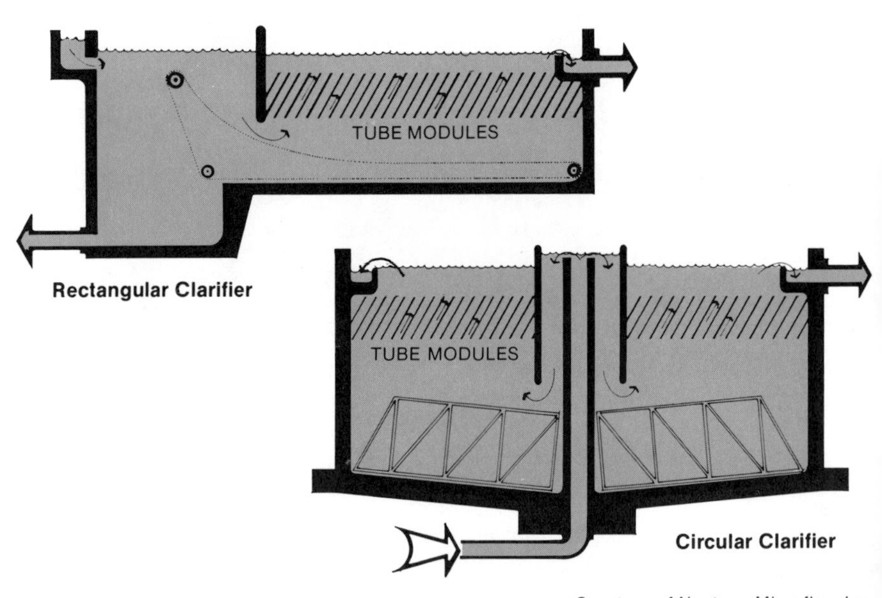

Courtesy of Neptune Microflox, Inc.

Figure 4-12. Retrofitting

settling solids move downward. This provides continuous sludge removal, which would not occur if the tubes were inclined at a lesser angle. Since they are modular, the tubes can be incorporated into new facilities or placed in existing basins. Retrofitting existing basins has resulted in improved settled-water quality and even higher flow rates. Examples of retrofit installations are shown in Figure 4-12.

Continuous mechanical sludge collection is desirable when tube settlers are

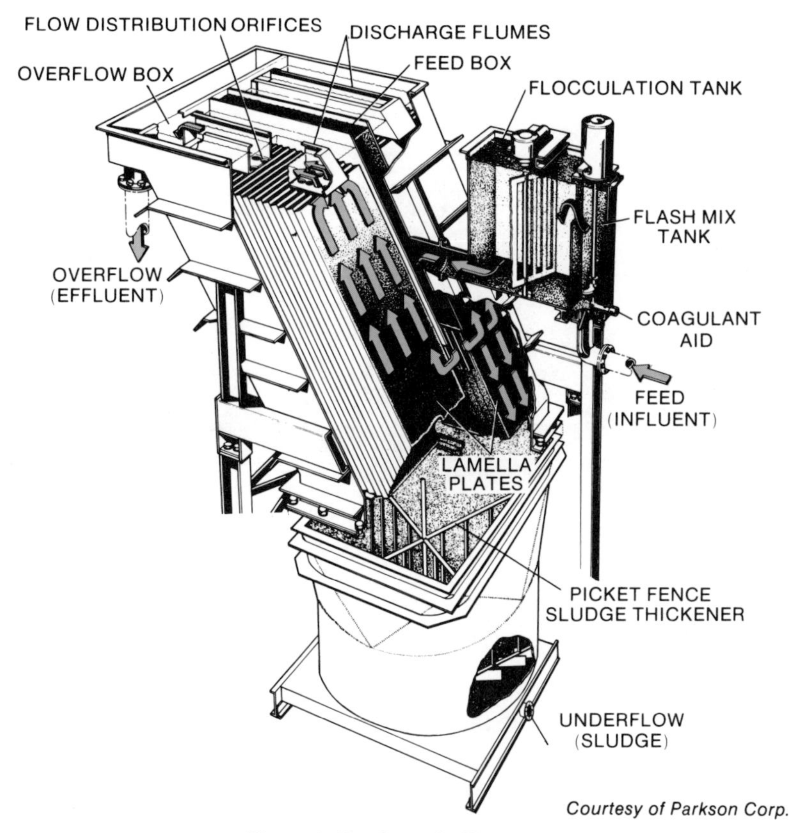

Figure 4-13. Lamella Plates

used, since more sludge will be created due to higher flow rates and more effective solids removal. If manual cleaning is used, the frequency of cleaning must be increased.

Another design of shallow-depth sedimentation uses LAMELLA PLATES (Figure 4-13), which are installed parallel at a 45-degree angle. In this case, the water and sludge flow in the same direction. The clarified water is returned to the top of the unit by small tubes.

Solids-Contact Basins

SOLIDS-CONTACT BASINS, also called UPFLOW CLARIFIERS or SLUDGE-BLANKET CLARIFIERS, are primarily used in the lime-soda ash process for water softening (described in Module 5, Softening) but are also used for turbidity and color removal. Several designs are available, but all are divided by baffles into two distinct zones: mixing and settling. Coagulation and flocculation take place in the mixing zone (which includes the reaction/flocculation area in the basin illustrated), where raw water and coagulant chemicals are combined and slowly agitated.

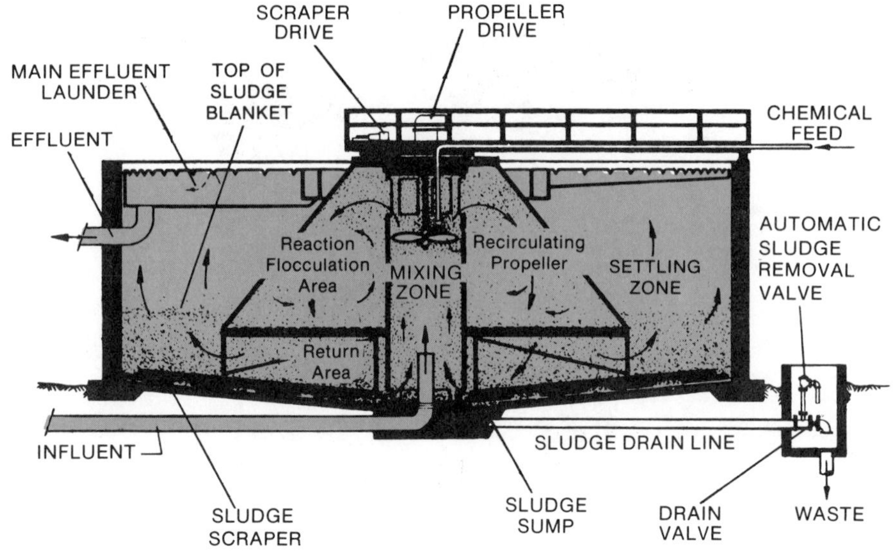

Courtesy of General Filter Company, Ames, Iowa

Figure 4-14. Sludge-Blanket Clarifier

Near the bottom of the basin, the flow is directed upward into the settling zone, which is separated from the reaction/flocculation area by baffles. A point exists at which the upflow velocity can no longer support the floc particles. This point defines the top of the sludge blanket. Essentially, the water has been "filtered" through this sludge blanket and the clarified water flows upward into the effluent troughs. The floc particles in the sludge blanket contact other particles and grow larger until they settle to the bottom. A portion is recycled to the mixing zone and the remainder settles in a concentration area for periodic disposal through the blow-off system. This must be done to maintain an almost-constant level of solids in the unit.

The advantage of this unit is that the chemical reactions in the mixing area occur faster and more completely due to the recycled materials from the sludge blanket. This allows much lower detention times than with conventional basins.

4-3. Operation of Sedimentation Basins

The primary function of sedimentation is to produce settled water with the lowest possible turbidity. For effective filtration, settled-water turbidity should not exceed 10 NTU (NEPHELOMETRIC TURBIDITY UNITS). An effective sedimentation process can also remove substantial amounts of organic compounds that cause color and trihalomethane formation.

Since effective sedimentation is closely linked with proper coagulation and flocculation, the operator must ensure that the best possible floc is being formed. Methods to do this are covered in Module 3, Coagulation/Flocculation.

Surface and Weir Overflow Rates

An important factor in proper sedimentation is the SURFACE OVERFLOW RATE—the flow rate through the basin divided by the basin's surface area.[1] The optimum surface overflow rate depends on the settling velocity of the floc particles. If the floc is heavy (as it is with lime softening), the overflow rate can be higher than with lighter, alum floc. A typical overflow rate for alum floc is 500 gpd/sq ft (0.24 mm/s).

A related factor, the WEIR OVERFLOW RATE, is the flow rate through the basin divided by the length of weir being used.[2] The higher the weir overflow rate, the more influence the outlet zone could have on the settling zone. To minimize this impact, it is recommended that the rate not exceed 20,000 gpd/ft (250,000 L/d/m). For light alum floc, the rate may have to be decreased to 14,400 gpd/ft (180,000 L/d/m), or 10 gpm/ft (125 L/min/m).

Detention time. The flow rate and basin volume determine the time it takes (theoretically) for water to move through the tank. This is known as the DETENTION TIME, also defined as the time it takes to fill the basin at a given flow rate.[3]

Conventional Basins

Conventional basins provide detention times from two to six hours. However, the actual detention period is often much different than the calculated period due to currents and other factors that cause the water to SHORT CIRCUIT through the basin. (Short circuiting, the flow of water through the basin in less time than the calculated detention time, is discussed later in this module.) For this reason, the surface and weir overflow rates are better factors to use for operation of sedimentation basins.

It may appear that little can be done to control the surface and weir overflow rates since an operator cannot change the size of the basins or the water demand. This is true to a certain extent, but the operator can experiment to determine the most effective overflow rates. If water demand makes it impossible to maintain these rates, the operator will at least know that adjustments elsewhere in the plant will have to be made to compensate for the poorer-quality settled water that may result. The operator must ensure that the total flow is divided evenly among the basins and that any changes in flow rate are made smoothly so that the basins do not receive surges that will break up the floc. Care must also be taken to make sure that the flow over the effluent weirs is uniformly and evenly distributed along the weir length.

Tube Settlers

If tube settlers are used, some additional operational considerations are necessary, since overflow rates used are two to three times those for conventional

---

[1] *Basic Science Concepts and Applications*, Mathematics Section, Surface Overflow Rate.

[2] *Basic Science Concepts and Applications*, Mathematics Section, Weir Overflow Rate.

[3] *Basic Science Concepts and Applications*, Mathematics Section, Detention Time.

basins. Due to these rates (and corresponding shorter detention times), it is essential that the floc have good settling characteristics. This may require some modification of the flocculation process, including addition of a coagulant aid. At times, the floc may bridge across the upper edge of the tube openings. This can result in a buildup of solids several inches thick. To dislodge this accumulation so that the floc particles can settle to the bottom, the water level of the basin must be dropped below the top of the tubes. If this is not possible, a gentle current of water must be directed across the top of the tubes using a hose or a permanent header (pipe) with openings along its length. Sludge withdrawal is likely to be required more frequently with tube settlers than with conventional basins.

Solids-Contact Basins

With solids-contact basins, the upflow rates used for turbidity removal are typically about 1 gpm/sq ft (0.7 mm/s), which results in detention times of only one to two hours. Proper coagulation/flocculation and control of the solids concentration in the slurry or sludge blanket (Figure 4-14) is essential for good results. To maintain a good sludge blanket, coagulant aids or weighting agents may be necessary. The solids concentration should be determined at least twice a day—more frequently if the water quality is always changing. This is done by taking samples from the taps provided on the basin and conducting settling tests prescribed by the manufacturer. Since solids-contact basins are used primarily in water softening, their operation is discussed more completely in Module 5, Softening.

Maintenance

Conventional basins should be drained at least once a year for a general inspection. The inspection will help determine if the sludge-removal equipment is operating properly. The inlet system should be examined closely to make sure all openings are clear of obstructions that could cause unequal flow distribution. The weirs must be level and should be routinely inspected. Inspection ensures that they are not being blocked by debris or chemical deposits, which could cause uneven flow and eddies that would disturb the settling zone. The baffles should be checked to determine if they are intact. Any algae or slime accumulations on basin walls, weirs, and baffles should be removed. Basins that are cleaned manually must be drained more frequently to wash the sludge deposits into the disposal system. Solids-contact basins should be drained at least annually to check the condition of the baffles and mixing equipment. Sludge pumping lines and equipment should be inspected routinely to ensure that plugging is not occurring. The lines should be flushed occasionally to prevent buildup of solids.

Sludge Disposal

Regardless of the type of basin used, sludge disposal is probably the most troublesome operating problem. The method used depends on the nature and volume of the sludge formed, which in turn depends on the raw-water quality and type of coagulants used. For example, different methods are typically used for lime and alum sludges. Methods of handling lime sludge are discussed in Module 5, Softening.

In the past, sludges were discharged into streams without treatment. This is no longer done because the sludge can form deposits in the receiving water that are harmful to aquatic life and can produce objectionable tastes and odors.

Alum sludge. Alum sludge is the most common form of sludge resulting from the sedimentation process, since alum is the coagulant most often used to remove turbidity. Alum sludge, a gelatinous, VISCOUS material, typically contains only 0.1 to 2 percent solid material by weight. However, it is extremely hard to handle and DEWATER because much of the water is chemically bound to the aluminum hydroxide floc.

One of the most common ways to handle alum sludge is to pump it into specially designed lagoons. When one lagoon is full, the sludge is diverted to a second lagoon. The sludge in the first lagoon is allowed to dry until it can be removed for final disposal. Water DECANTED from the top can be returned to the treatment plant. This process can take a year or longer, and the sludge can still be over 90 percent water, making it difficult to remove and unfit for disposal in a sanitary landfill. Therefore, it must be placed on land owned by the water utility. Because of the large land requirements, lagooning may not be possible for large plants or where land is not available. In areas where freezing temperatures are common, the freezing and thawing can provide excellent dewatering.

Sand drying beds can accomplish more efficient dewatering of sludge than lagoons. The sludge is spread in 2- to 4-ft (0.6- to 1.2-m) layers over 6 to 9 in. (150 to 230 mm) of sand overlying 12 in. (300 mm) of gravel and drain tiles to facilitate drainage. However, land requirements, difficulties with sludge removal, and poor performance during rainy periods are disadvantages of this method that must be considered.

Improved dewatering with far less land requirements can be accomplished using mechanical equipment. Vacuum filters, centrifuges, and filter presses can successfully dewater alum sludge to at least 20 percent solids by weight so that it can be placed in most landfills or used as a soil conditioner. However, mechanical dewatering usually has higher operation and maintenance costs than lagoons or sand drying beds.

Regardless of the mechanical equipment used, the sludge must undergo some pretreatment to make the subsequent dewatering more effective. This usually requires a sludge thickener, which is a circular tank much like a clarifier, equipped with a stirring mechanism (Figure 4-15). This mechanism breaks apart the floc particles in the sludge, allowing the water to escape and the solids to settle. Usually, a polymer is added to the sludge as it enters the thickener to enhance settling. A thickener can concentrate sludge up to about 5 percent solids by weight.

Because of the difficulties with dewatering and disposing of alum sludge, many operators reduce alum dosages through effective use of coagulant aids

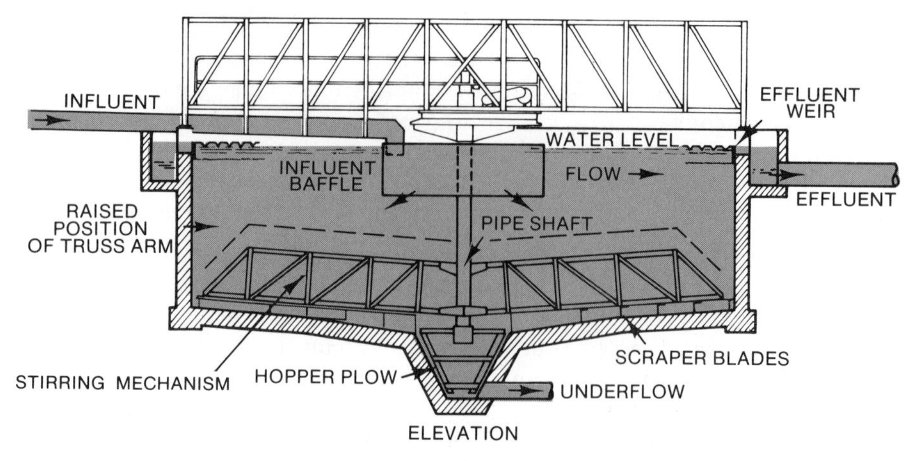

Figure 4-15. Sludge Thickener

such as polymers. This can greatly reduce the volume of alum sludge that must be handled.

Other sludges. Ferric chloride ($FeCl_3$), a common coagulant, produces a sludge that is difficult to dewater. This sludge is also difficult to thicken because it has a low density, which causes it to settle slowly.

Filter backwash water. Filter backwash water, discussed more thoroughly in Module 6, Filtration, is another source of sludge. Backwash water may be combined with the sludge from the sedimentation basins or treated separately.

4-4. Operating Problems

Poorly formed floc and short circuiting are the two most common problems in sedimentation. Both degrade the settled-water quality and increase turbidity loading on the filters.

Poorly formed floc, characterized by small or loosely held particles, does not settle properly and is carried out of the settling basin. This condition is the result of poor coagulation and flocculation. Inadequate rapid mixing, improper coagulants or dosages, and improper flocculation are the primary causes of poor coagulation/flocculation. JAR TESTS can provide the information necessary to find the specific problem. The solution may be to increase mixing energy, use a coagulant aid, or install baffles in the flocculation basin to prevent short circuiting.

Short circuiting—water bypassing the normal flow path through the basin and reaching the outlet in less than the normal detention time—occurs to some extent in every basin. It can be a serious problem in some installations, causing floc to be carried out of the basin due to shortened sedimentation time.

The major cause of short circuiting is poor inlet baffling. If the influent enters the basin and hits a solid baffle, strong currents and short circuits will result. A perforated baffle can be very successful in providing effective distribution at the

Figure 4-16. Perforated Baffles

Figure 4-17. Barrier Around an Uncovered Basin

inlet without causing troublesome currents (Figure 4-16). If short circuiting is suspected, TRACER STUDIES using salt or a dye tracer to determine the actual detention time can be conducted.

DENSITY CURRENTS can also be a problem. They occur because the influent contains more suspended solids, which cause it to have a greater density than the water in the basin. They can also occur when the influent is colder than the water in the basin. In either case, the denser influent sinks to the bottom of the basin, where it can create upswells of sludge and short circuits. If this is a problem, an engineering study should be conducted to determine the best solution. Many times the problem can be lessened by modifying the effluent weirs.

Wind can cause currents in open basins, which will result in short circuiting if the basins are not protected. A barrier should be constructed around uncovered basins in order to lessen the wind's effect and keep debris out of the water (Figure 4-17).

Another problem occurring in open basins located outside is algae and slime growth on the walls. These growths can cause tastes and odors. If the algae and slime detach from the walls, they can clog weirs or filters. They also create very slick, hazardous surfaces. The growths can be controlled by coating the walls with a mixture of 10 g of copper sulfate ($CuSO_4$) and 10 g of lime ($CaOH_2$) per litre of water. The basin should be drained and the mixture applied to the problem areas with a brush.

4-5. Control Tests and Record Keeping

The major test used to indicate proper sedimentation is the test for turbidity. The turbidity of samples taken from the raw water and outlet of each basin should be tested at least three times a day—more frequently if water quality is changing rapidly. By comparing these turbidities, an indication of the efficiency of the sedimentation process can be gained. For example, if the raw-water turbidity is 50 NTU and effluent turbidity is 40 NTU, very little turbidity is being removed. This indicates that better coagulation/flocculation is needed or that short circuiting is occurring. Visual examination of water samples from the effluent can also indicate if floc is being carried over onto the filters.

Turbidity of the settled water should be kept below 10 NTU. If turbidity is above 10 NTU, the coagulation/flocculation and sedimentation processes should be checked and their operation improved. Turbidity test procedures are discussed in Module 5 of Volume 4, *Introduction to Water Quality Analyses.*

As mentioned in Module 3, Coagulation/Flocculation, the temperature of the raw water should be measured and recorded at least daily. As water gets colder, it becomes more viscous, thus presenting more resistance to the settling particles. To compensate for this, the surface overflow rates may have to be reduced.

Record Keeping

Good records can be invaluable in helping the operator to solve problems and produce high-quality water. As raw-water quality or other conditions change, the operator can review past records to help determine what adjustments are needed. Since sedimentation is closely linked with coagulation/flocculation, records for both should be kept together.

Sedimentation records should include:

- Surface and weir overflow rates (calculated knowing the flow rate through each basin)
- Turbidity results for raw water and effluent from each basin
- Quantity of sludge pumped or cleaned from each basin
- Types of operating problems and corrective actions taken.

4-6. Safety and Sedimentation

Open sedimentation basins should be equipped with guardrails that will prevent falls into the basins. Walkways and bridges connected to basins should also have guardrails. Life rings or poles should be kept near the basins for rescue purposes.

Particular care must be taken when rain, ice, or snow is on walkways, since they become very slippery. Caution is also needed when cleaning or inspecting drained basins, since growths of aquatic organisms and sludge deposits can make the surfaces slippery.

Moving parts of all machinery should be equipped with guards to prevent the machinery from catching legs, fingers, or clothing. Guards must be left in place while the machinery is operating.

Selected Supplementary Readings

Basic Water Treatment Operation. Ministry of the Environment, Toronto, Ontario (1974). Chap. 3.

Cox, Charles R. *Operation and Control of Water Treatment Processes.* World Health Organization, Geneva, Switzerland (1969). Chap. 6.

Do Utilities Really Need Operators? *OpFlow*, 1:5:1 (May 1975).

Doe, P.W. Now It's Time to Rethink Sludge Disposal Ideas. *OpFlow*, 2:11:1 (Nov. 1976).

Doe, P.W. Classification, Collection, and Concentration of Sludge. *OpFlow*, 2:12:1 (Dec. 1976).

Doe, P.W. Thickening Methods and Treatment Processes for Sludge. *OpFlow*, 3:1:1 (Jan. 1977).

Doe, P.W. Ultimate Disposal of Sludge. *OpFlow*, 3:2:1 (Feb. 1977).

Manual of Instruction for Water Treatment Plant Operators. New York State Dept. of Health, Albany, N.Y. (1975) Chap. 8.

Manual of Water Utility Operations. Texas Water Utilities Assoc., Austin, Texas. (6th ed., 1975). Chap. 8.

Safety Practice for Water Utilities. AWWA Manual M3. AWWA, Denver, Colo. (1983).

Streicher, L. Can Organic Polymers Help Reduce the Total Volume of Sludge for Disposal? *OpFlow*, 5:2:1 (Feb. 1979).

Upgrading Water Treatment Plants to Improve Water Quality. AWWA Seminar Proc. Atlanta, Ga. (June 1980).

Water Quality and Treatment. AWWA Handbook. McGraw-Hill Book Company, New York (3rd ed., 1979). Chap. 4.

Water Works Operators Manual. Alabama Dept. of Public Health, Montgomery, Ala. (3rd ed., 1972). Chap. XV.

Glossary Terms Introduced in Module 4

(Terms are defined in the Glossary at the back of the book.)

Baffle
Clarification
Clarifier
Decanted
Density current
Detention time
Dewater
Effluent launder
Inlet zone
Jar Test
Lamella plate
NTU
Nephelometric turbidity unit
Outlet zone
Overflow weir
Plain sedimentation
Presedimentation
Radial flow
Rectilinear flow
Sedimentation basin
Sedimentation tank
Settling basin
Settling tank
Settling zone
Shallow-depth sedimentation
Short circuit
Sludge
Sludge-blanket clarifier
Sludge zone
Solids-contact basin
Solids-contact process
Surface overflow rate
Tracer study
Tube settler
Tube settling
Turbulence
Upflow clarifier
Viscous
Weir overflow rate

Review Questions

(Answers to Review Questions are given at the back of the book.)

1. What is the purpose of sedimentation?
2. What are the two basic types of sedimentation and how are they used?
3. What is the flow pattern through a rectangular sedimentation basin? What is the flow pattern through a circular clarifier?
4. List the four sedimentation zones and their function.
5. Define detention time.
6. What problems are associated with alum sludge? What can be done to reduce these problems?

7. A circular clarifier has a diameter of 45 ft and a depth of 10 ft. The overflow weir is mounted on the perimeter of the basin. Water is flowing through the basin at the rate of 1.5 mgd. Determine:
   (a) Detention time (in hours)
   (b) Surface overflow rate
   (c) Weir overflow rate.

8. Coagulated/flocculated water enters a sedimentation basin 25 ft wide and 100 ft long at a rate of 3.3 mgd. There are 200 ft of overflow weir and the basin has a 10-ft depth. Calculate:
   (a) Detention time (in hours)
   (b) Surface overflow rate
   (c) Weir overflow rate.

9. A sedimentation basin is treating water at a flow rate of 0.85 mgd. The basin is 57 ft long and 12 ft wide. The basin is 10 ft deep. Calculate the flow-through velocity (ft/min).

10. What is short circuiting and what causes it?

11. What is the key operational test to judge performance of the sedimentation process?

Study Problems and Exercises

1. Investigate the methods used for sludge disposal, reclamation, and reuse in your local area. Prepare a brief report describing these methods and what operating problems are being experienced.

2. From flow data and dimensions of individual sedimentation basins, calculate the following factors for one sedimentation unit at a local water treatment plant:

   (a) Surface overflow rate
   (b) Weir overflow rate
   (c) Detention time.

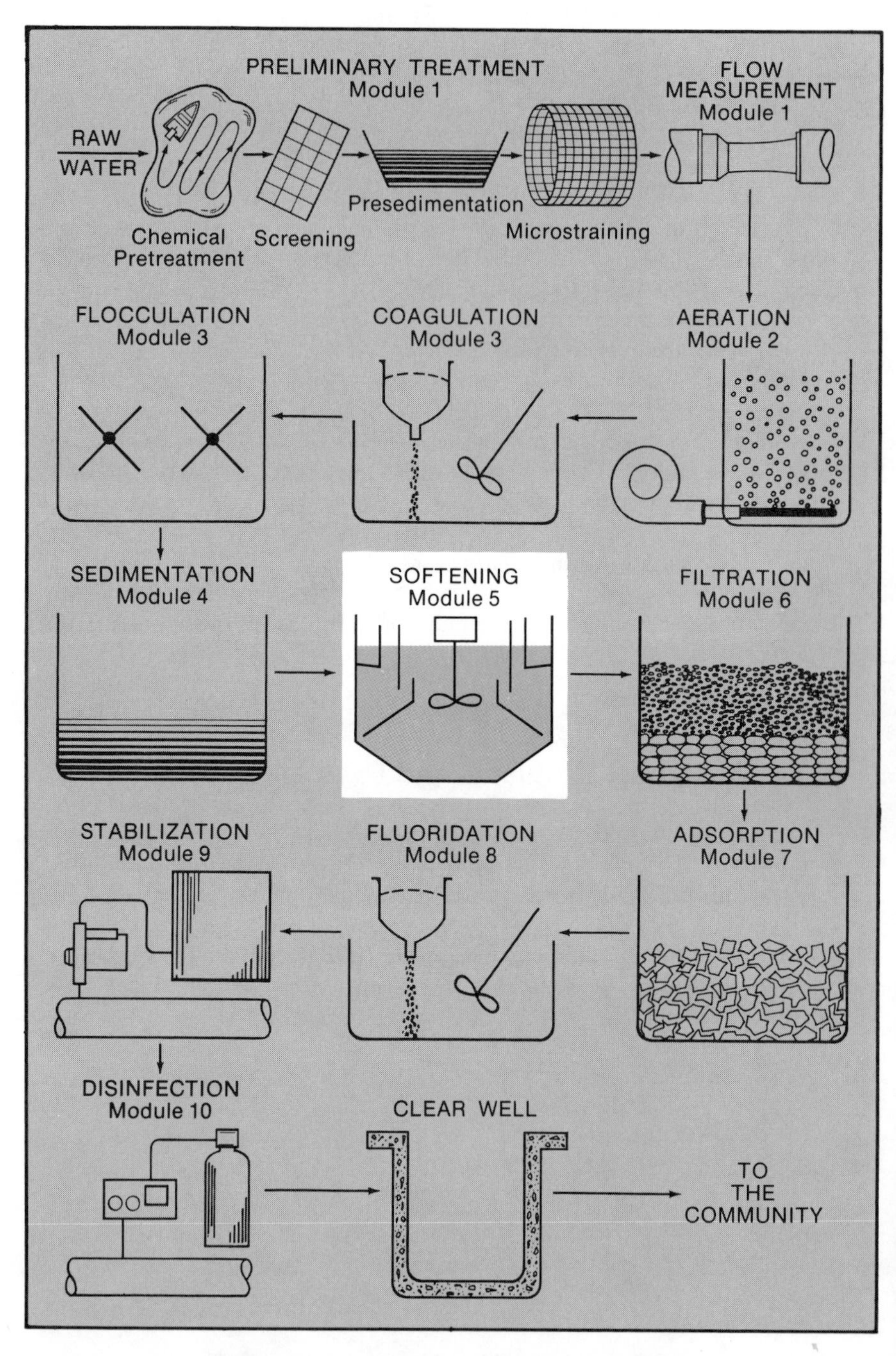

Figure 5-1. Softening in the Treatment Process

Module 5

Softening

Water contains various amounts of dissolved minerals. Some of these minerals, calcium and magnesium in particular, cause HARDNESS. Consumers frequently complain about the problems caused by hard water, such as the formation of scale in cooking utensils and hot-water heaters. The softening process (Figure 5-1) removes the hardness-causing minerals from water. The two most common water softening techniques are the LIME-SODA ASH METHOD and the ION-EXCHANGE PROCESS. To successfully perform either of these softening processes, an operator must have a clear understanding of the complex chemistry involved.

After completing this module you should be able to:

- Describe the purpose of softening water.
- Name the minerals causing water hardness.
- Identify the various types of water hardness.
- List the methods of softening water.
- Describe the lime-soda ash softening process.
- Describe the ion-exchange softening process.
- List the major operating variables of each process.
- Identify typical operating problems and solutions.
- Describe the commonly used laboratory control tests.
- Describe what safety precautions should be observed.

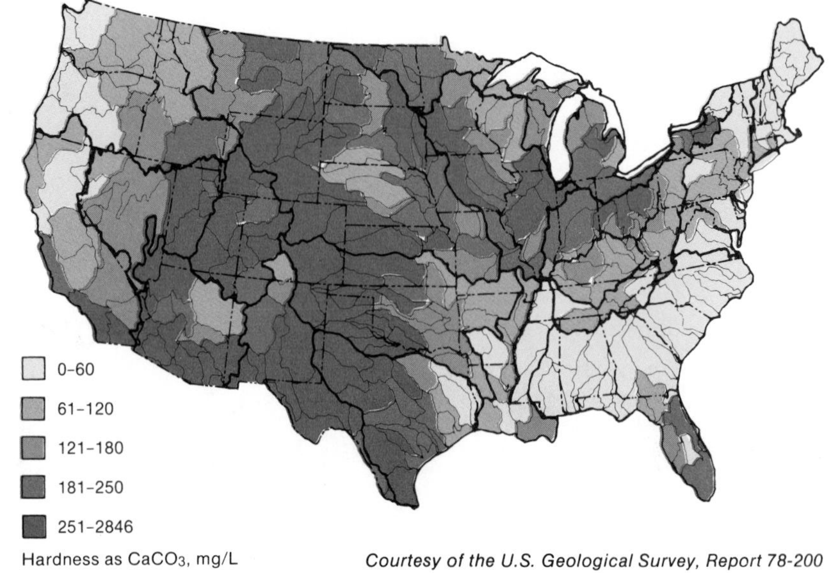

Courtesy of the U.S. Geological Survey, Report 78-200

Figure 5-2. Average Water Hardness in the Continental United States

5-1. Water Hardness and Softening Processes

The chemistry of water hardness and the related chemistry of water STABILIZATION are complicated and difficult subjects. This section describes the basic chemistry of hardness and its effects, then gives general descriptions of the two major treatment processes used to reduce hardness to an acceptable level. The following sections in this module describe the treatment processes in detail. The chemistry of stabilization is covered in Module 9, Stabilization.

Chemistry and Effects of Hard and Soft Water

Hard water is caused by soluble, DIVALENT, metallic CATIONS (positive ions having a valence of 2).[1] The two main cations that cause water hardness are calcium (Ca) and magnesium (Mg). Although strontium, aluminum, barium, iron, manganese, and zinc also cause hardness in water, they are not usually present in large enough concentrations to contribute significantly to total hardness.

As shown in Figure 5-2, water hardness varies considerably in different geographic areas due to the different geologic formations in each area and the length of time the water has been in contact with these formations. Calcium is dissolved as water passes over and through limestone deposits. Magnesium is dissolved as water passes over and through dolomite and other magnesium-bearing minerals. Because ground water is in contact with these geologic

---

[1] *Basic Science Concepts and Applications*, Valence, Chemical Formulas, and Chemical Equations—Valence.

Table 5-1. Comparative Classification of Water for Softness/Hardness

Classification	mg/L as $CaCO_3$*	mg/L as $CaCO_3$†
Soft	0-75	0-60
Moderately Hard	75-150	61-120
Hard	150-300	121-180
Very Hard	over 300	over 180

*Sawyer, C.N. *Chemistry for Sanitary Engineers*. McGraw-Hill Book Company, New York (1960).
†Briggs, J.C. & Ficke, J. F. Quality of Rivers in the United States, 1975 Water Year. USGS Open-File Rpt. 78-200. USGS, Reston, Va. (1977).

formations for a longer period of time than surface water, ground water is normally harder than surface water.

Expressing hardness concentrations. Water hardness is generally expressed as a concentration of CALCIUM CARBONATE[2] (in terms of mg/L as $CaCO_3$). The degree of hardness consumers consider objectionable will vary, depending both on the water and on the degree of hardness to which consumers have become accustomed. Consumers who are accustomed to soft water of 20 mg/L hardness as $CaCO_3$, for example, might consider 100 mg/L excessive; consumers who are accustomed to very hard water might consider 100 mg/L quite acceptable. Table 5-1 shows a comparative classification for softness and hardness in water.

Types of hardness. Hardness can be categorized by either of two methods:

- Calcium and magnesium hardness
- Carbonate and noncarbonate hardness.

Calcium and magnesium hardness is identified by the minerals involved. Hardness caused by calcium is called CALCIUM HARDNESS, regardless of the salts associated with it, which include calcium sulfate ($CaSO_4$), calcium chloride ($CaCl_2$), and others. Likewise, hardness caused by magnesium is called MAGNESIUM HARDNESS. Since calcium and magnesium are normally the only significant minerals that cause hardness, it is generally assumed that Total Hardness = Calcium Hardness + Magnesium Hardness.

Carbonate and noncarbonate hardness are identified both by the bicarbonate salts of calcium and by the normal salts of calcium and magnesium involved in causing the water hardness. CARBONATE HARDNESS is primarily caused by the bicarbonate salts of calcium and magnesium—that is, calcium bicarbonate [$Ca(HCO_3)_2$] and magnesium bicarbonate [$Mg(HCO_3)_2$]. Calcium and magnesium combined with carbonate (CO_3) also contribute to carbonate hardness. NONCARBONATE HARDNESS is a measure of calcium and magnesium salts other than carbonate and bicarbonate salts—that is, calcium sulfate, calcium chloride, magnesium sulfate ($MgSO_4$), and magnesium chloride ($MgCl_2$). Although it is a very rare condition, calcium and magnesium combined with nitrate (NO_3) may also contribute to noncarbonate hardness. For carbonate and noncarbonate hardness, Total Hardness = Carbonate Hardness + Noncarbonate Hardness.

---

[2] *Basic Science Concepts and Applications*, Chemistry Section, Solutions—Hardness.

When hard water is boiled, carbon dioxide (CO_2) is driven off. Bicarbonate salts of calcium and magnesium then precipitate (settle) out of the water to form calcium and magnesium carbonate PRECIPITATES. These hardness-causing precipitates form the familiar chalky deposits on teapots and other containers. Because it can be removed by heating, carbonate hardness is sometimes called TEMPORARY HARDNESS. Since noncarbonate hardness cannot be removed or precipitated by prolonged boiling, it is sometimes called PERMANENT HARDNESS.

Objections to hard water. Hard water forms scale, usually calcium carbonate, which causes a variety of problems. Left to dry on the surface of glassware and plumbing fixtures (shower doors, faucets, and sink tops), hard water leaves unsightly white scale, called water spots. Scale that forms on the inside of water pipes eventually reduces the pipes' carrying capacity. Scale that forms within appliances and water meters causes wear on moving parts. When hard water is heated, scale forms much faster. This creates problems inside boilers, water heaters, and hot-water lines. A coating of 0.04 in. (1 mm) of scale on the heating surfaces of a water heater creates an insulation effect that will increase heating costs by about 10 percent.

The historical objection to hardness has been its effect on soap. Hardness ions form precipitates with soap. These precipitates are unsightly and result in laundering or washing problems. The bathtub ring is a familiar example of the precipitate formed by hard water and soap. To counteract these problems, synthetic detergents have been developed and are used almost exclusively for washing clothes and dishes. Synthetic detergents have additives known as SEQUESTERING AGENTS that "tie-up" the hardness ions so they cannot form the troublesome precipitates. Although modern detergents counteract many of the problems of hard water, some consumers may seek softer, alternative water supplies—many of these supplies cannot provide the assurance of safe water that a closely monitored public water system guarantees.

Concern over soft water. Although a relatively soft water is preferable, in most cases, for consumer use, it does have several disadvantages. Excessively soft water can cause corrosion in pipes. This corrosion can shorten the service life of pipes and household appliances and can result in toxic materials, such as lead and cadmium, being dissolved in drinking water. The problems of corrosion and corrosion control are discussed in Module 9, Stabilization. In addition, if the ion-exchange process is used to soften water, the sodium concentration in the treated water will increase significantly. This is an important health consideration, particularly for people who are on a restricted sodium diet due to heart disease or hypertension.

The decision to soften a water supply depends completely on the community. Drinking water regulations do not generally require softening—what may be "hard" water in one area may be perfectly acceptable water in another area. Some industries (such as the soft drink and laundry industries) need soft water for their manufacturing processes. Therefore, the decision to soften a water supply should be based on an analysis of the benefits that will be gained against the costs of installing and operating softening equipment. In many cases, it is not economical to soften the entire water supply to meet the special needs of users who could more economically install their own softening equipment.

Softening Methods

The two methods most widely used for softening public water supplies are:

- Lime-soda ash softening
- Ion-exchange softening.

Each softening process has advantages and limitations. Other methods can also be used to soften water, such as electrodialysis, distillation, freezing, and reverse osmosis. Although each of these processes can produce softened water, they are complex and expensive processes, used only in unusual circumstances.

Lime-soda ash softening. In the lime-soda ash process, lime and soda ash are added to the water and react with various salts of calcium and magnesium to form two insoluble precipitates: calcium carbonate and magnesium hydroxide. These precipitates are then removed by conventional sedimentation and filtration processes. The lime-soda ash process is used in most larger treatment plants. With some modifications, the process can also remove turbidity and color; therefore it should be used when softening surface water.

Caustic soda (sodium hydroxide—NaOH) can be used in place of lime and soda ash to remove carbonate and noncarbonate hardness. The major advantage of caustic soda is that it produces less SLUDGE than lime and soda ash. However, since caustic soda is more expensive and increases the amount of total dissolved solids in the treated water, it is not used as widely as lime and soda ash.

Ion-exchange softening. The ion-exchange method removes hardness ions by exchanging or replacing them with sodium ions, which do not cause hardness. In this softening process, ion-exchange materials, such as POLYSTYRENE RESINS, are placed in an ion-exchange unit through which unsoftened water is passed. As the name suggests, ion-exchange involves the transfer of one ion for another. In this case, calcium and magnesium ions attach to the resin, which releases sodium ions in exchange. With use, the ion-exchange resins lose their ability to remove hardness-causing ions. The resins are then regenerated by passing brine (a salt-water solution) through the ion-exchange unit.

The ion-exchange process is used in household softeners and is favored in smaller municipal water treatment plants using ground-water supplies. When a large percentage of water hardness is in noncarbonate form, the ion-exchange softening process is used because the chemical costs of removing noncarbonates are lower with the ion-exchange process than with the lime-soda ash process. The ion-exchange process cannot be easily adapted to treat surface waters.

5-2. Description of Lime-Soda Ash Softening

In the lime-soda ash softening process, lime [$Ca(OH)_2$] and soda ash [Na_2CO_3] are added to hard water to remove the hardness-causing minerals from the water. When lime and soda ash are added, the hardness-causing minerals form nearly insoluble precipitates. These precipitates are removed by the conventional processes of coagulation/flocculation, sedimentation, and filtration. Because the precipitates are very slightly soluble, some hardness remains in the water—usually about 50 to 85 mg/L (as $CaCO_3$). This level of hardness is desirable to prevent the corrosion problems associated with waters having little or no hardness.

Chemical Reactions

The following equations represent the basic chemical reactions involved in the lime-soda ash process. The down arrow [↓] indicates precipitates formed by the reactions. (Note: Equations are numbered the same as in *Basic Science Concepts and Applications*, Chemistry Section, Chemistry of Treatment Processes—Lime-Soda Ash Softening.)

First, although carbon dioxide does not cause hardness, it reacts with and consumes the lime added to remove hardness; therefore, it must be considered in determining lime dosage. Sufficient lime must be added to convert carbon dioxide (CO_2) to calcium carbonate ($CaCO_3$) as shown in Eq 25. This reaction is complete when a pH of 8.3 is reached. (In some cases aeration is used to remove the CO_2 thereby reducing the lime requirement.)

$$\underset{\text{Carbon dioxide}}{CO_2} + \underset{\text{Lime}}{Ca(OH)_2} \rightarrow \underset{\text{Calcium carbonate}}{CaCO_3\downarrow} + \underset{\text{Water}}{H_2O} \qquad \textit{Eq 25}$$

The minerals that cause carbonate hardness are then precipitated out of the water either as calcium carbonate (Eq 16) or as magnesium hydroxide (Eq 17 and Eq 18). In Eq 16, enough lime is added to raise the pH to 9.4, at which point the calcium precipitate is formed. Eq 17 and Eq 18 show a similar process—additional (excess) lime is added to elevate the pH above 10.6 to form the magnesium precipitate, Mg $(OH)_2$.

$$\underset{\text{Calcium bicarbonate}}{Ca(HCO_3)_2} + \underset{\text{Lime}}{Ca(OH)_2} \rightarrow \underset{\text{Calcium carbonate}}{2CaCO_3\downarrow} + \underset{\text{Water}}{2H_2O} \qquad \textit{Eq 16}$$

$$\underset{\text{Magnesium bicarbonate}}{Mg(HCO_3)_2} + \underset{\text{Lime}}{Ca(OH)_2} \rightarrow \underset{\text{Calcium carbonate}}{CaCO_3\downarrow} + \underset{\text{Magnesium carbonate}}{MgCO_3} + \underset{\text{Water}}{2H_2O} \qquad \textit{Eq 17}$$

$$\underset{\text{Magnesium carbonate}}{MgCO_3} + \underset{\text{Lime}}{Ca(OH)_2} \rightarrow \underset{\text{Calcium carbonate}}{CaCO_3\downarrow} + \underset{\text{Magnesium hydroxide}}{Mg(OH)_2\downarrow} \qquad \textit{Eq 18}$$

The minerals that cause noncarbonate calcium hardness are precipitated out of the water by adding soda ash. The chemical reactions differ slightly, based on the types of noncarbonate calcium compounds causing the hardness. Eq 19 and Eq 20 show two examples of these chemical reactions.

$CaSO_4$ Calcium sulfate	+	Na_2CO_3 Soda ash	→	$CaCO_3\downarrow$ Calcium carbonate	+	Na_2SO_4 Sodium sulfate		*Eq 19*
$CaCl_2$ Calcium chloride	+	Na_2CO_3 Soda ash	→	$CaCO_3\downarrow$ Calcium carbonate	+	$2NaCl$ Sodium chloride		*Eq 20*

The minerals that cause noncarbonate magnesium hardness are precipitated out of the water by adding lime (Eq 21 and Eq 23). However, this process also results in the formation of noncarbonate salts (such as $CaSO_4$ and $CaCl_2$) that cause noncarbonate hardness. Therefore, soda ash must be added to the water to react with these salts to form $CaCO_3$ (Eq 22 and Eq 24).

$MgCl_2$ Magnesium chloride	+	$Ca(OH)_2$ Lime	→	$Mg(OH)_2\downarrow$ Magnesium hydroxide	+	$CaCl_2$ Calcium chloride	*Eq 21*
$CaCl_2$ Calcium chloride	+	Na_2CO_3 Soda ash	→	$CaCO_3\downarrow$ Calcium carbonate	+	$2NaCl$ Sodium chloride	*Eq 22*
$MgSO_4$ Magnesium sulfate	+	$Ca(OH)_2$ Lime	→	$Mg(OH)_2\downarrow$ Magnesium hydroxide	+	$CaSO_4$ Calcium sulfate	*Eq 23*
$CaSO_4$ Calcium sulfate	+	Na_2CO_3 Soda ash	→	$CaCO_3\downarrow$ Calcium carbonate	+	Na_2SO_4 Sodium sulfate	*Eq 24*

Since the softened water has a pH close to 11 and a high concentration of $CaCO_3$, it must be "stabilized" so the $CaCO_3$ will not precipitate on the filter media, on filter underdrains, or in the distribution system. The process of stabilizing water is called RECARBONATION. It involves adding carbon dioxide to the softened water (Eq 26). When carbon dioxide is added, soluble calcium bicarbonate is formed (resulting in a small amount of hardness), and the pH is reduced to about 8.6 or to a level at which the water is stabilized to prevent scale formation or corrosion.

Eq 27A shows that any residual $Mg(OH)_2$ will also be converted to a soluble compound—$MgCO_3$.

$CaCO_3$ Calcium carbonate	+	$CO_2\uparrow$ Carbon dioxide	+	H_2O Water	→	$Ca(HCO_3)_2$ Calcium bicarbonate	*Eq 26*

$Mg(OH)_2$	+	$CO_2\uparrow$	$\rightarrow$	$MgCO_3\downarrow$	+	H_2O	*Eq 27A*
Magnesium hydroxide		Carbon dioxide		Magnesium carbonate		Water	

Sulfuric acid is sometimes used to accomplish stabilization. Since this process adds noncarbonate hardness ($CaSO_4$) and the acid is more difficult to handle, carbon dioxide is generally used for stabilizing softened water.

Types of Lime-Soda Ash Processes

As indicated by the variety of chemical reactions, different lime-soda ash softening processes can be used. The best process to use should be determined by (1) the types and amounts of hardness in the water to be treated, and (2) the quality required of the finished water. For example, if only carbonate hardness is present in the water, only lime will need to be added to soften the water. However, if both carbonate and noncarbonate hardness are present, both lime and soda ash must be used. A discussion of some of the common lime-soda ash softening processes follows.

Conventional lime-soda ash treatment. The conventional process is used when there is only a small amount of magnesium hardness in the water. Treatment is typically performed using the single-stage process shown in Figure 5-3(a).

Excess-lime treatment. When the magnesium hardness of water is more than about 40 mg/L (as $CaCO_3$), magnesium-hydroxide scale will deposit in household hot-water heaters that are operated at normal temperatures of 140 to 150° F (60 to 66° C). To reduce this magnesium hardness, more lime than is used in the conventional process must be added to the water. The extra lime will raise the pH above 10.6 so that the magnesium hydroxide will precipitate out of the water (Eqs 17, 18, 21, and 23). This process is known as EXCESS-LIME TREATMENT. When this treatment process is used, soda ash is added to remove noncarbonate hardness, and recarbonation is performed to minimize or eliminate the formation of scale. Excess-lime treatment can be performed either in a single-stage process or in a double-stage process. The double-stage process allows greater removal of magnesium hardness and more control over the quality of the treated water. Figure 5-3(b) shows the double-stage excess-lime treatment process.

Split treatment. Split treatment is a modification of the excess-lime process. It is designed to lower the amount of lime and carbon dioxide required to soften water. As shown in Figure 5-3(c), only a portion of the water is treated with excess lime. A smaller remaining portion of the water bypasses the lime-treatment process. The amount of water bypassing lime treatment depends on the quality of the raw water and the desired quality of the finished water. After the larger portion of water has been treated with excess lime and the smaller portion of water has bypassed lime treatment, the two portions of water are recombined. The carbon dioxide and bicarbonate alkalinity in the untreated portion of water help stabilize the treated portion of water. Stabilizing the

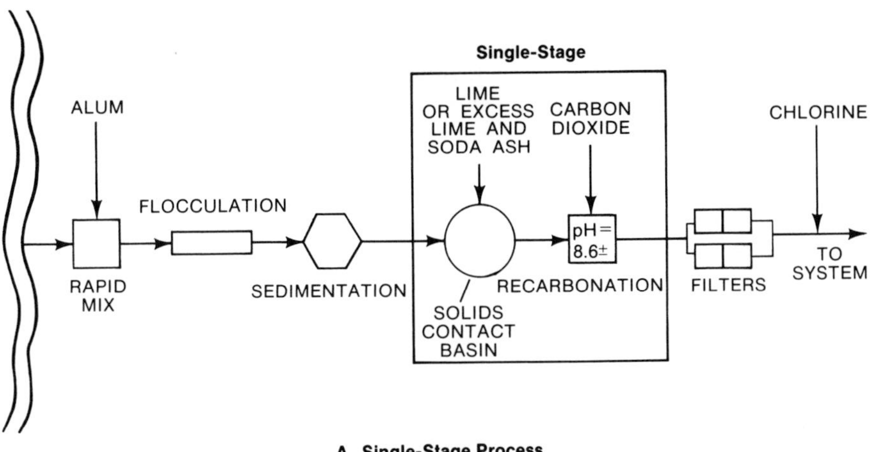

A. Single-Stage Process

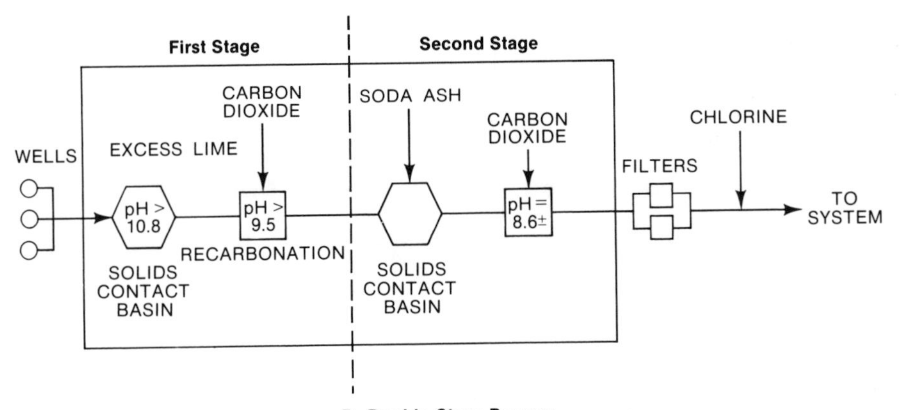

B. Double-Stage Process

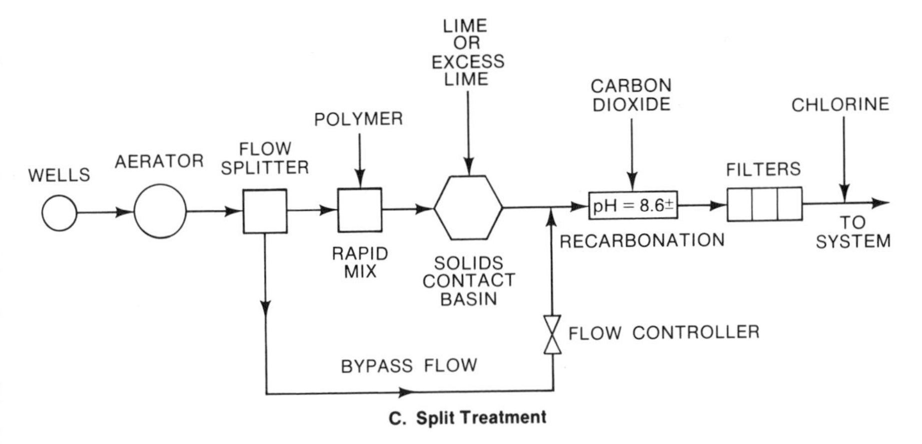

C. Split Treatment

Figure 5-3. Types of Lime-Soda Ash Softening Processes

treated portion of water minimizes or eliminates the addition of carbon dioxide for recarbonation.

Caustic-soda treatment. Caustic soda (sodium hydroxide—NaOH) can be used in place of lime and soda ash since it can remove both carbonate and noncarbonate hardness. When caustic soda is used, soda ash does not have to be added because it is formed during the other chemical reactions of sodium hydroxide with calcium and magnesium bicarbonate. Caustic soda is also easy to handle and results in less sludge formation. The major disadvantage of caustic soda is that it usually costs considerably more than lime and soda ash. Where sludge disposal is a problem, however, caustic soda may be economical to use.

Coagulation

The addition of coagulants is often necessary in the lime-soda ash softening process to help gather together the fine calcium and magnesium precipitates into particles that will readily settle out of the water. Adding coagulants will reduce the amount of suspended material that is deposited on the filters, which will result in longer filter runs, more efficient TURBIDITY removal, and less scale formation on the filter media. If the raw water is highly turbid or colored, a separate coagulation/flocculation step may be necessary before the softening step, since turbidity and color can adversely affect the softening process.

5-3. Lime-Soda Ash Softening Components

The lime-soda ash softening process involves the following major components:

- Chemical storage facilities
- Chemical feed facilities
- Rapid-mix basins
- Flocculation basins
- Sedimentation basins and related equipment
- Solids-contact basins (often used in place of the rapid-mix, flocculation, and sedimentation basins)
- Sludge recirculation, dewatering, and disposal equipment
- Recarbonation facilities
- Filtration facilities.

Chemical Storage Facilities

The size and number of facilities used for storing chemicals depend on the type and quantity of chemicals used. Lime can be purchased in two forms: (1) calcium hydroxide, $CaOH_2$, which is called either hydrated lime or SLAKED lime; and (2) calcium oxide, CaO, which is called either QUICKLIME or unslaked lime. Both calcium hydroxide and calcium oxide are dry chemicals, varying in texture from a light powder (Figure 5-4) to pebbles. Large treatment plants usually use

Courtesy of the National Lime Association

Figure 5-4. Pebble Quicklime

Courtesy of the National Lime Association

Figure 5-5. Chemical Storage Bins

Courtesy of the National Lime Association

Figure 5-6. Conveyor Used to Transport Calcium Oxide and Soda Ash

calcium oxide because it is less expensive. However, since calcium oxide does require special feeding equipment, small treatment plants may find calcium hydroxide less expensive to use. Soda ash (sodium carbonate—Na_2CO_3) is available in either powdered or granular form. All of these chemicals are available in bags, drums, or in bulk form (delivered in train or truck carloads).

Large quantities of chemicals are used in the water softening process. For example, in order to reduce the hardness of 1 mgd (4 ML/d) of water from 200 mg/L to 50–85 mg/L, more than 1200 lb (540 kg) of lime (in the form of either calcium oxide or calcium hydroxide) and 1200 lb (540 kg) of soda ash must be used every day. Chemical storage bins (Figure 5-5) are used to store these large quantities of chemicals. Since it is good practice to store a minimum of one month's supply of each chemical (depending on availability and transportation), chemical storage space requirements for each plant should be carefully calculated. Calcium oxide and soda ash are usually delivered either by rail car or by hopper-bottom truckload and are transferred by mechanical or pneumatic conveyor to storage bins (Figure 5-6). Calcium hydroxide is usually delivered and stored in 50- to 100-lb (23- to 45-kg) bags. When liquid caustic soda (50 percent NaOH) is used, it is stored and fed from lined-steel or plastic tanks.

Chemical Feed Facilities

Slaked lime and soda ash can be conveniently fed by conventional dry feeders located directly below the storage bins. Dispensing calcium oxide (unslaked lime) is a more complex process—in addition to the dry feeder, a lime slaker (Figure 5-7) and a solution feeder are required. The dry feeder dispenses the calcium oxide into the slaker, where the calcium oxide and water are mixed together to form a slurry known as MILK OF LIME [$CaO + H_2O \rightarrow Ca(OH)_2$]. The milk-of-lime slurry is then fed into the water by a solution feeder. Since the slurry will cake on any surface, slurry-moving equipment must be cleaned regularly. To allow for cleaning, open troughs or troughs with easily removable covers should be used to transport the slurry to the point where it is added to the water.

Tremendous heat is generated when lime is slaked, so adequate ventilation where lime slakers are located must be provided. Provisions must also be made to remove the grit produced during slaking. The grit consists of undissolved lime and impurities. The quantity of grit produced depends on the amount of lime slaked. Small amounts of grit can be collected in a wheelbarrow, but larger quantities are usually loaded directly into dump trucks and hauled to disposal sites daily.

In addition to dispensing calcium hydroxide and calcium oxide, conventional dry feeders and solution feeders are also used to dispense coagulants (such as alum) or coagulant aids (such as organic polymers) into the water. Dry feeders and solution feeders used for this purpose are illustrated in Module 3, Coagulation/Flocculation. Whenever lime, soda ash, or other dry powdered chemicals are used, chemical dust can pose a health hazard. Therefore, dust-control equipment should always be used when working with powdered chemicals. Such equipment is described in Module 8, Fluoridation.

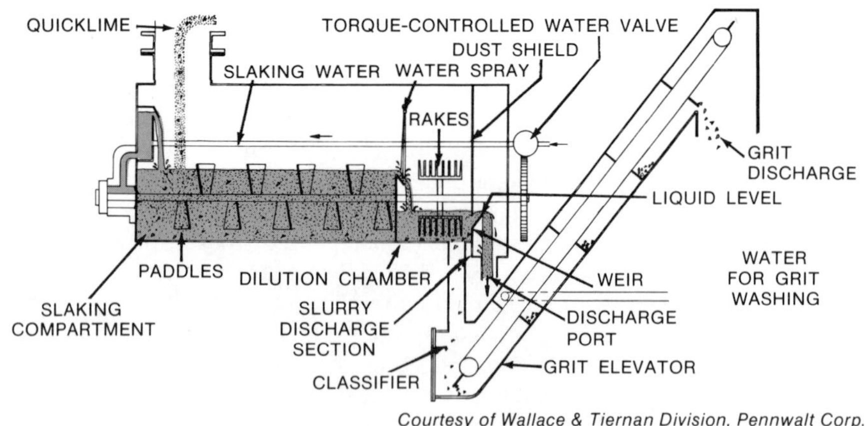

Courtesy of Wallace & Tiernan Division, Pennwalt Corp.

Figure 5-7. Lime Slaker

Rapid-Mix Basins

Rapid-mix basins used in lime-soda ash softening are similar to those described in Module 3, Coagulation/Flocculation, and serve two purposes:

- To thoroughly mix lime and soda ash with water
- To maintain mixing long enough for the chemicals to dissolve (this is especially critical for lime since it dissolves slowly).

Rapid-mix basins (having high-speed, mechanical mixers) are preferred for mixing chemicals over baffled basins because the mixing rate can be controlled independently of flow rate, making it possible to mix the chemicals more vigorously. In order to speed up the chemical reactions that occur in the lime-soda ash softening process, rapid-mix basins are often equipped to receive sludge that has been recirculated from the settling basin. Although coagulants can be added to rapid-mix basins, better results can be obtained by adding coagulants upstream in a separate rapid-mix basin.

Flocculation Basins

Flocculation is the process by which impurities are gathered together into solid particles (known as floc) that form and grow after coagulants have been added to the water. Flocculation basins used in the lime-soda ash softening process are similar to the flocculation basins used in conventional water treatment. They are equipped with gentle mixing devices (preferably rotating paddle wheels) so that the mixing of chemicals can be controlled. Variable-speed drives on the mixers provide the proper mixing speed needed for effective flocculation.

Sedimentation basins. Sedimentation (also known as settling) is the process by which suspended solid particles of floc settle out of water. Sedimentation basins used in the lime-soda ash softening process can be identical to those described in Module 4, Sedimentation. However, because of the larger amount of sludge formed in the lime-soda ash process, mechanical sludge collection and removal equipment in the basins is essential.

Solids-Contact Clarifiers

The solids-contact basin (or clarifier) described in Module 4, Sedimentation, was developed primarily for use in lime-soda ash softening. The basin is a single unit in which the processes of coagulation/flocculation, sedimentation, sludge collection, and sludge recirculation are performed (Figure 5-8). Several designs of solids-contact basins are available. All have two major zones:

- A *mixing zone* (which also includes the reaction/flocculation area in the basin illustrated) where lime (or lime with soda ash) is mixed with water and where previously formed slurry is recirculated to enhance chemical reactions and the formation of precipitates.
- A *settling or clarification zone* where the precipitates are allowed to settle as the clarified water passes upwards through the sludge blanket to the effluent troughs.

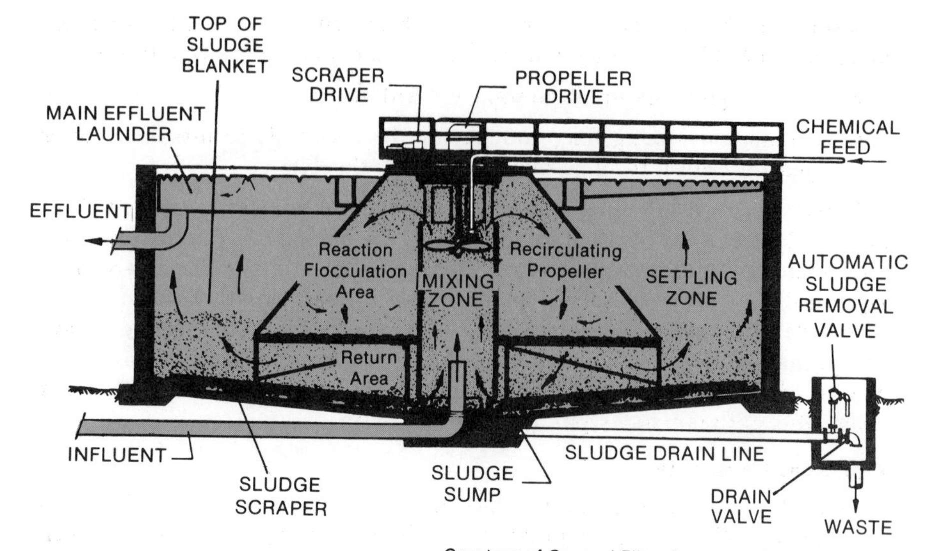

Courtesy of General Filter Company, Ames, Iowa

Figure 5-8. Solids-Contact Basin

One portion of the settling solids is returned to the mixing zone. The other portion of the solids is allowed to settle completely and is periodically withdrawn to maintain a fixed slurry concentration in the basin.

The major advantage of the solids-contact basin is that it saves construction costs by handling many softening processes in one component.

Sludge Recirculation, Dewatering, and Disposal Equipment

The flocculation and settling of calcium carbonate and magnesium hydroxide precipitates occur more quickly in the presence of previously formed floc particles. For this reason, settled sludge is commonly recirculated back to a rapid-mix basin in facilities where solids-contact basins are not used. Whether or not sludge recirculation is practiced, sludge must be removed from the settling basin. Although sludge removal can be accomplished by gravity drains, it is normally achieved with special sludge-handling pumps that provide for sludge removal and sludge recirculation.

Sludge-handling pumps are usually located in rooms adjacent to the settling basins. Centrifugal pumps with open impellers (to prevent plugging) are used to dispose of sludge, whereas slow-speed centrifugal pumps (or positive-displacement pumps) are used to recirculate sludge. The relatively gentle action of positive-displacement pumps minimizes the breakup of the flocculated solids, which helps the sludge retain good settling characteristics. Sludge recirculation equipment includes sludge flow-measuring devices, such as a flow meter (a magnetic meter, ultrasonic meter, or flow tube) or a progressing-cavity, variable-speed metering pump.

The large amount of sludge produced during lime-soda ash softening must be disposed of (or reclaimed). The disposal of sludge can be a difficult task. The best disposal method to use depends on many factors, such as the amount and type of sludge to be disposed of and the available disposal sites.

The process of dewatering sludge reduces the amount of water in the sludge. Most sludge is disposed of by trucking it to a land-disposal site, so the drier (and less voluminous) the sludge, the cheaper the disposal process. The more commonly used sludge-dewatering devices include:

- Drying beds
- Lagoons
- Sludge thickeners
- Vacuum filters
- Centrifuges.

Drying beds. A drying bed is a layer of sand placed over graded aggregate (gravel or stones) and an underdrain system. The sand is usually 4 to 9 in. (100 to 230 mm) deep. It is placed on top of 8 to 18 in. (200 to 460 mm) of graded aggregate. The aggregate rests on top of an underdrain system, consisting of perforated clay pipe or pipe laid with open joints. The underdrain system collects the water and either returns it to the water treatment plant or transports it to a waste disposal site. Drying beds are affected by weather. They perform best in areas having clear sunny days, warm weather, and little rain and snow.

Lagoons. Lagoons (large holding or detention ponds) provide the most economical way to dewater sludge; however, they cannot be used in many areas because the land required is not available. Lagoons are usually about 10 ft (3 m) deep and may be an acre (4000 m^2) or more in size. Once sludge is piped into the lagoon, the solids settle out of the sludge. Usually, the water on top of the sludge (DECANT) is either piped back to the treatment plant or discharged to a sewer or stream. When lagoons are used to dewater sludge, more than one lagoon is needed so that as the settled solids are being cleaned from one, the sludge from the plant can be diverted to the other.

Sludge thickeners. Sludge thickeners (Figure 5-9) are similar to circular basins. The rake arms, which rotate like a sludge collector, gently stir the blanket of sludge, opening channels through which the water is released. The channels allow sludge solids to settle (by gravity) and thicken better. To more effectively dewater sludge, the sludge is thickened before mechanical dewatering processes, such as vacuum filtration or CENTRIFUGATION (Figure 5-10), are performed. Sludge that has been thickened is about 20 to 40 percent solid particles.

Vacuum filters. A vacuum filter (Figure 5-11) consists of a cylindrical filter drum covered with a porous fabric woven from fine metal wire or natural or synthetic fibers (cotton, nylon, etc.). Sludge fills the feed tank and is pulled onto the filter fabric by a vacuum within the filter drum. This vacuum pulls water away from the sludge and into the drum, leaving a thin, dry cake of sludge on the filter fabric (Figure 5-12), which is taken to a waste-disposal site. The filter cake

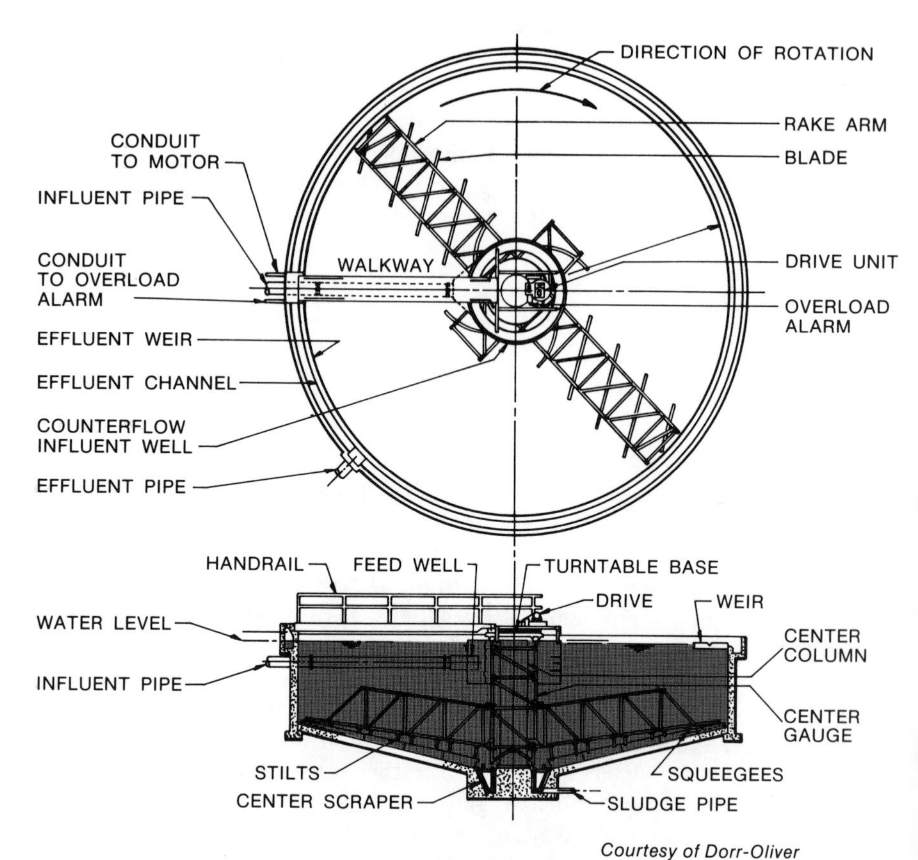

Courtesy of Dorr-Oliver

Figure 5-9. Sludge Thickener

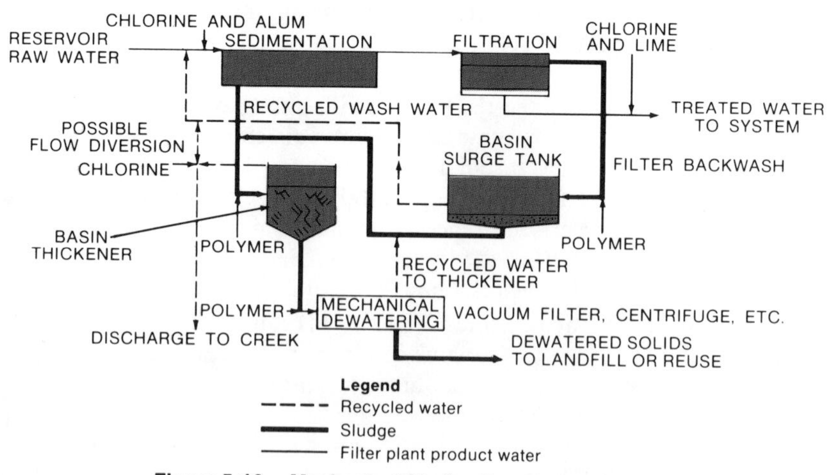

Figure 5-10. Mechanical Sludge-Dewatering System

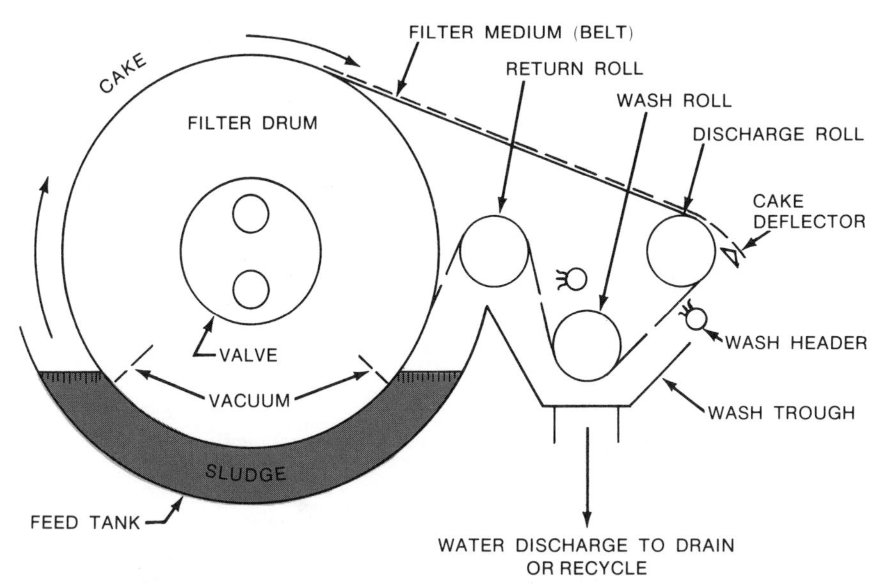

Figure 5-11. Sludge-Dewatering Vacuum Filter (Belt Type)

Courtesy of Komline-Sanderson Corp.

Figure 5-12. Dry Sludge on Filter Fabric

contains from 45 to 70 percent solid sludge particles. The clear water removed from the sludge, called filtrate, is returned to the thickener.

Centrifuges. A centrifuge is a sedimentation bowl that rotates sludge at high speeds to help separate sludge solids from the water. As shown in Figure 5-13, sludge enters the rotating bowl through a stationary feed pipe. The rotating bowl spins at high speeds, causing solids in the sludge to be thrown outward to the wall of the bowl at a force equal to 3000 to 10,000 times the force of gravity. The solids that settle against the wall of the bowl are scraped forward by a screw-type rotating conveyor and are discharged as shown in the figure. Clear water, called CENTRATE, discharges through controlled outlets and is returned to the thickener. The sludge cake produced by the centrifuge is 50 to 65 percent solid particles.

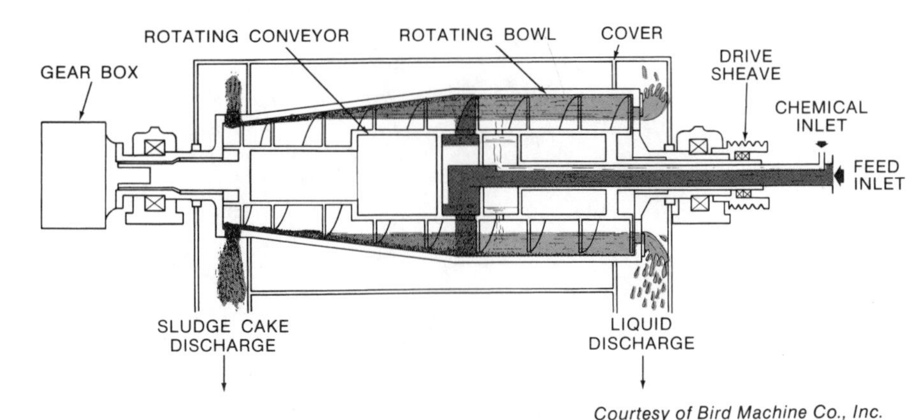

Courtesy of Bird Machine Co., Inc.

Figure 5-13. Sludge-Dewatering Centrifuge

Recarbonation Facilities

Recarbonation is usually accomplished by adding carbon dioxide (CO_2) gas to softened water to stabilize the water before it enters the distribution system. Recarbonation is performed in a basin that provides 15 to 30 min (or more) detention time.[3]

There are several ways to obtain carbon dioxide gas. In large, older plants, it is collected from furnace exhaust gases (as stack gas), scrubbed (by spraying water through the gas to remove impurities), moved by compressor to the recarbonation basin, and then diffused into the water. This technique for collecting carbon dioxide gas is not recommended because the presence of sulfur and phenolic materials in the gas may cause taste and odor problems in the finished water. In addition, the scrubbers, compressors, valves, and piping required to collect, clean, and move carbon dioxide gas need considerable maintenance because moist carbon dioxide gas is corrosive.

In large, newer plants, carbon dioxide gas is produced (1) by burning a mixture of natural gas and air under water (in a submerged combustion burner, such as the one shown in Figure 5-14), or (2) by burning a mixture of natural gas and air on the surface (in a forced-draft generator). Natural gas is used in both of the processes because it has few impurities. The carbon dioxide gas produced by either of these processes is then released near the bottom of the 10- to 15-ft (3.0-to 4.5-m) deep recarbonation basin. The amount of carbon dioxide gas generated is controlled by regulating the amount of natural gas burned. The equipment used in this process also produces carbon monoxide, which can reach dangerous concentrations in low, basin areas; therefore, good ventilation in recarbonation facilities is essential.

Small treatment plants and plants requiring only small amounts of carbon dioxide may purchase carbon dioxide either in the form of dry ice or liquid or in

[3] *Basic Science Concepts and Applications,* Chemistry Section, Chemistry of Treatment Processes—Recarbonation.

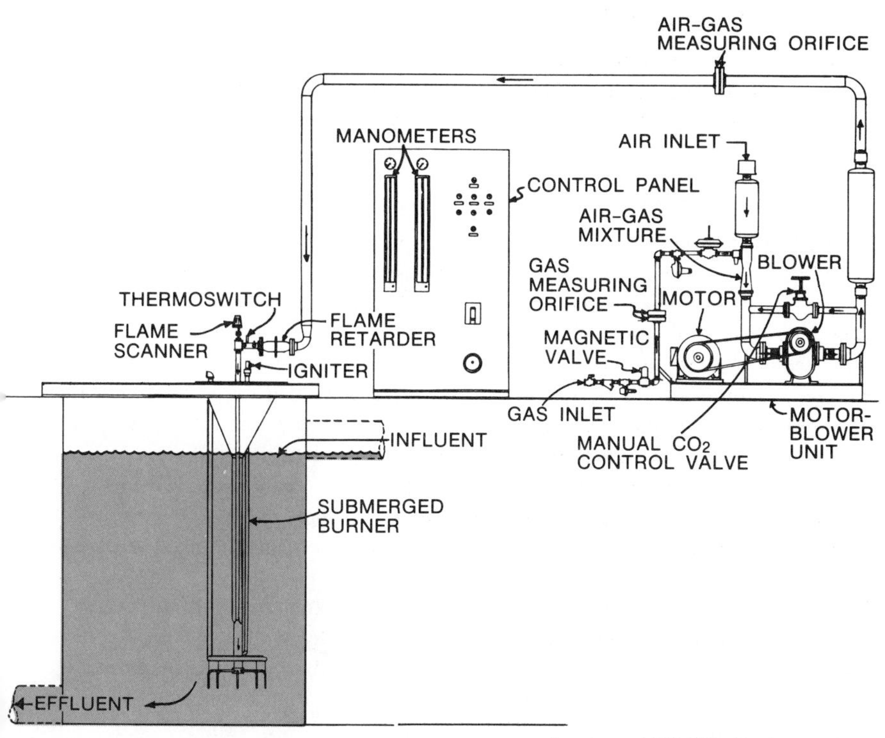

Courtesy of TOMCO$_2$ Equipment Co.

Figure 5-14. Submerged Combustion Recarbonation System

gas cylinders. If carbon dioxide is purchased in any of these forms, the need for combustion equipment and compressors and the danger of carbon monoxide are eliminated. If carbon dioxide is purchased in the form of dry ice or liquid, an evaporator is used to change the dry ice or liquid to a gas. The gas then passes through a pressure-regulating valve into diffusers in the recarbonation basin. The flow of carbon dioxide gas is controlled by a rotameter, which is similar to the meter on a gas chlorinator. Since the gas generated from dry ice or liquid carbon dioxide is pure carbon dioxide, pipes for the diffusers can be much smaller than those needed on-site for producing carbon dioxide gas by combustion. Carbon dioxide gas is heavier than air, and the escape of carbon dioxide gas into the atmosphere can result in a hazardous buildup of the gas in low areas. Therefore, ventilation systems are needed whenever carbon dioxide gas is generated.

Filtration Facilities

The filters used in a lime-soda ash softening plant are identical to those used in a conventional rapid-sand or multi-media filter plant. There are, however, special operational considerations that are discussed later in the module.

5-4. Operation of the Lime-Soda Ash Process

Operation of the lime-soda ash process involves storing and feeding large amounts of chemicals. It also involves processes used only for softening, as well as modifications to the operation of other processes. Finally, operation of the lime-soda ash process requires calculating chemical dosages.

Storing and Feeding Chemicals

Storing chemicals. Calcium oxide (quicklime) is usually purchased in pebble form to avoid the dust problems associated with the powdered form. However, some dust is created when any form of calcium oxide is used, and operators should always wear protective equipment when handling this chemical to avoid injury. Calcium oxide is shipped in bulk form by covered railroad-hopper car or by truck. Unloading can be done by screw conveyors, covered-belt conveyors, bucket elevators, or pneumatic conveyors. Calcium oxide is not corrosive and can be stored in concrete or steel silos or bins.

Calcium hydroxide (hydrated lime) is available in 50- and 100-lb (23- and 45-kg) bags. Care must be taken to keep the bags intact because calcium hydroxide will absorb carbon dioxide from moisture in the air and change to calcium carbonate.

Soda ash is purchased in bulk, bags, or barrels. It can be unloaded using the conveyors previously described for calcium oxide, although the fine, white soda-ash powder can pose chemical-dust hazards. Since soda ash is non-corrosive, it may be stored in ordinary steel bins, concrete bins, or silos.

It is a good practice to keep a 30-day supply of lime and soda ash on hand at all times to avoid running out of these chemicals in case unexpected problems (such as delays in transportation) arise. One way to keep a 30-day supply of lime and soda ash on hand is to have a chemical storage capacity of 60 days, taking delivery of a 30-day supply every month. Since chemicals lose effectiveness over a period of time, it is important to operate storage facilities in such a way that the oldest chemicals are used first. All chemicals used in the softening process should meet AWWA standards.

Feeding chemicals. The specific methods of operating chemical feed facilities will depend on the chemicals and feed equipment being used. Manufacturers' recommendations and information from chemical suppliers should be used to ensure proper operation of chemical feed facilities.

Calcium oxide (quicklime) is fed by a gravimetric feeder into a lime slaker. Two types of lime slakers are used:

- Detention type
- Paste type.

The detention-type lime slaker is the older of the two types of slakers and requires a water-to-lime ratio (by weight) of 4:1 to 5:1. Detention-type slakers take 20 to 30 min to slake lime. The water-to-lime ratio is adjustable and is controlled by the operator. Heat generated during slaking is transferred by a heat exchanger to incoming water in order to accelerate (speed up) the slaking process. A slaking temperature of 160° F (71° C) or higher should be maintained.

The paste-type lime slaker requires a water-to-lime ratio (by weight) of 2:1 to 2.5:1, and slaking is performed at near-boiling temperatures. These higher temperatures accelerate the slaking process. Paste-type slakers take 10 to 15 min to slake lime. The slaked lime is then diluted to approximately 10 percent $Ca(OH)_2$ in a solution chamber, forming a milk-of-lime slurry.

Calcium hydroxide (slaked lime) is fed directly into a solution chamber by a gravimetric feeder. (Since calcium hydroxide is already slaked, it does not need to pass from the gravimetric feeder into the lime slaker.) The resulting slurry is similar to the milk-of-lime slurry.

The process of feeding lime slurries presents many problems because the slurries will deposit (cake) on any surface. To cope with these problems, valveless proportioning pumps can provide a satisfactory method of pumping lime slurries. In addition, open troughs provide an effective means of transporting lime slurries from the feeder to the point of application since clogging and caking can be easily detected and removed. The basin into which lime slurry is fed should be as close to the slaker as possible to cut down on caking problems while the slurry is being transported.

Operation of Unit Processes Related to Lime-Soda Ash Softening

Rapid mix. A high-energy, rapid mixing process, called FLASH MIXING, is used to dissolve and thoroughly mix lime and soda ash into the water being treated. The optimum mixing speed is that speed which produces the best chemical reactions. The optimum mixing speed can best be determined from actual operating experience by running the mixers at different speeds and testing them to see which speed produces precipitates with the best settling characteristics.

A detention time of 5 to 10 min is required when mixing lime and soda ash with water since lime is slow to dissolve in water. All mixing equipment and the settling basin should be cleaned frequently to remove calcium carbonate and magnesium hydroxide precipitates. The concentration of these precipitates in the basin can be reduced by recirculating sludge from the settling basin. This process allows the precipitation to take place on the previously formed precipitate solids rather than in the basin.

Flocculation. Flocculation requires slow, gentle mixing to allow the chemical reactions to be completed and to promote the growth and precipitation of the floc. As such, flocculation basins usually require a detention time of 40 to 60 min. The speed of the mixing paddles must be adjusted to obtain the best floc. Once the floc has formed, local eddy currents and flow surges must be avoided to prevent breaking up the solid floc particles.

Sedimentation basins. The operation of sedimentation basins is discussed in Module 4, Sedimentation. To soften water, detention times of 2 to 4 hours are usually adequate. Surface overflow rates vary over a wide range, from 540 gpd/sq ft (0.25 mm/s) up to 1600 gpd/sq ft (0.75 mm/s) or higher. These surface overflow rates are higher than the rates in a conventional water treatment plant because the floc particles settle more readily in the softening process than the floc particles formed by alum in the conventional water treatment process. Due to the large amounts of sludge produced in the softening process, continuous mechanical collection of sludge in sedimentation basins is essential. The sludge draw-off lines should be backflushed periodically to remove any clogs or caking.

Solids-contact basins. In a solids-contact basin (Figure 5-8), the water passes upward through a blanket of flocculated material (called a sludge blanket). The purpose of the blanket is to entrap slowly settling particles that would otherwise escape the basin and end up in the filters. The mixing speed and the level of the sludge blanket are key operational factors with solids-contact basins. The mixing speed determines how well the chemicals and water are mixed. Increasing the mixing speed shortens the detention time of the initial mix, which may result in inefficient treatment because of incomplete chemical reactions. Inadequate mixing speed, on the other hand, may result in poor mixing of the chemicals and water. Settling tests and visual observations must be used to determine the best mixing speed. Many plants run the mixer continuously in order to keep the sludge blanket in a state of suspension at all times, including periods of time when the plant is shut down.

In addition to proper dosage and mixing, successful treatment in a solids-contact basin also depends on keeping the sludge blanket (in the settling zone) at desirable levels. If the sludge blanket drops below the "skirt" of the flocculation compartment (Figure 5-8), water will not pass through a full bed of solids, and some particles could pass through the basin and into the filters. The sludge blanket must not be allowed to build up or rise to such high levels that solids will pass over the overflow weirs and discharge into the filters. Usually, the top of the sludge blanket should be 4 to 8 ft (1.2 to 2.5 m) below the top of the overflow weirs. The height of the sludge blanket is adjusted according to the frequency, duration, and rate of sludge removal (known as SLUDGE BLOWDOWN) from the basin. Blowdown valves are operated intermittently in response to a timer or a water meter that signals a valve to open for a set period of time after a given number of gallons of water have passed through the solids-contact basin.

Sampling lines (pipes) should be provided at various levels in the settling zone to permit the operator to accurately determine the height of the sludge blanket and the degree of solids present in the water. In the lower level of the sludge blanket, the concentration of solids (by weight) should range from 5 to 15 percent. By maintaining a high concentration of solids while keeping the sludge blanket at a level low enough to avoid a loss of solids over the overflow weir, the amount of chemicals needed to soften the water can be kept to a minimum.

Solids-contact basins are designed for a flow rate of 1 to 1.75 gpm/sq ft (0.70 to 1.20 mm/s). Higher flow rates are sometimes possible, depending on the settling characteristics of the floc. Where multiple basins are provided, flow must be balanced among the basins so that no individual basin will be overloaded. This balance of flow may be accomplished either by flow meters or by adjusting valves or overflow weirs to provide flow rates that are proportional to the sizes of the basins.

Solids-contact basins must be drained in order to be cleaned. The frequency at which basins are drained varies from 3 to 12 months, depending on the types and amounts of chemicals being used in the basin. Sludge withdrawal lines and sampling lines should be vigorously flushed and cleaned on a routine basis to remove sludge residues and any scale that has built up in the basins. An alum or chlorine solution will effectively clean sludge withdrawal lines and sludge sampling lines.

Sludge collection and disposal equipment. Sludge collectors normally operate at a slow, constant speed (one revolution every 20 to 30 min). Usually, the operator does not need to vary this speed, except to stop the collector for routine maintenance and repair. Once the sludge has been scraped into the sludge hopper in the sedimentation basin, the hopper is emptied (usually by pumping), and the sludge is either recirculated or removed for disposal. Constant-speed centrifugal sludge pumps may be provided with a valve on the discharge side of the pump so that the operator can control the pumping rate by partially opening or closing the valve. Some settling basins are equipped with variable-speed, positive-displacement pumps so that the operator can control sludge flow by varying motor speed.

The frequency at which sludge is pumped varies significantly, based on the amount of hardness removed from the water and on the concentration of sludge solids. As a general rule, for each pound of lime used, 2.5 lb (1.1 kg) of sludge will be formed. Sludge concentration can vary from less than 5 percent solids to more than 15 percent solids by weight.

The operating procedures for dewatering sludge vary, depending on the type of dewatering process used. If a drying-bed process is used, the sludge should be pumped to the sand drying beds and spread to a depth of 8 to 12 in. (200 to 300 mm). In good weather, the sludge will dry and crack in about a week. The dried sludge, containing 50 to 60 percent solids, is removed either by hand raking or (preferably) by using a front-end loader. After being removed from the drying beds, the sludge is transported to a burial or refuse site. The success of the drying-bed process for dewatering sludge depends on clear, dry, warm weather. In cold, wet climates, the drying-bed process can be used if the drying beds are covered like greenhouses, heated, and ventilated.

If lagoons are used to dewater sludge, the sludge is pumped directly to the lagoons and allowed to settle. It is best to fill a lagoon with sludge at one end and decant (draw off) the water from the other end. The depth of sludge should be from 3 to 5 ft (0.9 to 1.5 m), and the lagoons should be operated on a "fill-and-dry" schedule. The water level should be kept as low as possible. The dewatered sludge, which should contain 50 percent solids, is removed by dragline, clamshell, or similar heavy equipment. It is sometimes spread on the ground near the lagoon to dry further before final disposal.

In some cases, the sludge solids are left to accumulate in the lagoon. When the lagoon is filled with sludge solids, it is abandoned, and a new lagoon is excavated. In cold climates, freezing temperatures help dewater sludge by separating water from the sludge solids. After thawing, the solids are small granular particles.

In softening plants that dewater sludge mechanically (by a vacuum filter or a centrifuge), sludge is usually pumped into a sludge thickener before going to the mechanical dewatering device. Sludge entering the thickener with 5 percent solids, for example, can be thickened to 15 to 20 percent solids. This reduced volume of sludge is easier and less costly to treat and dispose. The clear water removed during the thickening process is returned to the plant for conventional treatment. When sludge that has been thickened to 15 to 20 percent solids is dewatered either by a vacuum filter or by a centrifuge, a fairly dry, thin, cake-like sludge, having a solids concentration of 40 percent or more, is produced.

After the mechanical dewatering process is completed, the sludge is usually removed for disposal. Disposal can be achieved either by burying the sludge in a sanitary land fill or by spreading the sludge on the land to serve as a soil conditioner. If softening sludge is incinerated, lime can be reclaimed and reused; however, this procedure is economical only for large water treatment plants.

Recarbonation basins. Recarbonation basins are used to stabilize water after softening. In order to provide adequate time for the chemical reactions to take place, recarbonation basins usually provide 15 to 30 min detention time. If carbon dioxide is being injected by diffusers, the operator should routinely check the diffusers to see that they are not plugged. The ventilation system should be checked periodically, and the flow of carbon dioxide gas should be monitored at least every eight hours. Fluctuations in the flow rate of carbon dioxide gas or in the burner generating carbon dioxide gas can cause excessive scale to form downstream. Control of the recarbonation process should be based on (1) meeting water-stabilization goals (as determined by the Langelier Index or other analytical measures of water stability), and (2) using coupons in the distribution system (as described in the Control Tests section of this module).

Operation of the filtration process. The filtration process is described in Module 6, Filtration. However, there is one major difference between filtration in a plant that does not practice softening and filtration after softening that can cause serious operating problems. Since the water being filtered after the softening process contains calcium carbonate and, in some cases, magnesium hydroxide, precipitates of these chemicals can form on the filter media, causing the grains of filter media to stick (cement) together. If this happens, the effectiveness of the filtration process will be destroyed. To avoid this problem, the filter media should be inspected frequently for buildup of scale. If the grains of the filter media are coated with white scale (calcium carbonate) or black scale (magnesium hydroxide), recarbonation is probably inadequate. The presence of scale can be verified by adding a small quantity of hydrochloric acid to a porcelain dish containing a small quantity of filter media. If scale is present, the hydrochloric acid will react with the scale and will boil or foam.

Sometimes, scale can be removed from the filter media while it is in the filter bed by increasing the carbon dioxide feed rate to dissolve the scale. Care must be exercised during this operation to maintain water quality and to avoid damaging metallic underdrain components by prolonged exposure to corrosive water.

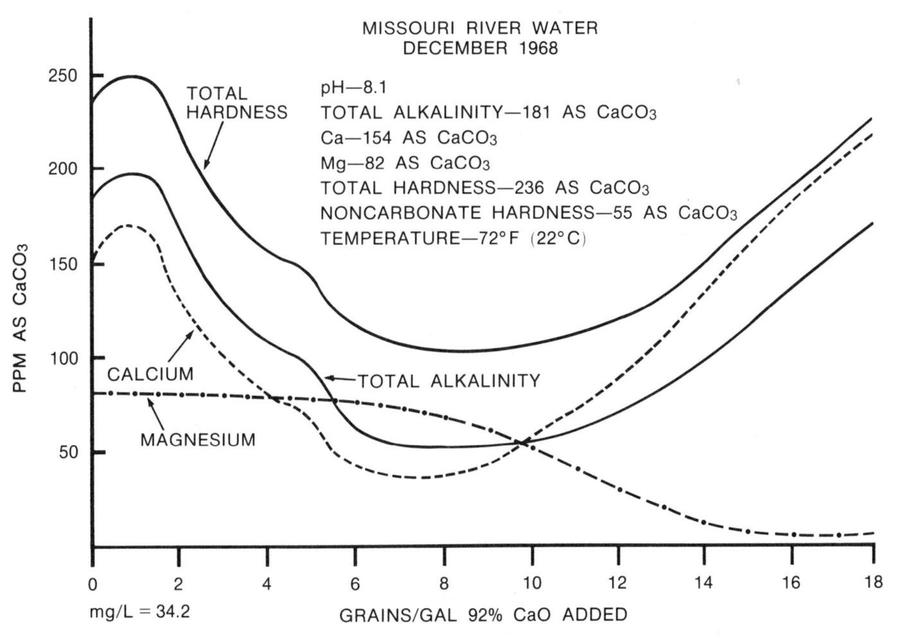

Figure 5-15. Softening Curve Used to Establish Most Effective Lime Dosage

Determination of Chemical Dosages

The amount of lime and soda ash required to soften water depends on (1) the hardness of the water, and (2) how much of the hardness is to be removed. The required dosage of lime is based on the combined amount of carbonate hardness, noncarbonate hardness, and carbon dioxide to be removed. The required dosage of soda ash is based only on the amount of noncarbonate hardness to be removed.

There are two methods for calculating lime and soda ash dosages: the conventional-dosage method and the conversion-factor method.[4] Initial dosages based only on these two methods can be computed for lime and soda ash; however, either sludge recirculation or the use of solids-contact basins will allow dosages to be reduced. Since adding chemicals in excessive amounts may not increase the amount of hardness removed from water, the operator must find the most effective dosages of lime and soda ash for the water being treated by using the computed figures as a base and then trying various dosages in the same range.

The use of a softening curve can be helpful in establishing the most effective lime dosage. This curve can be developed by using jar tests, followed by process testing in the plant. As shown in Figure 5-15, a softening curve is developed by plotting lime dosage against the concentration of carbonate and noncarbonate

[4] *Basic Science Concepts and Applications*, Chemistry Section, Chemical Dosage Problems—Lime-Soda Ash Softening Calculations.

hardness remaining in the water. At each point on the curve, pH should also be recorded. In the example shown in Figure 5-15, a total hardness of 100 mg/L can be achieved with a lime dosage ranging from 7.4 grains (127 mg/L) to 10 grains (171 mg/L). By using the lower dosage, a substantial amount of lime and money can be saved.

The dosage of soda ash needed to remove noncarbonate hardness should be based on the calculations of either the conventional-dosage method or the conversion-factor method. By monitoring finished water's noncarbonate hardness and alkalinity, the dosage of soda ash can be adjusted to the most efficient level.

Calculations for the theoretical feed rate for carbon dioxide in the recarbonation process are based on the amount of alkalinity that must be removed in order to stabilize the water. Any adjustments in the dosage of carbon dioxide should be based both on the Langelier Index and on coupons in the distribution system (described in the Control Tests section of this module).

5-5. Operating Problems of the Lime-Soda Ash Softening Process

There are several potential problems commonly associated with the operation of the lime-soda ash softening process:

- Excess calcium carbonate
- Magnesium hydroxide scale
- After-precipitation
- Carryover of sludge solids
- Unstable water
- Interference with other treatment processes.

Unless recognized and eliminated, these conditions can lead to high maintenance costs, failure of filtration processes, loss of distribution capacity, inefficient operation of water heaters and boilers, and ineffective treatment.

Excess Calcium Carbonate

Extremely small, almost colloidal particles of calcium carbonate can be formed during the coagulation/flocculation process. These particles can pass through the entire softening process to the filters, where they adhere very tightly to the individual grains of filter media (sand, anthracite coal, etc.). The filter grains become coated with calcium carbonate and cement together. In a similar way, calcium carbonate can pass through the media and plug filter underdrains. Clogged underdrains will usually cause visibly uneven backwash patterns. If these conditions are not corrected, they can cause problems that are serious enough to require complete replacement of the filter media and underdrains.

In addition, excess calcium carbonate can also precipitate in fine particles on the walls of pipelines. Eventually, as shown in Figure 5-16, the particles build up in thickness, reducing the pipeline's capacity. The restricted pipe capacity makes pumps work harder, which increases operating costs and shortens equipment life. Excess calcium carbonate is also the cause of scale deposits on filter walls and wash troughs. Such deposits are both unsightly and costly to remove. These calcium carbonate deposits result from incomplete coagulation/flocculation of lime, which may be caused by inadequate mixing, short detention times, or interferences from organic contaminants in the water during chemical precipitation.

The best solution to the problem is to eliminate whatever is causing the incomplete coagulation/flocculation by improving the mixing process, increasing detention times, or pretreating the water to remove organic contaminants. If the fine particles of calcium carbonate cannot be eliminated at the source, they must either be removed (dissolved) by recarbonation or controlled by applying a recommended dosage of 0.25 to 1.0 mg/L of sodium hexametaphosphate (a sequestering agent). A dosage of sodium hexametaphosphate in this range holds the calcium carbonate in solution, preventing precipitation onto the filter media.

Magnesium Hydroxide Scale

When water softened by the lime-soda ash process contains magnesium hardness in excess of 40 mg/L, magnesium hydroxide scale can form inside boilers and household water heaters operated at 140° F (60° C) or higher. This scale reduces the efficiency of heat transfer, thereby increasing heating costs. A magnesium hydroxide scale deposit of 0.04 in. (1 mm) thick can increase heating costs by more than 10 percent. To prevent scale buildup, magnesium hardness should be reduced to 40 mg/L or less by using the excess lime treatment.

After-Precipitation

Immediately after the lime-soda ash process is performed, the pH of the treated water is about 9.4—this is the best pH for removal of calcium carbonate. If the pH is left at this level, calcium carbonate will continue to precipitate out of

Courtesy of Johnson Controls, Inc.

Figure 5-16. Calcium Carbonate Buildup in Pipe

the treated water even after the water leaves the sedimentation basin or upflow basin. This effect, called AFTER-PRECIPITATION, can cause the same problems that occur when excess calcium carbonate precipitates out of treated water due to inadequate coagulation/flocculation. To prevent after-precipitation, carbon dioxide should be added to the softened, settled water in order to lower the pH to about 8.7, a level that provides chemical stability. This procedure converts any unsettled calcium carbonate to soluble calcium bicarbonate. Encrustation can also be prevented by adding sodium hexametaphosphate.

Carryover of Sludge Solids

The primary reasons that sludge solids and properly formed floc particles are carried over the weirs of a sedimentation basin are (1) improper hydraulic conditions and (2) sudden changes in water quality. Carryover is indicated by the unclear water above the sludge blanket, by sludge solids appearing in the basin effluent, and by short filter runs. The specific causes of carryover are listed in Module 4, Sedimentation.

Carryover of sludge solids is often most severe when a considerable portion of the settleable solids is magnesium-hydroxide floc particles, which are very light. To prevent carryover, either reduce the hydraulic loading on the sedimentation or upflow basin or improve the settling characteristics of the floc. Recirculation of previously formed calcium-carbonate sludge is effective in improving the settling characteristics of the floc.

Unstable Water

If softened water is not properly stabilized, one of two major operating problems will result:

- Scale deposits
- Corrosiveness.

Recarbonation and other stabilization techniques are used to control these operating problems. These techniques are discussed in Module 9, Stabilization.

Interference With Other Treatment Processes

Disinfection of water is essential to make water safe for consumption. Although the high pH associated with lime-soda ash softening contributes to disinfection, it also results in the formation of a chlorine residual that is primarily in the form of hypochlorite (OCl) when the water is chlorinated. The hypochlorite residual has considerably less disinfecting power than hypochlorous acid (HOCl), which exists at lower pH values. Therefore, higher doses of chlorine or longer contact times may be required to disinfect water softened by the lime-soda ash process. (See Module 10, Disinfection, for a further discussion of residuals and contact times.)

If the lime-soda ash process is used to treat surface water, the processes for removing taste and odor (particularly adsorption) may be hindered due to the increased solubility of taste-and-odor-causing compounds at elevated pH values. If this problem occurs, it can usually be controlled by removing these compounds during pretreatment.

Many surface-water treatment plants must treat raw water having a high color concentration due to organic compounds. For the most part, color must be removed prior to softening since a pH in the range of 4.0 to 5.5 is often required for alum coagulation to effectively remove color.

The formation of TRIHALOMETHANES (THMs) increases at pH levels used in the softening process. If THM problems exist, pretreatment processes may need to be improved, or disinfection processes may need to be modified. (For a more detailed discussion on THM problems, see Module 10, Disinfection.) Regulatory agencies or technically qualified consultants should be contacted prior to adopting a plan to minimize THM formation.

5-6. Description of the Ion-Exchange Softening Process

The ion-exchange process is a water-softening method that has been used extensively in smaller water systems and individual homes. This process depends on the ability of certain materials to exchange or transfer one ion for another. In this process, water to be softened is passed through a filter-like bed of ion-exchange material called RESIN. The ion-exchange resins exchange sodium ions (Na^+) that they hold for hardness-causing calcium and magnesium ions that are carried in the water. Water that has passed through ion-exchange resins has zero or near-zero hardness.

Chemistry of Ion-Exchange Softening

The following chemical reactions describe the chemistry of ion-exchange softening. In the equations, *X* represents the ion-exchange material. (Equations are numbered the same as in *Basic Science Concepts and Applications*, Chemistry Section, Chemistry of Treatment Processes—Ion-Exchange Softening.)

Carbonate Hardness

$$Ca(HCO_3)_2 + Na_2X \rightarrow CaX + 2NaHCO_3 \qquad \textit{Eq 31}$$

$$Mg(HCO_3)_2 + Na_2X \rightarrow MgX + 2NaHCO_3 \qquad \textit{Eq 34}$$

Non-Carbonate Hardness

$$CaSO_4 + Na_2X \rightarrow CaX + Na_2SO_4 \qquad \textit{Eq 32}$$

$$CaCl_2 + Na_2X \rightarrow CaX + 2NaCl \qquad \textit{Eq 33}$$

$$MgSO_4 + Na_2X \rightarrow MgX + Na_2SO_4 \qquad \textit{Eq 35}$$

$$MgCl_2 + Na_2X \rightarrow MgX + 2NaCl \qquad \textit{Eq 36}$$

These chemical reactions represent CATION EXCHANGE, since positive ions (Ca^{+2}, Mg^{+2}, Na^+) are involved. To replenish the sodium ions, the ion-exchange materials must be periodically backwashed with brine (salt solution). This process, called REGENERATION, allows the ion-exchange materials to be used over and over again. The regeneration process involves the following chemical reactions:

Regeneration

$$CaX + 2NaCl \rightarrow CaCl_2 + Na_2X \quad \text{Eq 37}$$

$$MgX + 2NaCl \rightarrow MgCl_2 + Na_2X \quad \text{Eq 38}$$

The ion-exchange softening process does not alter a water's pH or alkalinity. Stability of the water, however, is altered by the removal of calcium and by an increase in total dissolved solids (TDS). For each 1 mg/L of calcium removed and replaced with sodium, the TDS increases 0.15 mg/L. For each 1 mg/L of magnesium removed and replaced with sodium, the TDS increases by 0.88 mg/L.

The measurements used to express water hardness in the ion-exchange softening process are different from the measurements used to express hardness in the lime-soda ash softening process. Hardness in the lime-soda ash process is commonly expressed as mg/L (as $CaCO_3$); in the ion-exchange process, hardness is expressed in grains-per-gallon (usually referred to simply as *grains*).[5] The following conversion factors show the relationship between mg/L and grains.

1 grain = 17.12 mg/L
1 grain = 0.143 lb per 1000 gal
7000 grains = 1 lb per gal

Advantages of Ion-Exchange Softening

Compared with lime-soda ash softening, ion-exchange softening has certain definite advantages. Since both methods are effective, it may be helpful to understand why the ion-exchange process might be preferable to the lime-soda ash process.

Compactness and low cost. Ion-exchange units are relatively compact. They are also relatively inexpensive compared with the much bulkier equipment used in the lime-soda ash process.

Use of safe, easy-to-handle chemicals. The only chemical involved in the ion-exchange process is salt, which is safe and easy to handle. With proper backwashing and rinsing, there is no danger of excessive amounts of brine reaching the consumer.

Ease of operation. Whereas the lime-soda ash process requires constant monitoring and a high-level of operating skill, the ion-exchange process is relatively simple and can be almost completely automated. As with any automated process, routine inspection and maintenance is essential. In addition, the voluminous quantities of sludge common with the lime-soda ash process are not produced.

---

[5] *Basic Science Concepts and Applications*, Chemistry Section, Chemical Dosage Problems—Ion Exchange Softening Calculations.

Ability to remove all hardness. The ion-exchange process removes all water hardness. Therefore, by blending softened and unsoftened water, a water having any desired hardness can be produced.

Based on the advantages discussed here, many water utilities, particularly those using ground water, have found the ion-exchange softening process to be more cost-effective than the lime-soda ash softening process. If the ion-exchange process is used to treat surface water, the conventional treatment processes of coagulation/flocculation, sedimentation, and filtration will probably be necessary since the turbidity, algae, and color found in surface water will foul the ion-exchange resin. Because of this potential problem, utilities treating surface water may find the lime-soda ash softening process more cost-effective than the ion-exchange softening process.

5-7. Ion-Exchange Softening Components

The ion-exchange process requires the following basic components:

- Ion-exchange materials
- Ion-exchange unit
- Salt storage tanks
- Brine feeding equipment
- Devices for blending hard and soft water.

Ion-Exchange Materials

A number of materials, including some types of soil, can act as cation exchangers for softening water. For example, a natural green sand known as glauconite (composed of sodium aluminum silicate and also called zeolite) has very good ion-exchange capabilities and was once widely used for water softening. However, synthetic zeolites and organic polymers, known as polystyrene resins, have replaced natural zeolites in modern water treatment plants because their quality can be more easily controlled than natural zeolites and they have higher ion-exchange capacities.

Polystyrene resins are the most commonly used ion-exchange material since they have three to six times the ion-exchange capacities of other materials. As shown in Figure 5-17, beads of polystyrene resin look like small BB's. Figure 5-18 depicts the ion-exchange sites on a typical polystyrene resin. As Figure 5-18 indicates, the polystyrene resin is not used up in the ion-exchange process; instead, it serves as a kind of "parking lot" on which ions are traded. In practice, some of the resin is lost in the ion-exchange process as the resin particles rub against each other, particularly during backwashing procedures.

Ion-Exchange Units

The tanks containing the ion-exchange resins (Figure 5-19) resemble pressure filters. However, ion-exchange units differ from pressure filters in that the

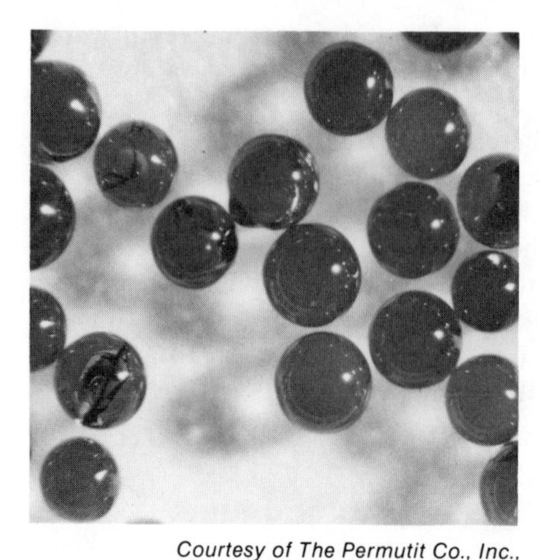

Courtesy of The Permutit Co., Inc., Subsidiary of Zurn Industries

Figure 5-17. Beads of Polystyrene Resin

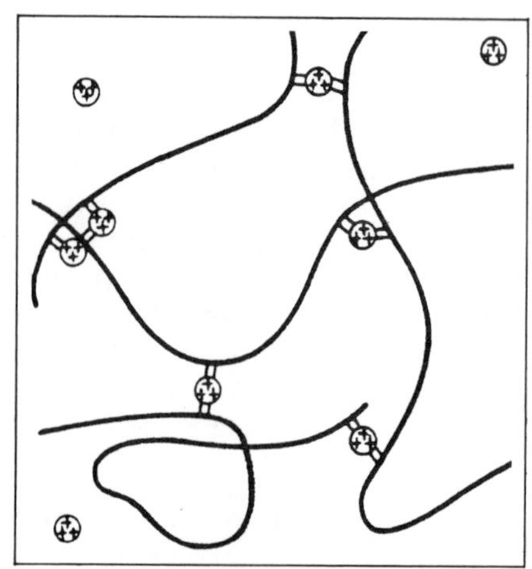

Figure 5-18. Ion-Exchange Sites on Polystyrene Resin

interior of an ion-exchange unit is coated with a special lining to protect it against corrosion from the brine used in regenerating ion-exchange resins. Figure 5-20 shows a cutaway view of a vertical-downflow ion-exchange unit. The unit is usually provided with the following components:

- Hard-water inlet
- Soft-water outlet
- Wash-water inlet and collector
- Brine inlet and distribution system
- Brine and rinse-water outlet
- Rate-of-flow controllers
- Sampling taps
- Underdrain system (which also distributes backwash water)
- Graded gravel to support ion-exchange resins.

Upflow types of ion-exchange softening units are available for use, but vertical-downflow ion-exchange units are more commonly used because they remove some sediment as they soften the water. Both upflow and downflow units can be equipped with automatic controls.

The size of the ion-exchange unit and the volume of ion-exchange resin needed to soften water is determined by the hardness of the raw water and the desired length of time between regenerating the resin. Regardless of these factors, the minimum recommended depth for resin in the ion-exchange unit is 24 in. (0.6 m). The resin is supported in the ion-exchange unit either by the underdrain system or by 15 to 18 in. (0.40 to 0.45 m) of graded gravel. The underdrain system must be capable of both distributing the wash water evenly

Courtesy of The Permutit Co., Inc., Subsidiary of Zurn Industries

Figure 5-19. Ion-Exchange Pressure Tanks

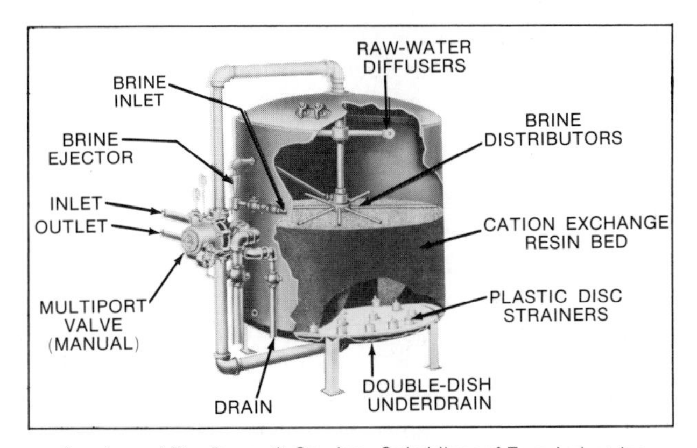

Courtesy of The Permutit Co., Inc., Subsidiary of Zurn Industries

Figure 5-20. Vertical-Downflow Ion-Exchange Unit

and draining the ion-exchange unit completely so that corrosive brine solution will not be trapped within the unit.

In vertical-downflow ion-exchange units, a brine distribution system is used to direct the flow of heavy brine downward through the unit. The system must distribute the brine evenly so that all of the resin comes into contact with it. The rinse water can also be distributed through the same system.

Salt Storage Tanks

Salt is used in the water-softening process to form brine, which regenerates the ion-exchange resins. The amount of salt used in creating the brine ranges from 0.25 to 0.45 lb (0.11 to 0.20 kg) for every 1000 grains of hardness removed. The salt is usually stored in two-bin concrete tanks. The tanks provide both storage and a container for dissolving the salt. Salt storage tanks are usually designed to hold enough salt and brine either for a 24-hour period of operation or for three regenerations of the resins, whichever is greater. Since salt attacks and wears away concrete, the walls of storage tanks should be painted with a salt-resistant, protective coating.

The salt used for resin regeneration should meet the requirements of AWWA B200, Standard for Sodium Chloride. Rock salt or pellet-type salt is the best type of salt for preparing the brine. Road salt is not suitable for this use because it often contains impurities. Block salt is not suitable because it has a low exposure of surface area per unit volume. Fine-grained (granular) salt is not desirable because it packs too tightly.

Salt storage tanks should be covered to prevent dirt and other foreign material from entering the tanks. In addition, a raised curb should be provided at each access hole in the tanks to prevent dirty or polluted water from entering the tanks.

Filling a salt storage tank and preparing the brine is accomplished by pumping water into the tank and filling the tank with rock salt. More water is added to submerge the rock salt so that it will dissolve, forming the brine. An air gap must be provided between the top of the tank and the bottom of the water line to prevent the brine from siphoning back into the water supply. An excess of undissolved rock salt should be kept in the tank to ensure a constant supply of brine.

Since brine is heavier than water, the highest concentration of brine will be in the lower part of the storage tank. The brine is usually drained or pumped from the bottom of the tank and pumped to the ion-exchange units.

Brine feeding equipment. Concentrated brine contains about 25 percent salt. To be most effective, the brine should be diluted to contain 10 percent salt. An hydraulic ejector or metering pump is used to dilute the concentrated brine from the salt-storage basin before it is applied to the resin bed in the ion-exchange unit. Brine is very corrosive, and special salt-resistant pumps and piping (such as plastic pipe) should be used to transport the brine.

Usually, brine pumps are specially fitted with corrosion-resistant materials and operate at low capacity and low head, thereby requiring low-horsepower motors. The base of a brine pump must be painted with a protective coating to protect it from corrosion due to salt leakage. Sometimes hardwood is installed around the base of a brine pump to eliminate corrosion problems.

The solubility of salt decreases at low temperatures, forcing the salt out of solution. The water that remains after the salt has separated out of solution is subject to freezing. Therefore, the piping that transports concentrated brine should be protected from freezing.

Devices for Blending Hard and Soft Water

A properly operated ion-exchange unit will produce water with zero hardness. However, both to reduce corrosion problems and because it is less expensive, it is desirable to produce water having a hardness in the range of 85 to 100 mg/L. To accomplish this, a bypass or some other device is used to blend a portion of the hard water with the softened water from the ion-exchange unit to produce water having the desired hardness. Blending hard and soft water is commonly accomplished and controlled by providing water meters both on the softener influent (or effluent) line and on the bypass line. Throttling valves blend the water to the desired hardness.

5-8. Operation of the Ion-Exchange Process

There are four basic cycles involved in the ion-exchange water softening process. As shown in Figure 5-21, the cycles are:

- Softening
- Backwash
- Regeneration
- Slow and fast rinse.

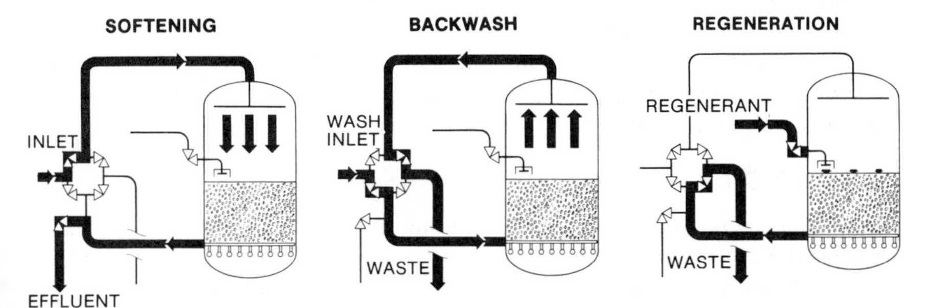

Softening. Influent water passes downward through the bed of ion-exchange material to the effluent.

Backwash. Influent water is passed upward through the bed of ion-exchange material to loosen the bed and remove suspended solids that may have been deposited in the bed during operation.

Regeneration. Regenerant solution is passed through the bed to waste at a controlled concentration and flow.

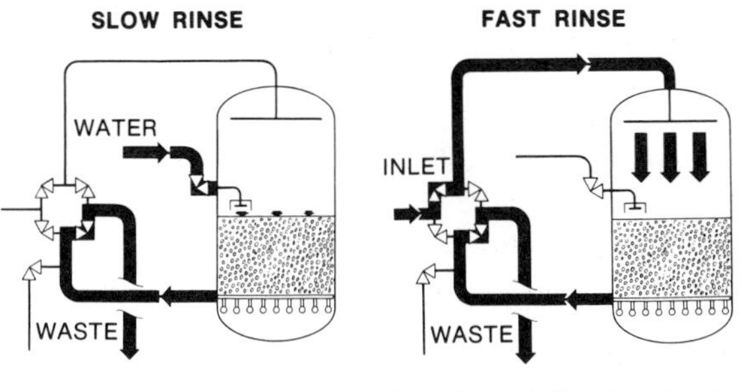

Slow rinse. Water is passed through the bed to displace the regenerant solution to waste.

Fast rinse. Influent water is passed through the bed to waste to remove the last traces of regenerant chemicals.

Courtesy of Infilco Degremont, Inc.

Figure 5-21. Four Cycles in Ion-Exchange Process

Specific operating procedures will vary, depending on the type of equipment used. Therefore, it is very important to follow the manufacturer's or engineer's recommendations for each piece of equipment used in the softening process. General guidelines for the operation of the ion-exchange softening process follow. Calculations used in the process are given in *Basic Science Concepts and Applications*.[6]

Softening Cycle

The softening cycle involves feeding hard water into the ion-exchange unit until the unit no longer produces water having a hardness of zero or near-zero. The cycle ends when water having a hardness of 1 to 5 mg/L first appears in the effluent. Almost all ion-exchange units include a meter for measuring water flow. The meter is usually equipped with an alarm so that when a volume of water corresponding to the rated capacity of the resin bed has been treated, the operator will be warned of the need to regenerate the resin. The loading rate for ion-exchange units using polystyrene resin is in the range of 10 to 15 gpm/sq ft (6.8 to 10.2 mm/s). This loading rate is high compared to other types of resin and zeolites (2 to 6 gpm/sq ft, or 1.4 to 4.1 mm/s), but it is quite acceptable because the rate at which ions are exchanged is so rapid.

Backwash Cycle

Once hardness "breaks through" (is found in the effluent), the ion-exchange unit must be removed from service, and the resin must be regenerated. In vertical-downflow ion-exchange units, the resin must be backwashed before being regenerated. Backwashing loosens the resin that has compacted during the softening cycle, randomly mixes the resin, and removes any foreign material that has filtered or precipitated from the water. The backwash rate (usually 6 to 8 gpm/sq ft, or 4.1 to 5.4 mm/s) is based on the flow rate needed to expand the resin bed by at least 50 percent of its depth.

Backwash and rinse water are usually discharged to a box containing orifice plates that establish the flow rates. These flow rates are usually controlled with float-operated butterfly valves. The butterfly valves are calibrated to throttle the flow of water at a desired rate for each operation.

Distributor pipes at the top of ion-exchange units provide uniform distribution of raw water and uniform collection of backwash water. An underdrain collector/distributor in the units maintains uniform flow-distribution patterns. These components help create a flow pattern through the ion-exchange unit that achieves close contact between the ion-exchange media and hard water or brine.

Regeneration Cycle

To regenerate ion-exchange resins, concentrated brine is drawn from the salt-storage basin, diluted by an ejector to a solution containing 10 percent salt (by

[6] *Basic Science Concepts and Applications*, Chemistry Section, Chemical Dosage Problems—Ion Exchange Softening Calculations.

weight), and passed slowly and continuously through the resin bed. Regeneration usually requires a contact time of 20 to 35 min for the brine to pass through the resin bed at application rates of about 1 gpm/cu ft (2.6 $L/s/m^3$) of resin. NOTE: The rate at which ion-exchange resins are regenerated (the REGENERATION RATE) is measured in gallons-per-minute per cubic feet of resin (or litres-per-second per cubic metre of resin); the rate at which water is introduced into the ion-exchange softening unit while the unit is in operation (the LOADING RATE) is measured in gallons-per-minute per square feet of bed surface (or millimetres per second). Adequate contact time for the brine to pass through the resin bed is extremely important in order to return proper ion-exchange capacity to the resin. During the contact time, sodium ions from the brine are exchanged for calcium and magnesium ions in the resin bed.

Rinse Cycle

The regeneration cycle must be followed by a thorough rinsing cycle to remove all of the unused salt. Rinsing is accomplished by first running unsoftened water through the brine distribution line. Rinse water is then added to the ion-exchange unit at a rate of about 2 gpm/sq ft (1.4 mm/s) through the raw-water distribution line until the water leaving the unit has a chloride concentration nearly equal to that of the water entering the unit. As shown in Figure 5-21, some ion-exchange units use a slow rinse to remove the bulk of the brine, followed by a fast rinse to remove the last remaining traces of brine. The amount of water needed to perform the rinse cycle is about 20 to 35 gal/cu ft (2.6 to 4.7 kL/m^3) of resin.

Disposal of Brine

One of the major problems with the ion-exchange process is the disposal of the spent brine used in the regeneration cycle. Proper disposal techniques must be thoroughly evaluated before selecting the ion-exchange process, since the costs involved in disposing of the brine might make other softening processes more cost-effective.

The total amount of wastewater (spent brine) will usually vary from 1.5 to 7 percent of the amount of water softened. The wastewater contains calcium chloride, magnesium chloride, and sodium chloride from the regeneration cycle. Even with the dilution provided by the backwash and rinse water, the total concentration of dissolved solids in the wastewater will probably be from 35,000 to 45,000 mg/L. Since these concentrations can upset sewage treatment operations, cause serious corrosion, seriously harm aquatic life, make soil unusable for agricultural purposes, and make water unusable for almost any purpose, proper disposal of the brine is essential. Brine may be disposed of by (1) the use of lagoons, (2) dilution and discharge to surface water, and (3) pumping wastewater through an injection well into deep saline aquifers. The method used will depend on each utility's needs and the requirements of the state pollution control agency.

5-9. Operating Problems of the Ion-Exchange Softening Process

There are several conditions that cause operating problems in the ion-exchange softening process, including:

- Resin breakdown
- Iron fouling
- Turbidity, organic color, and bacterial-slime fouling
- Unstable water.

Resin Breakdown

Polystyrene resins can generally be expected to serve for 15 to 20 years before needing replacement. However, certain conditions can cause the resin used in the ion-exchange unit to break down (degrade) sooner, thereby causing the ion-exchange process to malfunction. Oxidation by chlorine is the primary cause of resin breakdown. The chlorine dosages normally used in disinfecting water have only a minimal effect on resin. However, the residuals of high chlorine dosages, such as those occurring when chlorine is used for removing iron from water, can cause severe resin breakdown. When chlorine is used for iron removal, the excess chlorine should be removed before the ion-exchange softening process is performed.

Iron Fouling

Iron in raw water causes problems in the ion-exchange softening process. Iron can seriously affect the capacity of the resin to exchange sodium ions for calcium and magnesium ions. Ferrous iron is oxidized and precipitates as iron oxide within the resin bed. After this chemical reaction occurs, no amount of brine will remove the iron oxide. The accumulation of iron oxide can continue until the resin bed completely loses its ion-exchange capacity. If iron oxide is formed before the raw water enters the ion-exchange unit, it is deposited on the resin bed and can be removed during normal backwashing procedures. Resin beds containing or covered by iron oxide precipitates develop a dark red (rusty) color and will show a definite loss in ion-exchange capacity. Pretreatment to remove the iron will prevent this problem.

Turbidity, Organic Color, and Bacterial-Slime Fouling

Turbidity, color from organic chemicals (particularly humic compounds), and bacterial slimes adversely affect the ion-exchange softening process by coating resin particles. This results in a loss of ion-exchange capacity and excessive head loss. Although backwashing helps prevent this problem, it cannot remove a large volume of the material coating the resins because the particles are tightly held on the resins.

The best solution for this problem is to treat the water by conventional processes (coagulation/flocculation, sedimentation, filtration) prior to softening the water in the ion-exchange unit. If this procedure is not followed, the resin will have to be replaced or restored frequently—a procedure that can be costly.

Unstable Water

Water that has been softened by the ion-exchange process is usually corrosive. Softened water that is corrosive should be blended (as previously discussed), or chemicals must be added to the softened water to provide stabilization and corrosion control (as discussed in Module 9, Stabilization).

5-10. Operational Control Tests and Record Keeping

Both the lime-soda ash process and the ion-exchange process require careful and continuous monitoring to effectively soften water. Monitoring is accomplished by performing a series of control tests.

The lime-soda ash process is a complex process, requiring the following control tests:

- Alkalinity
- Total-hardness
- Carbon dioxide
- pH
- Jar test
- Langelier Index determination for water in the distribution system
- Monitoring of coupons in the distribution system.

The ion-exchange process is a relatively simple process, requiring only the following control tests:

- Total hardness
- Langelier Index determination for water in the distribution system
- Chloride tests (performed periodically to determine if rinsing after regeneration is adequate).

Most of these tests are described in Volume 4, *Introduction to Water Quality Analyses*, Physical/Chemical Tests Module. The key to successful control of the softening process is using the test results correctly.

Alkalinity Test

The results of alkalinity tests are used in calculating dosages for both lime and soda ash.[7] Once the alkalinity test has been completed and both phenolphthalein alkalinity and total alkalinity (also called methyl orange [MO] alkalinity) have been determined, the results must be converted to the specific type of alkalinity

[7] *Basic Science Concepts and Applications*, Chemistry Section, Acids, Bases, and Salts—Alkalinity.

Table 5-2. Calculation of Alkalinity from Alkalinity Titration Results*

	Alkalinity—mg/L as $CaCO_3$		
Titration Result	*Hydroxide*	*Carbonate*	*Bicarbonate*
$P = 0$	0	0	T
P less than $1/2T$	0	$2P$	T-$2P$
$P = 1/2\ T$	0	$2P$	0
P more than $1/2\ T$	$2P$-T	$2P$-$2T$	0
$P = T$	T	0	0

*P = phenolphthalein alkalinity (mg/L); T = total alkalinity (mg/L), also called methyl-orange (MO) alkalinity.

actually present. There are three specific types of alkalinity: (1) hydroxide alkalinity, (2) carbonate alkalinity, and (3) bicarbonate alkalinity. Table 5-2 shows how to calculate the three types of alkalinity.

As a sample calculation, if the phenolphthalein alkalinity is zero ($P = 0$ from the *Titration Result* column of Table 5-2), then the total alkalinity is all bicarbonate alkalinity. However, if phenolphthalein alkalinity is less than half of the total alkalinity (P less than ½T), then both carbonate and bicarbonate alkalinity are present. The carbonate alkalinity will be two times the phenolphthalein alkalinity ($2P$), and the bicarbonate alkalinity will be that number subtracted from the total alkalinity (T-$2P$). The most common alkalinity conditions found in raw water are represented by the first two conditions shown in Table 5-2. The last three conditions in the table are found in lime-softened water before recarbonation procedures have been performed. Figure 5-22 is a graphical representation of Table 5-2, showing the relative pH values associated with each of the five possible alkalinity combinations.

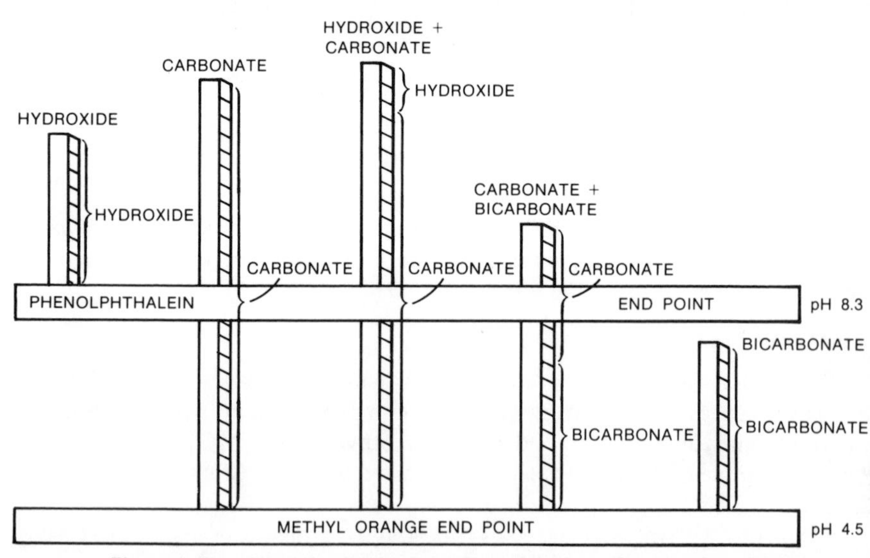

Figure 5-22. Alkalinity Determined From Alkalinity Titration Results

Table 5-3. Finding Carbonate and Noncarbonate Hardness

Laboratory Results *TH = Total Hardness* *TA = Total Alkalinity*	*Noncarbonate Hardness (Lime and Soda Ash Used)*	*Carbonate Hardness (Lime Only Used)*
1. TH less than TA	0	TH
2. TH = TA	0	TH
3. TH greater than TA	TH-TA	TA

Total-Hardness Test

The total-hardness test indicates the need for softening and provides information vital to the proper operation of the softening process. The total-hardness test determines the total-hardness concentration of a water. The total-hardness concentration is used together with total alkalinity to calculate carbonate and noncarbonate hardness.[8] The procedure for performing this calculation is summarized in Table 5-3.

If noncarbonate hardness is present, the total-hardness concentration and the bicarbonate alkalinity are used to calculate the dosage of soda ash needed to remove the noncarbonate hardness. If noncarbonate hardness is absent, no soda ash is needed, and the total-hardness concentration is used in place of bicarbonate alkalinity to calculate the correct lime dosage.

The total-hardness test should also be conducted routinely on raw and finished waters to determine how effectively the softening processes are removing hardness from the water.

Magnesium Test

The magnesium test is used to determine how much magnesium hardness is in the water, which, in turn, is used in determining lime dosages.[9] Since the results of magnesium tests indicate the magnesium hardness present, they can be used to calculate calcium hardness when the total hardness is also known.[10] The magnesium test also provides valuable insight into whether or not the water being treated contains enough magnesium to cause scale in water heaters and boilers. Whenever magnesium hardness exceeds 40 mg/L (as $CaCO_3$), there is a strong possibility that magnesium hydroxide scale will form within hot water heaters and boilers that are operated at temperatures of 140° F (60° C) or higher. If magnesium hardness is removed to levels at or below 40 mg/L (as $CaCO_3$), there should be no trouble with magnesium hydroxide scale.

Carbon Dioxide Test

The carbon dioxide test is used to determine the concentration of free carbon dioxide in the water. In turn, the free carbon dioxide concentration is used to

[8] *Basic Science Concepts and Applications,* Chemistry Section, Solutions—Hardness.

[9] *Basic Science Concepts and Applications,* Chemistry Section, Chemical Dosage Problems—Lime-Soda Ash Softening Calculations.

[10] *Basic Science Concepts and Applications,* Chemistry Section, Chemical Dosage Problems—Lime-Soda Ash Softening Calculations.

calculate lime dosages.[11] Although carbon dioxide does not cause hardness, it reacts with lime, precipitating as calcium carbonate. A sufficient amount of lime must be added to the water to allow this chemical reaction to occur and still have enough lime remaining in the water to complete the softening process.

pH Test

Measurements of the water's pH should be made before, during, and following lime-soda ash softening. As shown in Figure 5-22, a pH value of 8.3 denotes the phenolphthalein end point in the alkalinity test. Waters having a pH greater than 8.3 contain phenolphthalein alkalinity (hydroxide), whereas waters having a pH below 8.3 do not.

During the softening process, pH measurements provide valuable information about the success or completeness of the chemical reactions that are occurring. For example, when excess-lime treatment is used, the lime dosage is designed to raise the pH of the water to 10.6—the best pH for magnesium hydroxide precipitation. The pH is then reduced by recarbonation to 9.4, which is the best pH for calcium carbonate precipitation.

After soda ash is added and softening reactions have been completed, the final pH is adjusted by recarbonation to whatever pH is required to achieve stabilization, usually about 8.7. This process converts any residual calcium carbonate into soluble calcium bicarbonate, thus preventing after-precipitation (the precipitation of calcium carbonate after softening processes have been completed) either on the filter (filter encrustation) or in the mains (pipeline scaling).

By monitoring pH with an accurate, calibrated pH meter throughout the softening processs, an operator can determine whether or not softening and recarbonation dosages are working successfully. In addition, pH and total alkalinity can be used to determine the concentration of carbon dioxide in the water.

Jar Test

The jar test is a good way to test the performance of lime and soda ash dosages before they are used in the plant. A dosage correctly calculated from the results of the alkalinity test, total-hardness test, and carbon dioxide test is not necessarily the best dosage. There are numerous other factors, not easily defined or tested, that can greatly influence the selection of the "best" dosage. By performing the jar test, an operator can check the overall performance of the lime and soda ash dosages, such as the size and condition of floc particles, the ease and speed with which floc particles settle, and the resultant final pH. One very useful result of the jar test is that a softening curve, such as the one shown in Figure 5-15, can be developed from test results. The softening curve can help identify the lime dosage that will soften water to the desired point at the least cost.

---

[11] *Basic Science Concepts and Applications*, Chemistry Section, Chemical Dosage Problems—Lime-Soda Ash Softening Calculations.

Langelier Index and Coupons in the Distribution System

The stability of softened water should be monitored to assure that the water will neither cause corrosion nor form scale. On a daily basis at the plant, the Langelier Index should be determined and evaluated in connection with coupons placed in the distribution system. The Langelier Index and coupons are described in Module 9, Stabilization.

Record Keeping

Keeping accurate, up-to-date records on softening procedures is an important part of a successful softening operation. The type of records kept will depend on the softening process used. For the lime-soda ash process, records of the following information should be kept on a daily basis:

- Pounds (kilograms) of lime, soda ash, and carbon dioxide fed
- Chemical feeder settings
- Amount of water treated, in million gallons per day (megalitres per day)
- Hardness, alkalinity (P and MO), and magnesium/calcium concentrations in raw and treated water
- Free carbon dioxide in raw water
- The pH of the raw and treated water and the pH at those stages in the softening process when it is critical to chemical reactions
- Amount of sludge pumped to disposal
- Dosage calculations and results of jar tests (softening curves)
- Depth of the sludge blanket and results of settling tests if a solids-contact basin is used.

For the ion-exchange process, records of the following information should be kept on a daily basis:

- Hardness, alkalinity (P and MO), magnesium/calcium concentrations, and pH of the raw and treated water
- Amount of water treated, in million gallons per day (megalitres per day)
- Gallons (litres) of water treated in each softening cycle
- Gallons (litres) of backwash water, rinse water, and brine used
- Pounds (kilograms) of salt added to the storage tank.

These records will be useful in keeping track of the softening process and for ordering chemicals. However, each operator should develop additional records that are the most useful for the treatment being used.

5-11. Safety and Softening

The lime-soda ash softening process exposes operators to daily contact with large quantities of lime and soda ash. These chemicals pose hazards if not handled properly. As shown in Figure 5-23, operators should be adequately protected when handling chemicals. Since the ion-exchange process only uses salt, there are fewer dangers; however, brine can cause severe irritation if splashed on the skin or eyes.

Lime

Handling and storage. Calcium oxide (quicklime—CaO) is a strong, caustic chemical compound and an irritant. Handling calcium oxide creates quicklime dust, which is irritating to the eyes, mucous membranes, and lungs, and which can cause lung damage if exposure is prolonged. Operators exposed to this dust should wear goggles and a dust mask specifically suited for quicklime dust.

Skin contact with quicklime dust can cause dermatitis or skin burns. Skin contact is especially a problem when the dust mixes with perspiration on the skin. When handling quicklime, operators should wear heavy cotton clothing with long sleeves, gloves that reach to the elbow (gauntlets), bandanas, and trousers tied around the shoe tops. If clothing becomes covered with quicklime dust or spattered with lime slurry, clothes should be changed, and the lime-soiled clothing should be thoroughly laundered. Operators should always shower immediately after handling quicklime, even if quicklime dust is not visibly present. Dust collection equipment should be in proper operating condition. Chemical spills and dust accumulations should be cleaned up with a dry vacuum or similar dust-pickup system, not with a broom.

Calcium oxide is strongly attracted to water. When the two come into contact, the heat that is generated can start fires in nearby flammable materials. Therefore, calcium oxide must be stored in a dry place.

Dry calcium oxide and dry alum should never be mixed. The heat that results when they are mixed can generate hydrogen and result in an explosion. To avoid accidental contact of these two chemicals, use separate loading, handling, and storage equipment for each chemical.

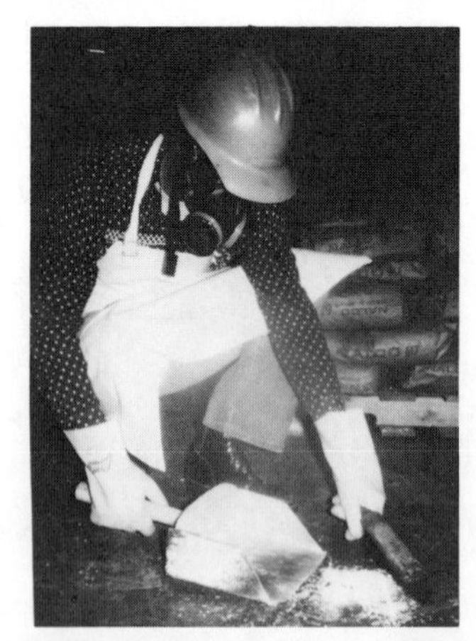

Figure 5-23. Worker Properly Dressed for Handling Chemicals

Lime slakers should be equipped with temperature override devices that will stop the lime feed before dangerous temperatures are reached.

Calcium hydroxide [hydrated lime—$Ca(OH)_2$] is less caustic and, therefore, less irritating to the skin than calcium oxide. Nonetheless, it is still a dangerous chemical, particularly to the eyes. When working with calcium hydroxide, exercise the same care and take the same precautions as with calcium oxide.

First aid. To treat skin burns from lime, wash the affected area alternately with a mild solution of boric acid and large quantities of soap and water. If lime gets into the eyes, first flush eyes with large amounts of water and then flush the eyes with boric acid solution. Consult a physician immediately. If lime slurry is splattered into the eyes, flush the eyes immediately because serious injury can occur in a matter of minutes. If throat or nasal passages become irritated, particularly because of prolonged exposure to lime, consult a physician.

Soda Ash

Handling and storage. Moisture will cause soda ash to cake, making it difficult and hazardous to handle. To prevent this problem, store soda ash in a cool, dry place. All soda ash equipment should be fitted with properly operating dust-collection systems.

Operators working with soda ash should wear protective equipment, including safety goggles for eye protection, a close-fitting dust respirator, and clothing that protects against skin contact. Skin areas that will be exposed unavoidably should be treated with a protective cream or petroleum jelly to minimize the harmful effects of soda ash dust. Some brands of soda ash contain ammonia. Operators allergic to ammonia should wear sufficient protective clothing and avoid the dangers of skin contact by applying an ointment or petroleum jelly to exposed areas.

Pumps and equipment that handle soda ash solutions should be equipped with spray or splash guards to protect personnel working in the area. Potential danger areas and storage rooms should be marked with warning signs.

First aid. Soda ash dust and the mist from soda ash solutions are very irritating to the respiratory system, mucous membranes, and eyes. Prolonged exposure can damage the nasal passages.

If soda ash dust or solution gets into the eyes, flush the eyes with warm water for at least 15 min. Consult a physician immediately. If skin contact with soda ash solution occurs, flush the affected area with large quantities of water. In the event that soda ash dust or mist is inhaled, consult a physician immediately. Gargling or spraying the nasal passages and throat with warm water will reduce the irritation.

Carbon Dioxide

Equipment safety. Carbon dioxide is a colorless, odorless gas, which is heavier than air. If it accumulates, it will displace the air; if it is inhaled, the carbon dioxide can result in an oxygen deficiency, causing asphyxiation. Since carbon dioxide is usually generated on-site, preventive maintenance of the generating equipment is essential to prevent leaks and to ensure that safety devices and alarms will function. The generating equipment should be located in a well ventilated area. Explosion-proof lighting should be used, and smoking should be strictly prohibited when carbon dioxide is being used.

Carbonation basins. Since generation of carbon dioxide can also produce carbon monoxide, a highly toxic gas, good ventilation of the carbonation basin is essential, particularly if submerged-combustion carbon-dioxide-generating units are used. Extreme care should be taken when cleaning carbonation basins, particularly if they are enclosed. They should be thoroughly ventilated by portable blowers before and during the cleaning operation. An approved carbon monoxide tester should be used to make sure the air is safe to breathe.

Machinery. In addition to storing and handling chemicals safely, general safety precautions should be followed when working around machinery. Moving parts of machinery should be equipped with protective guards. In addition, all basins should have guardrails.

Selected Supplementary Readings

Babbitt, H.E.; Cleasby, J.L.; & Doland, J.J. *Water Supply Engineering*. McGraw-Hill Book Company, New York (6th ed., 1962). Chap. 24.

Cox, Charles R. *Operation and Control of Water Treatment Processes*. World Health Organization. Geneva, Switzerland (1969) pp. 219–239.

Hirsch, A.A. *Manual for Water Plant Operators*. Chemical Publishing Company, Inc., New York (1945). Chap. 11.

Manual of Instruction for Water Treatment Plant Operators. New York State Dept. of Health, Albany, N.Y. (1975). pp. 171–178.

Manual of Water Utility Operations. Texas Water Utilities Assoc., Austin, Texas (6th ed., 1975). pp. 242–253.

The Nalco Water Handbook. Nalco Chemical Company. McGraw-Hill Book Company, New York (1979). Chap. 12.

Water Quality and Treatment. AWWA Handbook. McGraw-Hill Book Company, New York (3rd ed., 1971). Chap. 9 and 10.

Water Supply and Treatment. National Lime Association, Washington, D.C. (1962) pp. 84–114.

Water Treatment Plant Design. AWWA. Denver, Colo. (1969) Chap. 11 and 12.

Water Treatment Plant Design. Robert C. Sanks (ed.), Ann Arbor Science Publishers, Inc. Ann Arbor, Mich. (1978) Chap. 24.

Glossary Terms Introduced in Module 5

(Terms are defined in the Glossary at the back of the book.)

After-precipitation
Calcium carbonate
Calcium hardness
Carbonate hardness
Cation
Cation exchange
Centrate
Centrifugation
Decant
Divalent
Excess-lime treatment
Flash mixing
Hardness
Ion-exchange process
Lime-soda ash method
Loading rate
Magnesium hardness
Milk of lime
Noncarbonate hardness
Permanent hardness
Polystyrene resin
Precipitates
Quicklime
Recarbonation
Regeneration
Regeneration rate
Resin
Sequestering agent
Slaked
Sludge
Sludge blowdown
Stabilization
Temporary hardness
Trihalomethanes
Turbidity

Review Questions

(Answers to Review Questions are given at the back of the book.)

1. What causes water hardness?

2. How do the two major hardness-causing cations get into the water?

3. When does hardness become objectionable to consumers?

4. List the various types of water hardness.

5. Write the formula that describes the relationship between total hardness, carbonate hardness, and noncarbonate hardness.

6. List several specific disadvantages for both hard water and soft water.

7. List the two methods most widely used for softening public water supplies.

8. Name the compounds and write the chemical formulas for the two principal hardness-causing precipitates that settle out of water when the lime-soda ash softening process is used.

9. How does carbon dioxide enter into the chemical reaction of the lime-soda ash softening process?

10. List the major components used in the lime-soda ash softening process. Do not include the filtration component since that process is normally used whether or not softening is practiced.

11. What are the two forms of lime that can be used in the lime-soda ash softening process?

12. Which form of lime requires slaking?

13. It is good operating practice to keep a ________-day supply of lime and soda ash on hand at all times.

14. What are typical detention times in the lime-soda ash process for flocculation and sedimentation?

15. A recent sample of raw water had the following water quality characteristics:

    Calcium—177 mg/L as Ca
    Magnesium—39 mg/L as Mg
    Bicarbonate alkalinity—348 mg/L as HCO_3
    Carbon dioxide—11 mg/L as CO_2

    In treating this water, the best results were obtained when an excess-lime dosage of 35 mg/L as $CaCO_3$ was used. Calculate the total lime and soda ash dosage in mg/L. The lime used was 89-percent pure; the soda ash used was 100-percent pure.

16. In question 15, find the dosages in:
    - Pounds per million gallons
    - Pounds per day (assuming a flow rate of 1,825,000 gpd)

17. A recent laboratory result shows the following hard water characteristics:

    Calcium—224 mg/L as $CaCO_3$
    Total hardness—245 mg/L as $CaCO_3$
    Bicarbonate (total) alkalinity—150 mg/L as $CaCO_3$
    Carbon dioxide—10 mg/L as CO_2

    Determine the amount of quicklime (92.5-percent pure) and soda ash (100-percent pure) required to treat this water. Express the answer in pounds per million gallons. (Note: Excess lime is not added.)

18. How many pounds of chemicals would be required each day if the flow rate in question 17 was 736,000 gpd?

19. List six operating problems of the lime-soda ash process.

20. Describe two ways to control excess calcium carbonate.

21. What problem can occur if magnesium concentration in the treated water exceeds 40 mg/L as $CaCO_3$?

22. A laboratory returned the following test results. Re-express each chemical concentration as $CaCO_3$.

    Calcium—135 mg/L as Ca
    Total Hardness—360 mg/L as $CaCO_3$
    Bicarbonate alkalinity—300 mg/L as HCO_3
    Carbon dioxide—25 mg/L as CO_2

23. What is the concentration of magnesium ions in question 22? Will such a concentration present a scale problem for hot-water heaters?

24. At what pH do magnesium hydroxide and calcium carbonate precipitate?

25. How does ion-exchange work?

26. List the five major components of the ion-exchange softening process.

27. What is the name of the most commonly used ion-exchange material?

28. List the four cycles in the ion-exchange softening process.

29. The diameter of a vertical ion-exchange pressure tank is 2.5 ft. The softener loading rate is 5 gpm per sq ft. How long will it take to treat 156,000 gal of water?

30. An ion-exchange unit contains 11 cu ft of polystyrene resin, having a rated ion-exchange capacity of 30,000 grains per cu ft. What is the total-hardness removal capacity? Give the answer in grains and milligrams.

31. The ion-exchange unit in question 30 treats water that has a total hardness of 237 mg/L. How many gallons of water can be softened before the ion-exchange material must be regenerated?

32. The resin in question 30 requires 0.31 lb of salt to regenerate every 1000 grains of removal capacity. Once exhausted, how much salt is needed to regenerate this resin?

33. How much brine solution is needed in question 32 if the brine solution contains 0.5 lb of salt per gallon?

34. List four operating problems of the ion-exchange softening process.

35. A relationship between two laboratory tests is used to establish the concentrations of carbonate and noncarbonate hardness. List the two tests and describe the relationship.

Study Problems and Exercises

1. Assume that you have just been hired as the chief operator for a community of 5000 people in Arizona whose water supply comes from ground-water sources. The water has a total hardness of 420 mg/L, of which 200 mg/L is noncarbonate hardness. The water also has 0.6 mg/L of iron and 8 picocuries/L of radium 236. The city council asks for your recommendations.
   (a) Should the water be softened or should other types of treatment be performed?
   (b) If you recommend softening the water, which softening process should be used and why? Diagram the softening process you recommend.
   (c) What will be done with wastes from the treatment process?

2. You are the operator for a surface-water system that is softened by the lime-soda ash softening process. In addition, you prechlorinate and use alum coagulation and sedimentation to remove turbidity. You have seen the first test results for trihalomethanes in the treated water. The test results indicate an average concentration of 250 mg/L. Discuss what may be the causes for these high levels of trihalomethanes and what changes in treatment procedures you would propose to meet the MCL.

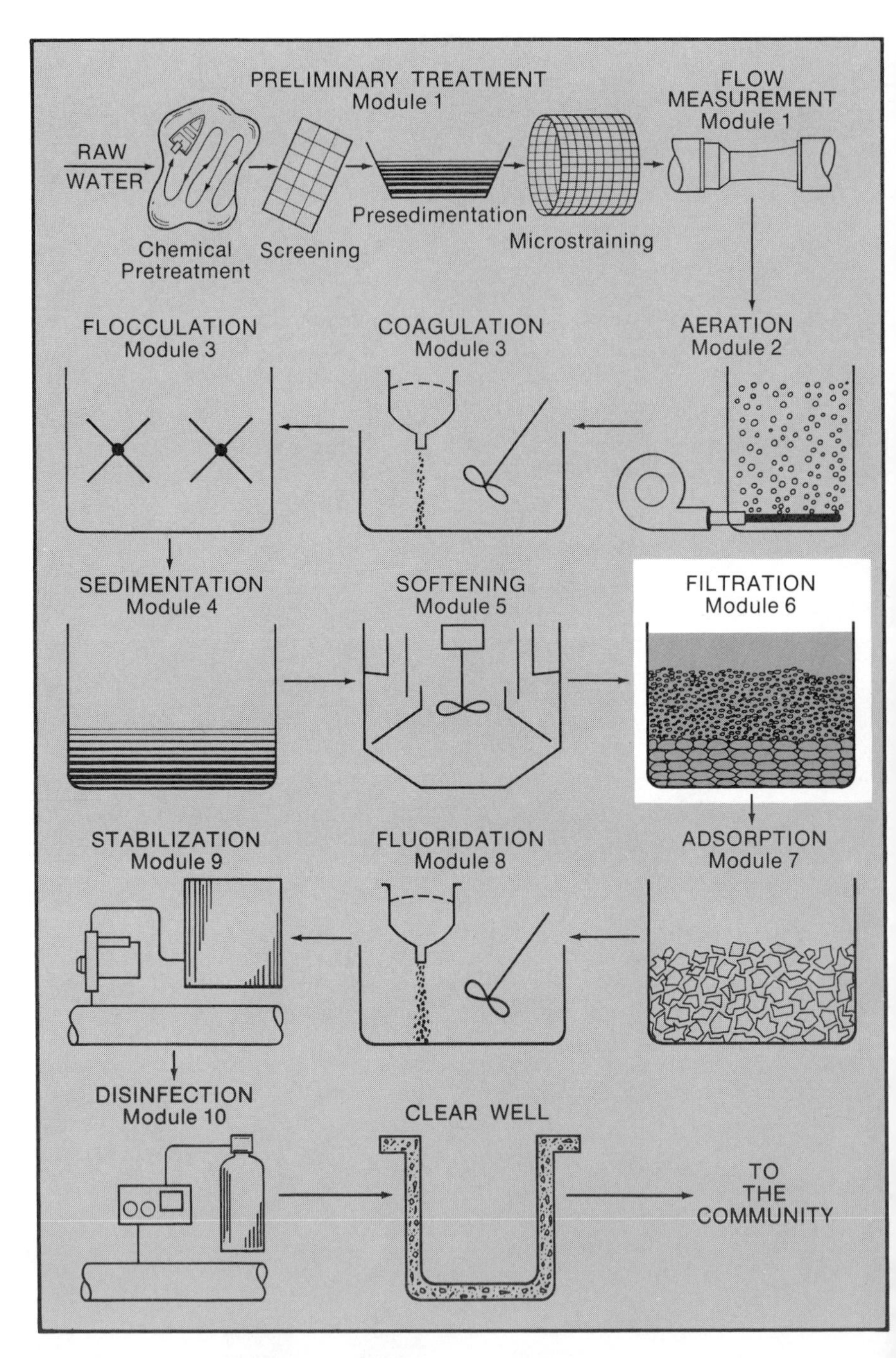

Figure 6-1. Filtration in the Treatment Process

Module 6

Filtration

The removal of suspended solids by filtration plays an important role in both the naturally occurring purification of ground water and the artificial purification of water in treatment plants. The location of the filtration process in a typical treatment plant is shown in Figure 6-1.

Filtration takes place naturally as water moves through porous layers of soil. This natural process, called PERCOLATION, removes most suspended material from ground water. (If ground waters are treated to remove hardness, iron, and manganese, filtration at the treatment plant may be needed to remove the chemical precipitates.) Since surface waters are subject to runoff and other sources of contamination, and do not undergo filtration in nature, filtration must be provided at the water treatment plant. At the plant, suspended material is removed and kept from interfering with later treatment processes (primarily disinfection) and from reaching the consumer. This suspended material (measured as turbidity) includes soil, oxidized metals, and microorganisms. Many of these microorganisms are extremely resistant to chlorination and other disinfection methods; however, they can be effectively removed with filtration.

To help assure effective turbidity removal, a treatment plant operator must understand the fundamentals of filtration. This module describes the types of filtration and common operating procedures used in water treatment. Typical operating problems are presented with discussions of how they can be avoided or solved. The module concludes with a discussion of operational control tests and important safety practices associated with the filtration process.

After completing this module you should be able to

- Discuss the role of filtration in water treatment, particularly concerning public health protection and drinking water standards.
- Diagram and discuss the major parts of a typical gravity filter.

- Discuss the advantages and disadvantages of slow-sand, rapid-sand, high-rate, and pressure filtration.
- Describe typical operating problems and indicate how they can be avoided or solved.
- Describe control tests that are used to monitor the filtration process.
- Describe safety precautions to be taken during the filtration process.

6-1. Description of Filtration

The major purpose of filtration is to remove suspended material (measured as turbidity) from water. This suspended material can include floc from the coagulation/flocculation or coagulation, flocculation, and sedimentation processes; microorganisms; PRECIPITATES such as calcium carbonate that remain after the lime or lime-soda ash softening of ground or surface waters; and iron and manganese precipitates from ground-water sources.

These suspended materials are removed when water passes through a bed of granular material known as the filter media. The filter media is usually sand or some combination of sand, anthracite coal, garnet, or similar substances manufactured specifically for water filtration. As shown in Figure 6-1, filtration follows coagulation, flocculation, and sedimentation in the treatment process.

Turbidity

Lowering turbidity is important in protecting public health and preventing operational problems in the distribution system. Turbidity interferes with the disinfection process because the suspended particles shield microorganisms from the disinfectant. The particles also combine chemically with the disinfectant and leave less disinfectant to combat the microorganisms. Turbidity also causes deposits in the distribution system that create tastes, odors, and bacterial growths. Therefore, lower turbidity results in more effective disinfection and better public health protection.

For these reasons, the USEPA drinking water regulations include a maximum contaminant level (MCL) of 1 NTU (NEPHELOMETRIC TURBIDITY UNIT) and the American Water Works Association has set a goal of 0.1 NTU for turbidity in drinking water (Appendix A). A well-designed and well-operated treatment plant should consistently meet both limits with few problems. To assure public health protection, surface-water sources should always be filtered even if the untreated water has turbidity levels less than the MCL of 1 NTU.

Filtration Process

The filter used in the filtration process is commonly thought of as a sieve or microstrainer that traps suspended material between the grains of filter media (Figure 6-2A). However, straining is the least important process in filtration since most suspended particles can easily pass through the spaces between the grains of the filter media.

A. Mechanical

RAW WATER

Large particles become lodged and cannot continue downward through the media.

B. Adsorption

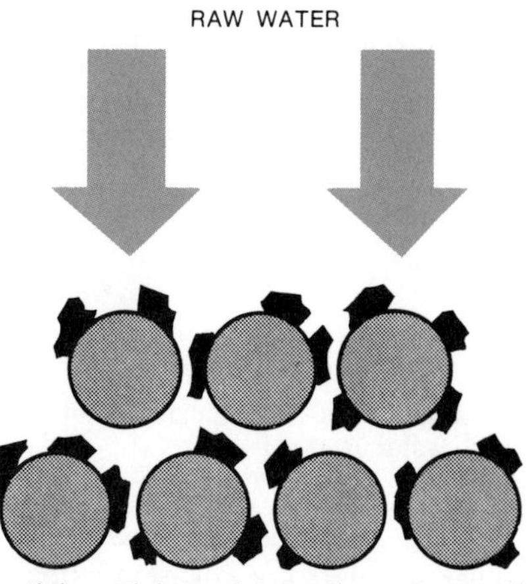

Particles stick to the media and cannot continue downward through the media.

Figure 6-2. The Two Removal Mechanisms

Filtration primarily depends on a combination of complex physical and chemical mechanisms, the most important being ADSORPTION (Figure 6-2B). As water passes through the filter bed, the suspended particles contact and adsorb (stick) onto the surface of the individual filter grains or onto previously deposited material. The forces that attract and hold the particles to the grains are the same as those at work in coagulation and flocculation. In fact, some flocculation and sedimentation occurs in the filter bed. This points out the importance of good chemical coagulation before filtration. Poor coagulation can cause serious operating problems with filters as discussed later in this module.

6-2. Filtration Facilities

Types of Filters

Filters can be classified by the filtration rate, type of filter media, or type of operation. For purposes of this module, the following classification, which combines the above factors, will be used:

I. Gravity Filters
- A. Slow-Sand
- B. Rapid-Sand
- C. High-Rate
  1. Dual-Media
  2. Multi-Media

II. Pressure Filters
- A. Sand or Multi-Media
- B. Diatomaceous Earth.

Table 6-1. Comparison of Slow-Sand, Rapid-Sand, and High-Rate Filters

Characteristic	*Slow-Sand Filters*	*Rapid-Sand Filters*	*High-Rate Filters*
Filtration rate	0.05 gpm/sq ft	2 gpm/sq ft	3–8 gpm/sq ft
Media	Sand	Sand	Sand and coal or sand, coal, and garnet
Media distribution	Unstratified	Stratified—fine to coarse	Stratified—coarse to fine
Filter runs	20–60 days	12–36 hours	12–36 hours
Loss of head	0.2 ft initial to 4 ft final	1 ft initial to 8 or 9 ft final	1 ft initial to 8 or 9 ft final
Amount of back-wash water used	Backwash not used	2–4% of water filtered	6% of water filtered

There are other filter types, such as upflow and biflow filters. Because these types of filters are not used extensively in the United States, they will not be discussed here. Gravity filters will be discussed in detail because they are the most common. Discussion of pressure filters will be limited to a description in the Filtration Facilities Section.

Gravity filters. In gravity filters, the force of gravity moves the water through the filter media. Slow-sand filters were the first type of gravity filter used in the United States; rapid-sand filters and high-rate filters came into use later. Gravity filters are compared in Table 6-1.

Slow-sand filters. Slow-sand filters, first introduced into the United States in 1872, are the oldest type of municipal water filtration. As shown in Figure 6-3, a slow-sand filter consists of a 3 ½-ft (1-m) layer of fine sand supported by 1 ft (0.3 m) of graded gravel.

Slow-sand filters effectively remove turbidity, but a number of significant disadvantages have led to their eventual replacement. First, they depend on fine sand and a sticky mat of suspended matter (a SCHMUTZDECKE) on the sand surface for filtering action. It is often necessary to waste the filtered water for as long as two weeks before a schmutzdecke forms that will effectively remove turbidity.

Secondly, the fine sand that is used has small void spaces, which fill quickly. Consequently, chemical coagulation generally is not used to help form the schmutzdecke because the void spaces will be filled and clogged by the floc formed by the coagulant. For this same reason, slow-sand filters are not used with waters that consistently have turbidities above 10 NTU.

Finally, to allow proper formation of the schmutzdecke, flow rates must be kept very low. As shown in Table 6-1, the filtration rate is only 0.05 gpm/sq ft (0.03 mm/s)[1]. Filtration rates are expressed in terms of gallons per minute per square foot of filter surface area. Consequently, from 0.5 to 1.0 acres (0.20 to 0.40 ha) of land are required for each 1 mil gal (4 ML) of water filtered. Because such a large land area is required, slow-sand filters are not BACKWASHED. Instead

[1] *Basic Science Concepts and Applications*, Mathematics Section, Filter Loading Rate.

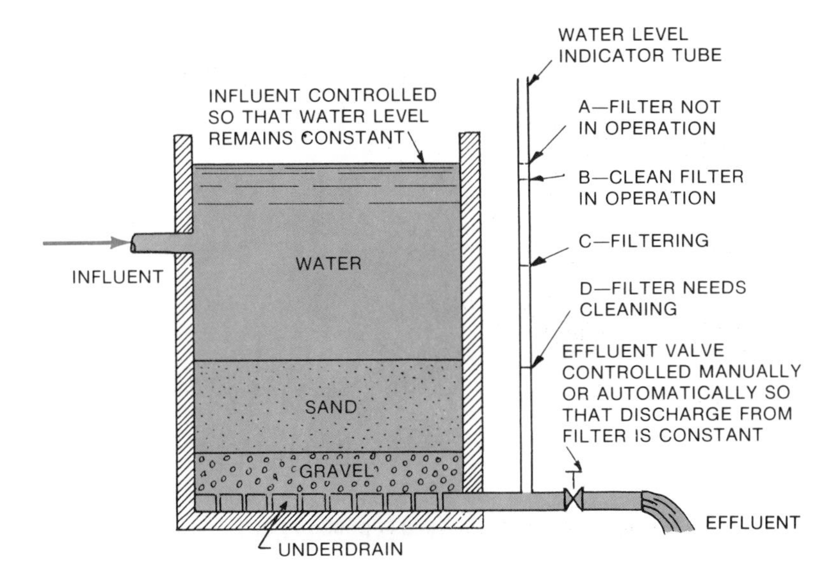

Courtesy of the New York State Department of Health

Figure 6-3. A Slow-Sand Filter

they are cleaned by scraping off the schmutzdecke and the top 6 in. (150 mm) of sand. The sand is washed and replaced after cleaning has reduced the sand depth to about 2 ft (0.6 m).

These disadvantages, combined with the relatively high turbidities of surface waters in the United States, have made the slow-sand filter uneconomical. This has resulted in development of rapid-sand filters.

Rapid-sand filters. The rapid-sand filter can accommodate filter rates 40 times those of the slow-sand filter. Because of the rapid-sand filter's increased filter rate and capability of being backwashed, chemical coagulation is practical, resulting in an effective means of treating waters with high or variable turbidities.

Figure 6-4 shows the basic components of a rapid-sand filtration system. They include:

- Filter tank
- Filter sand
- Gravel support bed
- Underdrain system
- Wash-water troughs
- Filter-bed agitators
- Control equipment (not shown in Figure 6-4).

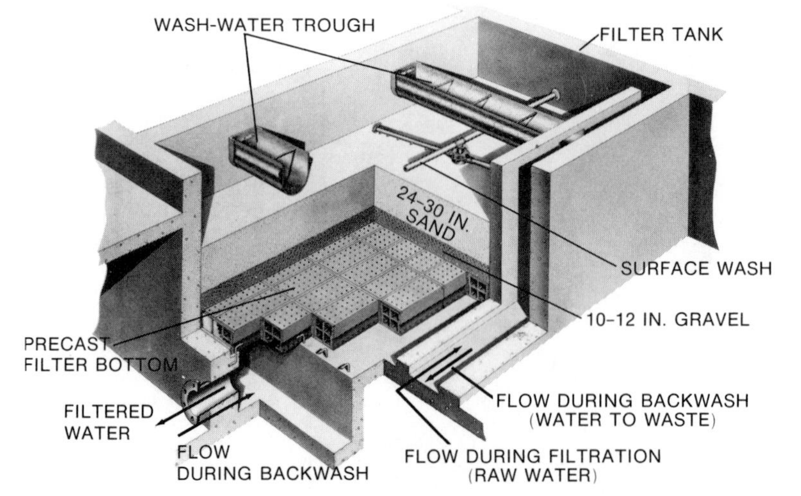

Courtesy of Leopold Company, Division of Sybron Corporation

Figure 6-4. A Rapid-Sand Filtration System

The watertight FILTER TANK contains all system components except for the control equipment. The tanks are usually rectangular and made of concrete; however, tanks in small plants, particularly "package plants," are often made of steel. The filter tanks are usually constructed side by side on either side of a central pipe gallery to minimize piping. Figure 6-5 shows such an arrangement with only one row of filters.

FILTER SAND is specifically manufactured for filtration as are all media used for this purpose. A rapid-sand filter usually uses a 24- to 30-in. (0.6- to 75-m) bed of fairly uniform sand grains, 0.4 to 0.6 mm in diameter. This sand is much coarser

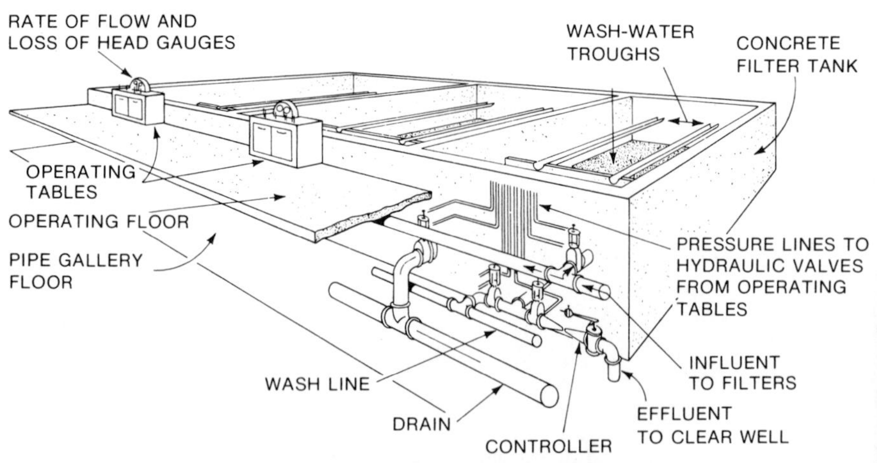

Figure 6-5. Filter Tank Construction

than that used for slow-sand filters. This accounts for many of the advantages of rapid-sand filters. One major advantage is that the coarser sand has larger void spaces, which do not fill and seal off quickly.

Three to five layers of graded gravel are placed between the sand and underdrain system. The total GRAVEL BED is usually 6 to 18 in. (150 to 450 mm) thick, depending on the underdrain system. This gravel bed prevents sand from entering the underdrains and helps distribute the backwash water evenly.

The filter UNDERDRAINS serve two functions. The most important function is to evenly distribute the backwash water so that the sand and gravel beds will not be disturbed. The other function is to uniformly collect the filtered water. The water then flows into the piping system for further treatment and then to the clearwell.

Types of underdrain systems commonly used are:

- Pipe-lateral
- Leopold bottom
- Wheeler bottom
- Porous plates or strainer nozzles.

The PIPE-LATERAL SYSTEM uses a control manifold with several perforated laterals on each side. Piping materials include cast iron, asbestos cement, and polyvinyl chloride (PVC). The perforations are usually placed on the underside of the laterals to help prevent them from being plugged with sand. Also, the force of the backwash water is directed against the filter floor, which helps keep the gravel and sand beds from being disturbed by high-velocity water jets.

Figure 6-6 shows a LEOPOLD FILTER BOTTOM, consisting of perforated, vitrified clay blocks with channels inside to carry and distribute the water. Figure 6-4 shows how these blocks are placed in a filter.

A WHEELER BOTTOM (Figure 6-7) consists of concrete conical depressions filled with porcelain spheres. Each cone has an opening in the bottom and contains 14

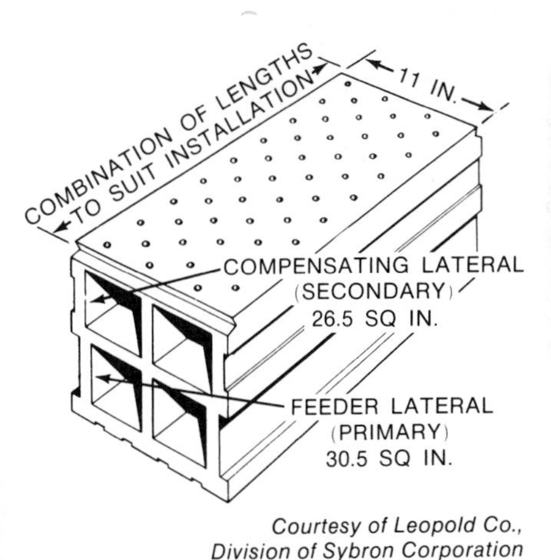

*Courtesy of Leopold Co.,
Division of Sybron Corporation*

Figure 6-6. A Leopold Filter Bottom

*Courtesy of BIF,
a Unit of General Signal*

Figure 6-7. A Wheeler Bottom

spheres ranging in diameter from 1 ⅜ to 3 in. (30 to 80 mm). The spheres are arranged to lessen the velocity of the wash water and distribute it evenly.

The above three systems need a gravel support bed to prevent the sand from being disturbed during backwash and to prevent plugging of the underdrains. Systems using POROUS PLATES support the filter media directly without intermediate gravel layers. The porous plates are made of ceramic and are supported by long bolts, concrete piers, or clay saddles, as shown in Figure 6-8. Plastic or stainless steel nozzles mounted in hollow, vitrified clay blocks are also used without a gravel bed. The nozzles have fine slots designed to keep the media from leaking into the underdrains.

WASH-WATER TROUGHS, placed above the filter media, collect the backwash water and carry it to the drain system. Proper placement is very important to ensure that filter media does not wash into the troughs and that the water travels away uniformly, which maintains an equal head on all parts of the underdrain system. The troughs are usually constructed of concrete, plastic, or other corrosion-resistant materials.

During operation of a rapid-sand filter, the top 6 to 10 in. (150 to 250 mm) of sand removes most of the suspended matter. It is important that this layer is thoroughly cleaned. Frequently, normal backwashing does not get the sand layer sufficiently clean, therefore an additional cleaning system is used. This system is known as FILTER AGITATION, AUXILIARY SCOUR, or SURFACE WASHING. The system consists of nozzles attached to either a fixed pipe (Figure 6-9A) or rotary pipe (Figure 6-9B) system installed just above the filter media. Water or an air–water mixture is pumped through the nozzles producing high velocity jets, which help break up and wash away the accumulated suspended matter in the top 6 to 10 in. (150 to 250 mm) of the filter bed.

Good control of the filtration process is essential to consistently produce low-turbidity water. Filtration is often the most automated process in the

Courtesy of the Norton Company, Industrial Ceramics Division

Figure 6-8. Porous Plates Supported by Clay Saddles

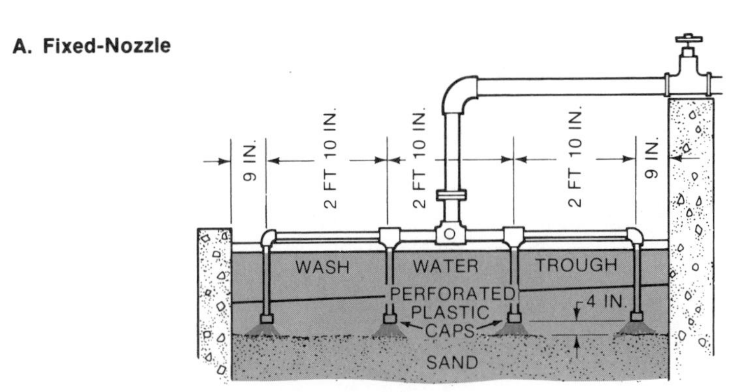

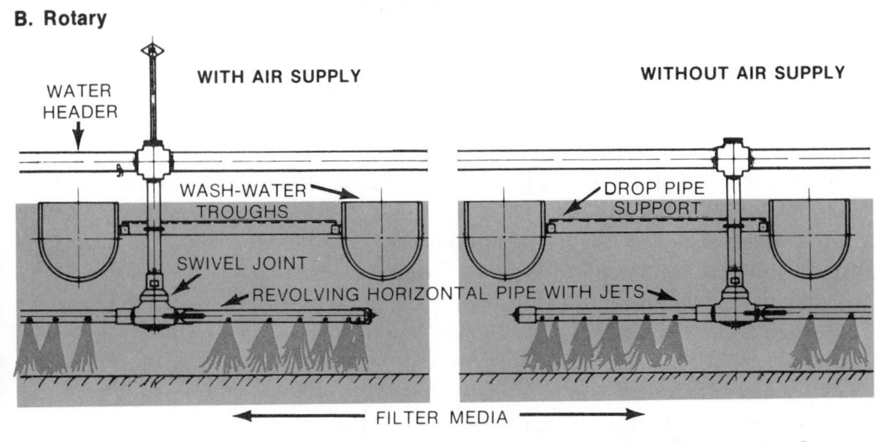

Courtesy of Roberts Filter Manufacturing Company

Figure 6-9. Filter Bed Agitators

treatment plant. Regardless of the degree of automation, rapid-sand filters usually require the following control equipment:

- Rate-of-flow controller
- Loss-of-head indicator
- On-line turbidimeters.

The RATE-OF-FLOW CONTROLLER maintains a fairly constant flow through the filter so that flow surges do not occur. These surges would force suspended particles through the filter. The controller generally consists of a flow measurement device (usually a Venturi tube), a throttling valve on the effluent line, and a means to automatically or manually set the throttling valve to maintain a fixed flow rate (Figure 6-10).

As filtration proceeds, more and more pressure is needed to overcome the HEAD LOSS (frictional resistance) that results from the buildup of suspended matter in the filter. The head loss should be continuously measured to help determine when the filter should be backwashed. Usually, the difference in the

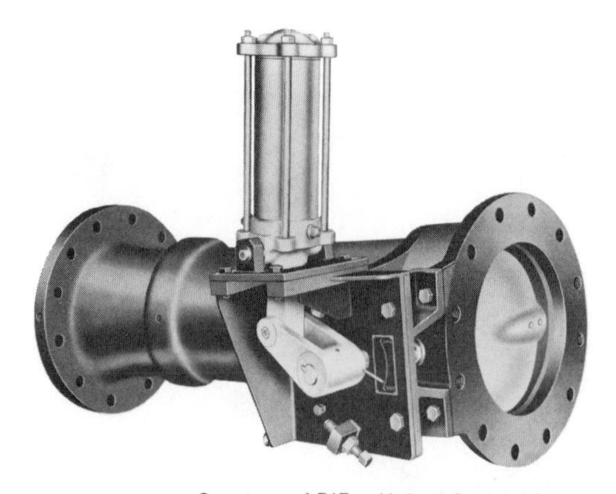

Courtesy of BIF, a Unit of General Signal

Figure 6-10. Rate-of-Flow Controller

height of water or mercury in tubes (PIEZOMETERS)[2] connected to the filter and the effluent line is measured, or the head is sensed by air pressure as shown in Figure 6-11. Recorders can be attached to provide a continuous record. Another control method that is becoming common is variable declining-rate filtration, discussed under Operation of Gravity Filters.

Turbidity of the filtered water is another key factor in filtration control. At some point, the suspended material will begin to break through the filter bed, causing turbidity to increase in the filtered water. The filter should be backwashed at this time. ON-LINE TURBIDIMETERS (Figure 6-12) can continuously monitor and record the effluent turbidity. This continuous record provides valuable information that can be used to schedule backwashing and control chemical addition ahead of the filters.

Rapid-sand filters effectively filter waters that have variable and relatively high turbidities. The major disadvantage is that only the top 6 to 10 in. (150 to 250 mm) of the sand does most of the filtering. Therefore, about 80 percent of the bed is not being used effectively. To overcome this problem, high-rate filters were developed.

High-rate filters. As shown in Table 6-1, high-rate filters can operate at rates three to four times those of rapid-sand filters. These filters use a combination of filter media, not just sand; all other components are the same as for rapid-sand filters. Dual-media filters use sand and anthracite coal. Multi-media (mixed-media) filters use three or more media—usually sand, anthracite coal, and garnet. Figure 6-13 compares the filter media used in high-rate filters with that used in a rapid-sand filter.

In rapid-sand filters, finer sand grains lie on top of the coarse sand grains. This is known as a fine-to-coarse gradation. As a result, most suspended material is

[2] *Basic Science Concepts and Applications*, Hydraulics Section, Piezometric Surface and Hydraulic Grade Line (Piezometric Surface); also, Hydraulics Section, Head Loss.

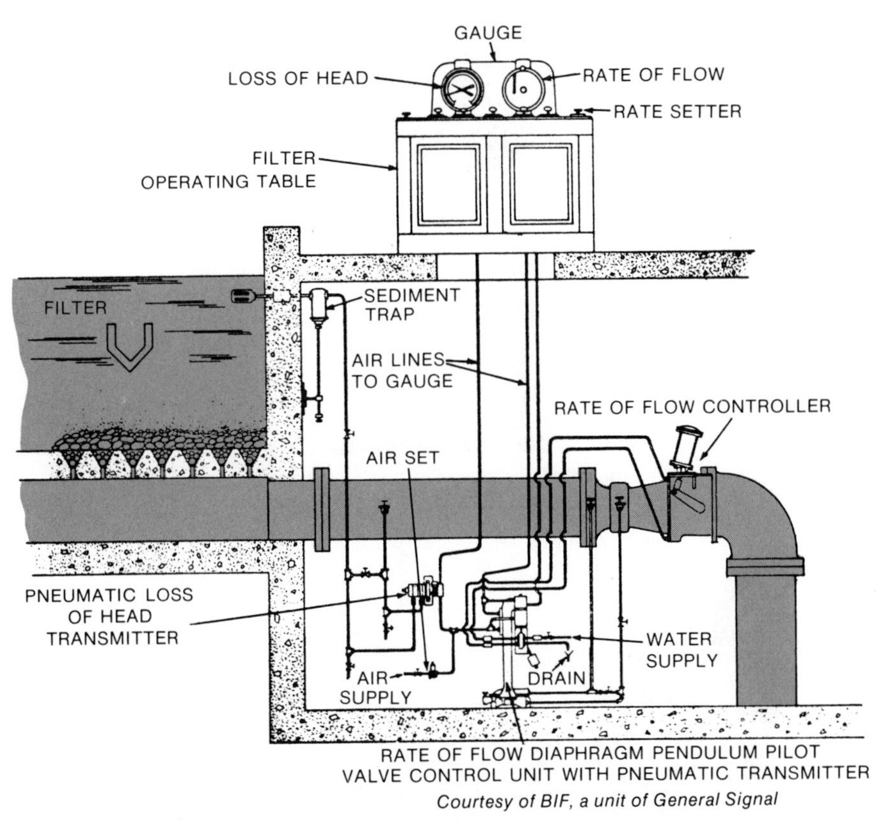

Courtesy of BIF, a unit of General Signal

Figure 6-11. Measurement of Head in the Filtration Process

removed in the top few inches of the filter, causing high head loss and shorter filter runs.

Dual- and mixed-media filters have a coarse-to-fine gradation of filter media. By using fine media with a high specific gravity and coarse media with a lower specific gravity, the layers of media approximately maintain their respective positions in the filter bed even after backwashing. Some mixing does occur, as illustrated in Figure 6-14. Mixing actually makes the filter more effective by providing more contact area for the suspended particles.

In operation, the coarse layer on top removes the larger suspended particles. The finer particles pass through this layer and are removed by finer media below. As a result, most of the filter bed is used to remove suspended particles. This allows for longer filter runs and higher filtration rates because head loss does not build up as quickly as with a rapid-sand filter. Multi-media filters are becoming popular because they can greatly increase a treatment plant's capacity without loss in water quality.

The type of filter media used depends on many factors including raw-water quality, raw-water quality variation, and chemical treatment. Pilot tests using

Courtesy of Hach Chemical Company

Figure 6-12. On-Line Turbidimeter

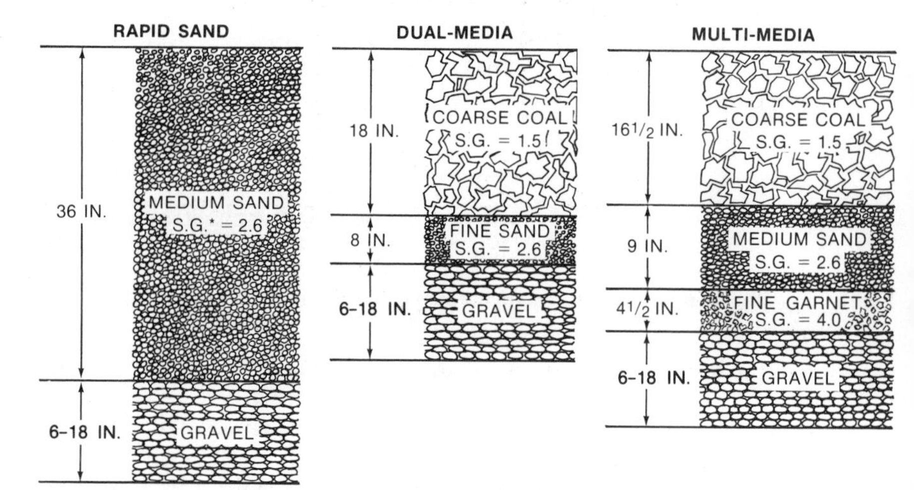

*Specific Gravity, see *Basic Science Concepts and Applications*, Hydraulics Section, Specific Gravity.

Figure 6-13. Comparison Between High-Rate Filter Media and Rapid-Sand Filter Media

different combinations of media are usually conducted to determine which media combination performs best for a particular water.

Pressure filters. Pressure filters can be divided into two broad categories:

- Pressure-sand filters
- Diatomite filters.

Both types have been used extensively in filtering water for swimming pools. Some small communities also use these filters because installation and operating costs are low.

Courtesy of Neptune Microfloc, Inc.

Figure 6-14. Mixing of Filter Media, Left: As Laid; Right: After Backwash

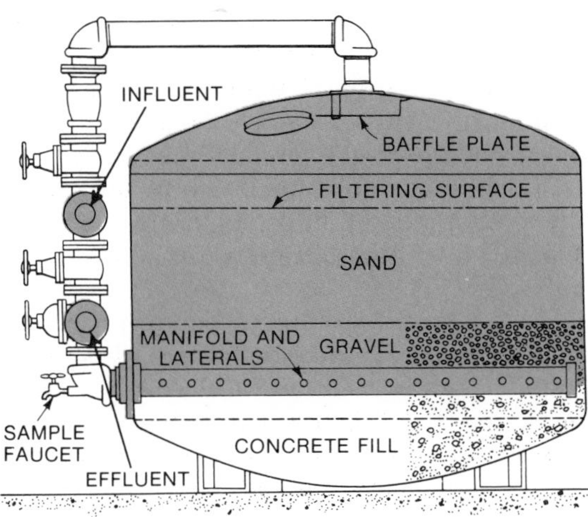

Figure 6-15. Vertical Pressure Sand Filter

Pressure-sand filters. PRESSURE-SAND FILTERS operate on the same principle as gravity-sand or rapid-sand filters. However, with pressure-sand filters the filtration process takes place in a cylindrical steel tank. The tank is either vertical as in Figure 6-15 or horizontal. As with gravity filters, sand or a combination media is used, and filtration rates are about the same as for gravity filters. Pumping forces the water through the media at the selected rate.

Commonly, these filters are used for iron and manganese removal from ground water. The water is usually aerated or treated with potassium permanganate to OXIDIZE the iron and manganese. The resulting chemical precipitates are then removed by filtration. If these filters are to be used for treating surface water, chemical addition and other treatment processes must be provided before filtration.

Because the water is under pressure, air binding will not occur. Pressure filters do have a major disadvantage—the filter bed cannot be observed during operation. Consequently, filter cracks, media disruption, or mudballs will not be found until a more obvious problem results. Because of this, continuous turbidity monitoring is essential to prevent serious degradation of the filtered water and to provide a warning when bed problems occur.

Diatomaceous earth filters. DIATOMACEOUS EARTH FILTERS (DE FILTERS) were developed by the military during World War II to remove the organisms causing amoebic dysentery. Today they are used extensively in swimming pools and they are starting to be used in small towns for potable water treatment. The filter media is diatomaceous earth—the skeletal remains of microscopic aquatic plants called DIATOMS. Diatoms, ranging in size from 5 to over 100 μm (0.005 to 0.1 mm), have always flourished in the world's waters. Prehistoric deposits are mined and processed to produce the diatomaceous earth media.

Diatomaceous earth filters can be operated either as pressure or vacuum filters. Regardless of the design, the components and the operation are similar (Figure 6-16). Filtration rates are usually between 0.5 to 2 gpm/sq ft (0.34 to 1.4 mm/s).

The first step in operating a DE filter is to apply a precoat of diatomaceous earth to the septum, which provides a filter surface (Figure 6-17A). A thickness of ⅛ in. (3 mm) is normally provided. Only previously filtered water should be used for PRECOATING. The untreated water is then applied to the filter. Because of possible cracking and clogging of the precoat, a BODY FEED of diatomaceous earth must be added constantly (Figure 6-17B). Once the filter run is completed,

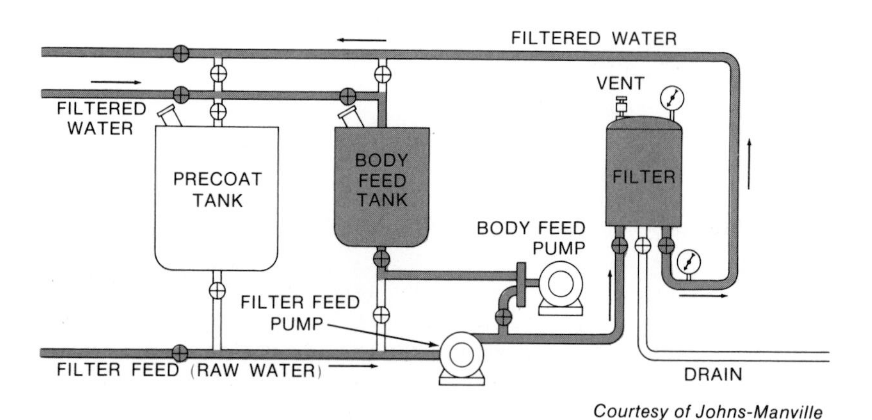

Courtesy of Johns-Manville

Figure 6-16. Components of a DE Filter

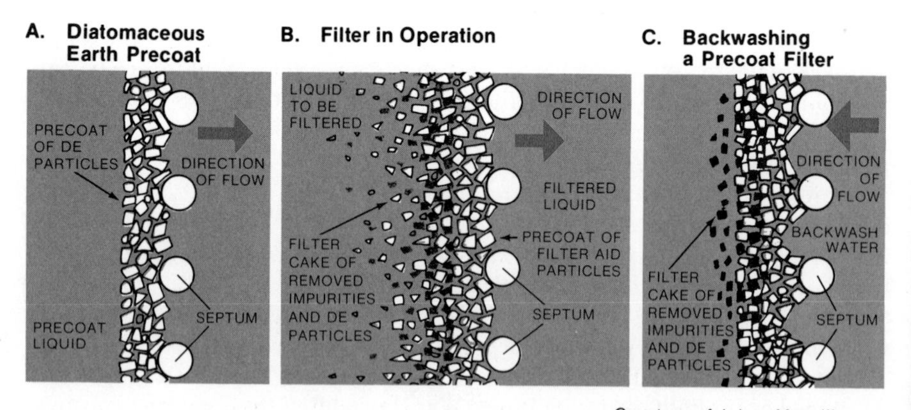

Courtesy of Johns-Manville

Figure 6-17. Diatomaceous Earth Filter

the filter cake is washed from the septum by reversing the water flow (Figure 6-17c). The spent filter cake is usually discharged to a lagoon for disposal.

The use of DE filters for potable water treatment has been limited because of the difficulty in maintaining an effective cake of diatomaceous earth for filtration. If these filters are to be used for treating surface water, proper treatment of the water before filtration and continuous monitoring of the filtered water must be provided.

Filtration Plants

There are two basic types of filtration processes currently used in the United States. Conventional filtration, shown in Figure 6-18A, has been the traditional design for many years and can provide effective treatment for practically any range of raw-water turbidity. Its success is primarily due to the sedimentation step, which removes most of the suspended material. After sedimentation, the water going through the filters usually has a turbidity of 10 to 15 NTU. For this reason, conventional filtration can be used regardless of raw-water turbidity and color levels. Historically, rapid-sand filters were used in the conventional process; however, conversion to dual- or multi-media filters in an effort to increase treatment capacity is increasing.

Direct filtration, as shown in Figure 6-18B, does not have a sedimentation process. This process is designed to treat raw waters with average turbidities

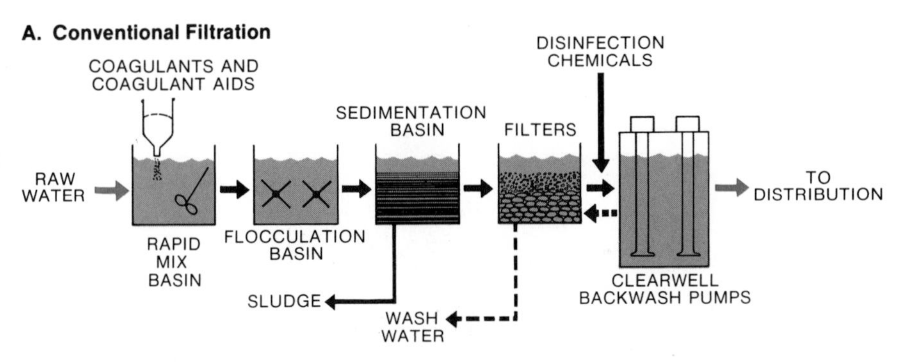

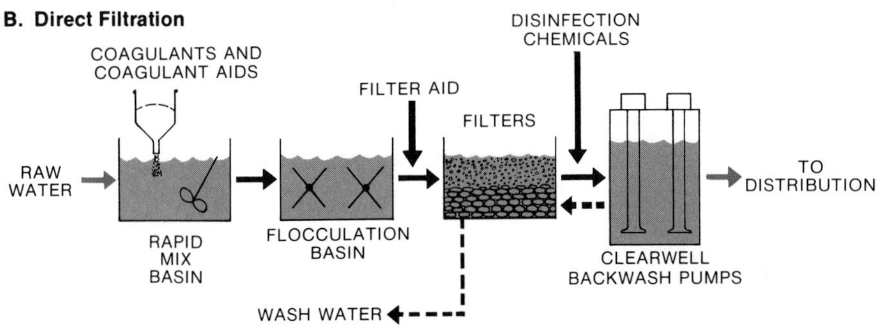

Figure 6-18. Filtration Processes

below 25 NTU and color levels below 25 units. However, they have been used when turbidity and color levels were higher. Dual- or multi-media filters should always be used with direct filtration because of their capability to remove more suspended solids than rapid-sand filters before backwashing. The major advantage of direct filtration is its low construction cost compared to conventional plants and ability to consistently produce high-quality water. However, because of the short time between coagulant addition and filtration and the greater load applied to the filters, this type of system must be carefully monitored.

6-3. Operation of Gravity Filters

Filtration has three steps: filtering, backwashing, and filtering to waste. Figure 6-19 indicates the key valves used in the filtration process and their positions during the filtration steps.

Filtering

During filtering, water is applied to the filter to maintain a constant depth of 4 to 5 ft (1 to 1.5 m) over the media. Initially, the media is clean and head loss is very low. The filtration rate is kept at the desired level by using a control. A

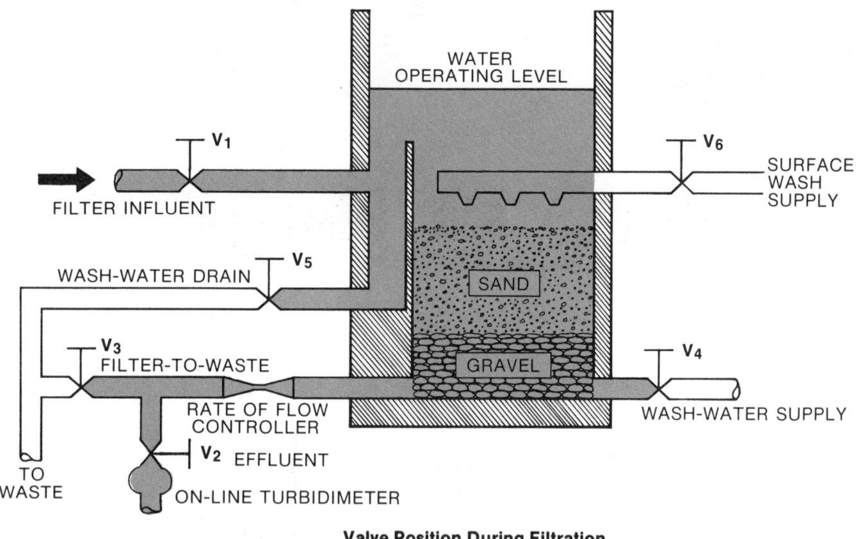

Valve Position During Filtration

Valve	*Filtering*	*Backwashing*	*Filtering to Waste*
V_1—Influent	Open	Closed	Open
V_2—Effluent	Open	Closed	Closed
V_3—Filter to Waste	Closed	Closed	Open
V_4—Wash-Water Supply	Closed	Open	Closed
V_5—Wash-Water Drain	Closed	Open	Closed
V_6—Surface-Wash Supply	Closed	Open	Closed

Figure 6-19. Key Valves Used in Filtration

control is also important in preventing harmful surges, which can disturb the media and force floc through the filter.

Rate-of-flow controller. Commonly, the rate-of-flow controller is used to maintain a constant desired filtration rate—usually 2 gpm/sq ft (1.4 mm/s) for rapid-sand filters and 4 to 6 gpm/sq ft (2.8 to 4 mm/s) for high-rate filters. To control flow at the beginning of the filter run, the flow controller valve is almost completely closed. This produces the necessary head loss and maintains the desired flow rate. As filtration continues and suspended material builds up in the bed, head loss increases. To compensate for this increase, the controller valve is gradually opened. When the valve is fully opened, the filter run must be ended, since further head loss cannot be compensated for and the filter rate will drop sharply. The major disadvantage of a rate-of-flow controller is its need for frequent regular maintenance. A malfunctioning controller can damage the filter bed and degrade the quality of the filtered water by allowing sudden changes in the filtration rate.

Variable declining-rate filtration. Another control method is variable declining-rate filtration. As shown in Figure 6-20, the filtration rate is not kept constant with this method. The rate starts high and gradually decreases as the filter gets dirty and head loss increases. This method is often preferred because it does not require a rate-of-flow controller or constant attention by the operator, and harmful rate changes cannot occur.

With this method, each filter accepts the proportion of total flow that its bed condition can handle. For example, as a filter gets dirty, the flow through it decreases. Thus, flow redistributes to cleaner filters and total plant capacity does not decrease. To prevent excessive flow rates from occurring in clean filters, a flow-restricting orifice is placed in the effluent line of each filter.

Regardless of the control method used, as filtration progresses, suspended matter builds up within the filter bed. At some time, usually after 15 to 36 hours of operation, the filter must be cleaned by backwashing.

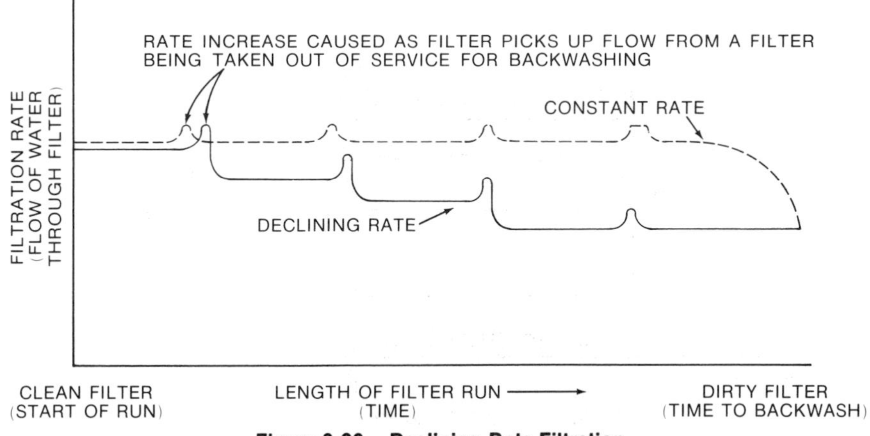

Figure 6-20. Declining-Rate Filtration

Backwashing

Backwashing is a critically important step in the filtration process, and inadequate backwashing causes most operating problems associated with filtration. For a filter to operate efficiently, it must be cleaned thoroughly before the next filter run begins. In addition, properly backwashed filters require far less maintenance. Treated water is always used for backwashing so that the bed will not be contaminated. The treated water comes from elevated storage tanks designed for this purpose, or it is pumped from the clear well.

During filtration, the voids between the media grains fill with the filtered material (floc). The grains also become coated with the floc and become very sticky, making the filter bed difficult to clean. To clean the filter bed, the media grains must be agitated violently and rubbed against each other to dislodge the sticky coating. Therefore, the backwash rate must be high enough to completely suspend the filter media in the water. The backwash causes the filter bed to expand as shown in Figure 6-21. The expansion, however, should not be so large that the media flows into the wash-water troughs. Since normal backwash rates are not sufficient to thoroughly clean the media, auxiliary scour (surface wash) equipment is recommended to provide the extra agitation needed. Auxiliary scour is a must with high-rate filters, since the filtered material penetrates much deeper into the bed.

Head loss, filter-effluent turbidity, and length of the filter run must all be considered when deciding when a filter needs backwashing. Generally, a filter should be backwashed when:

- Head loss is so high that the filter no longer produces water at the desired rate. This is known as TERMINAL HEAD LOSS and is usually about 8 ft (2.4 m).
- Floc starts to break through the filter bed causing the filter effluent turbidity to increase.
- A filter run reaches 36 hours.

Figure 6-22 shows the relationship of these three factors in a typical filter run.

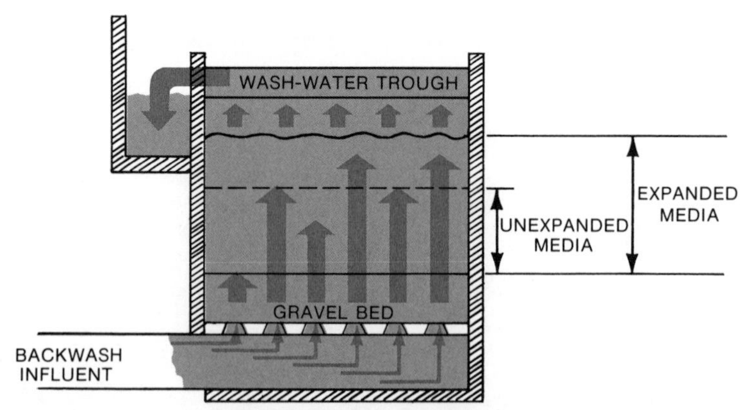

Figure 6-21. Filter Bed Expansion

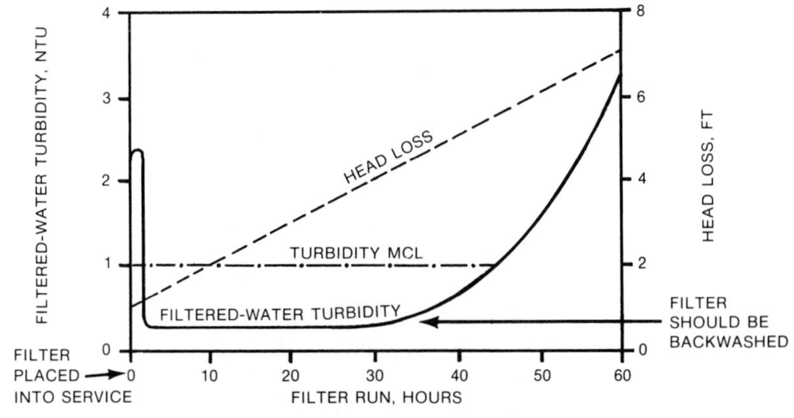

Figure 6-22. Typical Filter Run

The decision to backwash should not be based on only one of these three factors. This can lead to operating problems. For example, if a filter is not washed until terminal head loss is reached, a serious increase in filtered-water turbidity could occur well before the filter is backwashed. Likewise, depending on effluent turbidity alone can cause head loss to reach terminal conditions well before BREAKTHROUGH has occurred. This can cause the pressure in the filter bed to drop below atmospheric pressure, which causes operational problems.

A filter should definitely be backwashed when the effluent turbidity begins to increase as indicated in Figure 6-22. The turbidity should never be allowed to increase to 1 NTU before backwashing. In fact, filtration tests have shown that microorganisms start passing through the filter rapidly once breakthrough starts, even though the turbidity may be well below 1 NTU.

If the water applied to the filter is of very good quality, filter runs, based on head loss or effluent turbidity, can be very long. In fact, filters may be washed only once a week. It must be remembered, however, that long filter runs can cause gradual buildup of organic materials and bacterial populations within the filter bed. This, in turn, can lead to tastes and odors in the treated water as well as difficult-to-remove slime growths within the filter. On the other hand, short filter runs decrease finished-water production due to the time the filters are out of production and the increased quantity of backwash water needed. Usually, the amount of backwash water used should not exceed 4 percent of the amount of water treated for rapid-sand filters, and 6 percent for dual- and multi-media filters. Filter runs should range between 15 and 24 hours and should never exceed 36 hours. Very little is gained in net water production if filter runs are greater than 24 hours.

A typical backwash sequence begins by draining the water in a filter down to a level about 6 in. (150 mm) above the media. The surface washers are turned on and allowed to operate for 1 to 2 min (Figure 6-23). This allows the high-velocity water jets (usually having 40 to 75 psig [280 to 500 kPa, gauge] of water pressure) to break up any surface layers of filtered material. The backwash valve is opened

Courtesy of Roberts Filter Manufacturing Company

Figure 6-23. Surface Washer

part-way to allow the bed to expand to just above the level of the washers. This provides violent scrubbing of the top portion of the media, which has the greatest accumulations of filtered material. Intense scrubbing is particularly important with rapid-sand filters because the top 8 in. (200 mm) of media does most of the suspended-solids removal. After a few minutes, the backwash valve is fully opened to allow a filter bed expansion of 20 to 30 percent—the actual amount of expansion depends on how much agitation is needed to suspend the coarsest grains of media in the bed. With multi-media filters, the bed must be expanded so that the surface washers can scrub the area (interface) between the coal and sand layers where most of the filtered material has penetrated. A backwash rate of 15 to 20 gpm/sq ft (10 to 14 mm/s) is usually sufficient to provide the expansion needed.[3] The expanded bed is washed for 5 to 15 min depending on how dirty the filter is. The clarity of the wash water as it passes into the wash-water troughs can be used as an indicator of when to stop washing. The surface washers are usually turned off about 1 min before the backwash flow is stopped. This allows the bed to restratify into layers, which is particularly important for multi-media filters.

If surface wash equipment is not available, a two-stage wash should be used. The initial wash velocity should be just enough to slightly expand the top portion of the bed (around 10 gpm/sq ft [7 mm/s]). Although not as effective as surface washing, this method will provide the scrubbing action needed to clean the surface media grains. After the surface has been cleaned, the full backwash rate is applied.

Turning on the backwash too quickly can severely damage the underdrain system, gravel bed, and media. The time from starting the backwash flow to

[3] *Basic Science Concepts and Applications*, Mathematics Section, Filter Backwash Rate.

reaching the desired backwash flow rate should be from 30 to 45 sec. To prevent accidents, the backwash valve controls should be set to open slowly.

Disposal of backwash water. To avoid water pollution, backwash water must not be returned directly to streams or lakes. This water is usually routed to a lagoon or basin for settling. After settling, the water is usually recycled to the treatment plant as shown in Figure 6-24. The settled solids are combined with the sludge from the sedimentation basins and treated as discussed in Module 4, Sedimentation. This conserves most of the backwash water and relieves the operator of the need to obtain a discharge permit from the state water pollution control agency.

Since backwash water usually does not contain many suspended solids, the water is sometimes returned to the plant influent. The water must be placed in a storage tank so that it can be returned to the plant without causing flow surges.

Filtering to Waste

Once backwashing is completed, the water applied to the filter should be filtered to waste until the turbidity of the effluent drops to an acceptable level. As shown in Figure 6-22, at the beginning of the filter run, the effluent turbidity remains high for a certain time period. Depending on the type of filter and treatment processes used before filtering, this period may last from 2 to 20 min. This is probably due, in part, to filtered material that remains in the bed after backwashing. In addition, the initial high filtration rates continue to wash fine material through the filter until the media grains become sticky and more effectively adsorb the suspended material.

If filtering to waste cannot be done, a slower filtration rate can be used for the first 15 to 30 min of each filter run to prevent breakthrough of filtered material.

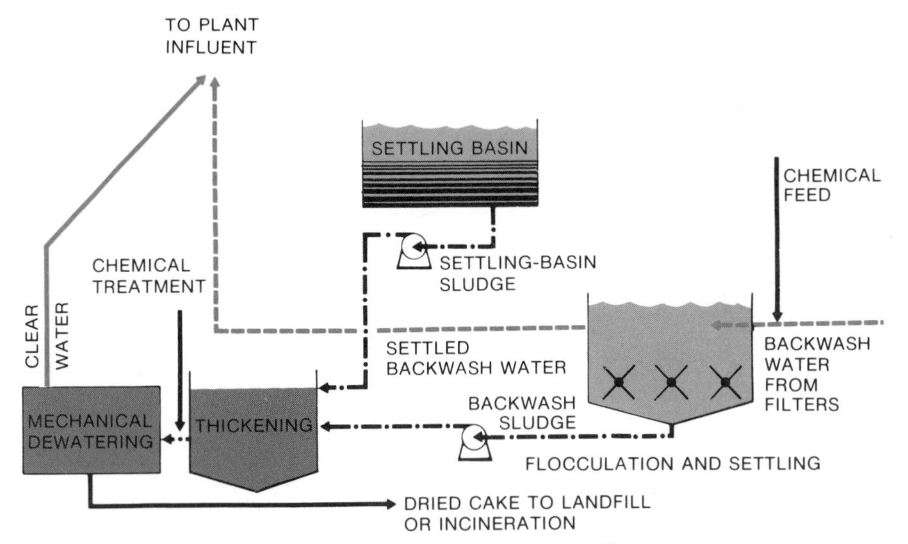

Figure 6-24. Disposal of Backwash Water

The filtered material can include large quantities of microorganisms, so breakthrough must be prevented.

Filter Aids

As water passes through the filter, the floc can be torn apart, resulting in very small particles, which can penetrate the filter. In this instance, turbidity breakthrough occurs well before the terminal head loss is reached (Figure 6-25A), resulting in short filter runs and high volumes of backwash water. This is particularly true of high-rate filters since loading in terms of suspended material and the filter rate are much higher than with rapid-sand filters.

To help solve this problem, POLYMERS (also called POLYELECTROLYTES) can be used as filter aids. Polymers are water-soluble, organic compounds and come in either dry or liquid form. The dry form is more difficult to use because it does not dissolve easily. Mixing and feeding equipment are required as indicated in Figure 6-26.

Polymers have high molecular weights and can be NONIONIC, CATIONIC, and ANIONIC, and as discussed in Module 3, Coagulation/Flocculation, they are also used as primary coagulants or coagulant aids. Usually nonionic or slightly anionic polymers are used as filter aids.

When used as a filter aid, the polymer strengthens the bonds between the filtered particles and coats the media grains to improve adsorption. The floc then holds together better, adheres to the media better, and resists the shearing forces exerted by the water flowing through the filter. For best results, the polymer should be added to the water just ahead of the filters. In order to select the proper

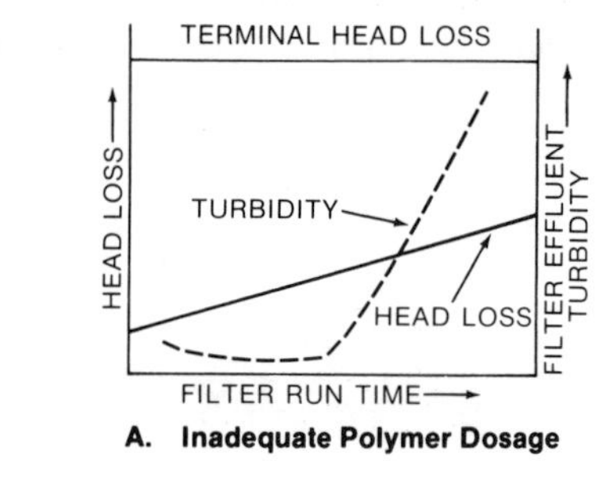

A. Inadequate Polymer Dosage

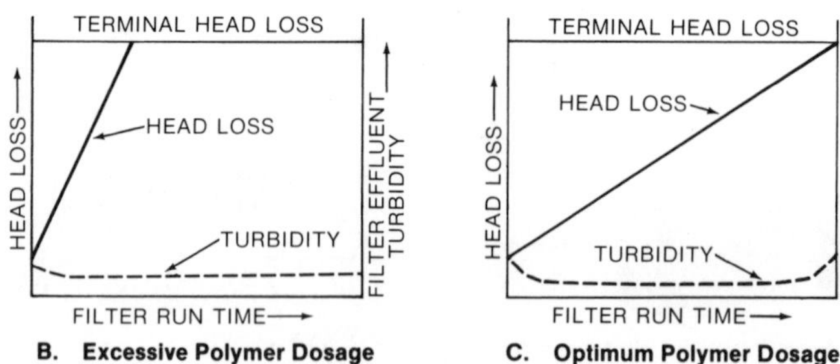

B. Excessive Polymer Dosage **C. Optimum Polymer Dosage**

Figure 6-25. Effect of Polymer Dosage on Turbidity and Head Loss

polymer for a particular treatment plant, several kinds should actually be used in the plant, and their performance and relative cost should be compared.

The dose required is normally less than 0.1 mg/L, but the proper dosage must be determined through actual use. Continuous turbidity monitoring of the filtered water is essential to ensure that the proper dose is being applied. Figure 6-25A illustrates the effect of an inadequate polymer dose. Figure 6-25B illustrates the effect of an excessive polymer dose. Too much polymer makes the filtered material too strong and causes it to stick in the upper few inches of the filter bed, creating a rapid head loss. The optimum dosage causes maximum head loss to be reached just before turbidity breakthrough occurs (Figure 6-25C).

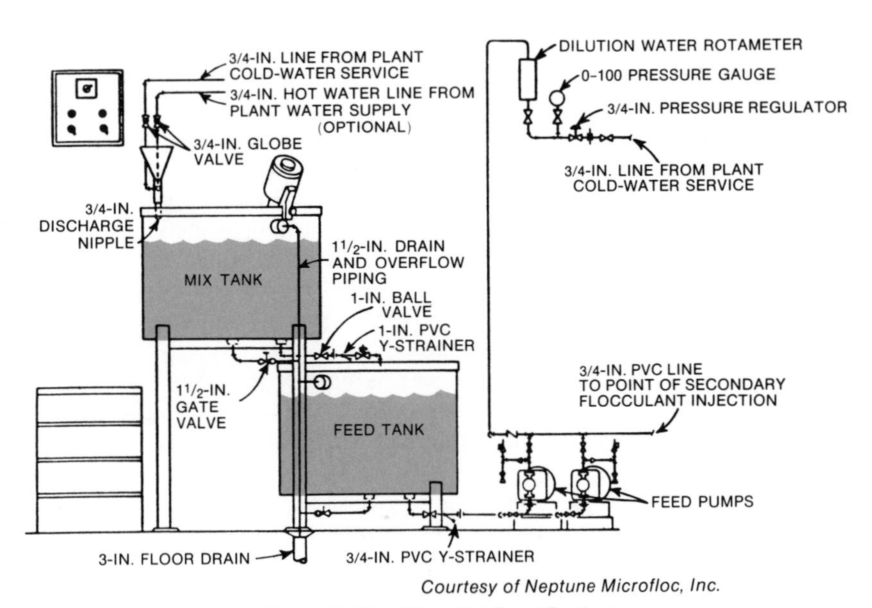

Courtesy of Neptune Microfloc, Inc.

Figure 6-26. Filter Aid Feed System

6-4. Operating Problems with Gravity Filters

The three major areas in which most filtration problems occur are:

- Chemical treatment before the filter
- Control of filter flow rate
- Backwashing the filter.

If these three procedures are not operated effectively, the quality of filtered water suffers and additional maintenance problems occur.

Chemical Treatment Before the Filter

The importance of proper coagulation and flocculation and the advantages of using a filter aid have already been discussed. Adjustments to these processes are possible only if the filtration process is closely monitored. Since many raw-water characteristics such as turbidity and temperature are not constant, dosage changes during filtration will be necessary. Consequently, continuously-recording instruments that keep track of turbidity, head loss, and flow rate are very important.

If short filter runs are occurring because of turbidity breakthrough, perhaps more coagulant, better mixing, or less filter aid is needed. If short runs are due to rapid buildup of head loss, perhaps less coagulant or less filter aid is required. It is the operator's job to recognize these types of problems and choose the proper solution.

Control of Filter Flow Rate

Rate increases or rapid fluctuations in the flow rate can force previously deposited filtered material through the media. The dirtier the filter, the more problems rate fluctuations can cause. These fluctuations can be caused by an increase in total plant flow, a malfunctioning rate-of-flow controller, flow increase when a filter is taken out of service for backwashing, or operator error.

Obviously, as demand increases filter rates may have to be increased to meet the demand, particularly if there is inadquate treated-water storage. If an increase is necessary, it should be made gradually over a 10-min period to minimize the impact. Filter aids can also reduce the harmful effects of a rate increase.

When a filter is backwashed, the filters remaining in operation must pick up the nonoperating filter's load. This can create an abrupt surge through the filters, particularly if rate-of-flow controllers are used. This problem can be avoided if a clean filter is kept in reserve. When a filter is taken out of service for backwashing, the clean filter is placed in service to pick up the extra load. This will probably not work for plants with less than four filters, since all filters are generally needed to keep up with the demand.

In many plants, particularly smaller ones that operate for only part of a day, the filtered material remaining on the filters can be shaken loose by the momentary surge caused by start-up when filters are placed into service the next day. This can be avoided by backwashing the filters before they are placed into service.

If rate-of-flow controllers are used, they should be well-maintained so that they function smoothly. Malfunctioning controllers will "hunt" for the proper valve position causing harmful rate fluctuations.

Backwashing the Filter

Effective backwashing is essential to consistent production of high-quality water. Ineffective backwashing can cause the following problems:

1. *Mudball formation.* During filtration, grains of filter media become covered with sticky, floc material. Unless backwashing removes this material, the grains clump together and form MUDBALLS. As the mudballs become larger, they can sink into the filter bed during backwashing and clog those areas where

they settle. These areas become inactive, causing higher than optimum filtration rates in the remaining active areas and unequal distribution of backwash water. Additional problems, such as cracking and separation of the media from the filter walls, may also result (Figure 6-27). Mudballs are usually seen on the surface of the filter after backwashing, particularly if the problem is severe, as shown in Figure 6-28. A periodic check for mudballs should be made; a checking procedure is described in the next section. Mudballs can be prevented by using adequate backwash flow rates and filter agitation (surface wash), which scrubs the filtered material from the media grains. Filter agitation is essential for dual- and multi-media filters, since mudballs can form deep within the bed.

2. *Filter bed shrinkage.* Bed shrinkage or compaction can result from ineffective backwashing. Clean media grains rest directly against each other with little compaction even at terminal head loss. However, dirty media grains are kept apart by the layers of soft filtered material. As the head loss increases, the bed compresses and shrinks, resulting in cracks (Figure 6-29) and separation of the media from the filter walls. The water then passes rapidly through the cracks and receives little or no filtration.

3. *Gravel displacement.* As mentioned earlier, if the backwash valve is opened too quickly the supporting gravel bed can be washed into the overlying filter media. This can also happen if part of the underdrain system is clogged, causing unequal distribution of the backwash flow. Eventually, the increased velocities displace the gravel and create a SAND BOIL as shown in Figure 6-30. When this occurs, the media starts washing into the underdrain system.

Since some gravel movement always occurs, filters should be probed at least once a year to locate the gravel bed. This can be done with a ¼-in. (6-mm) metal

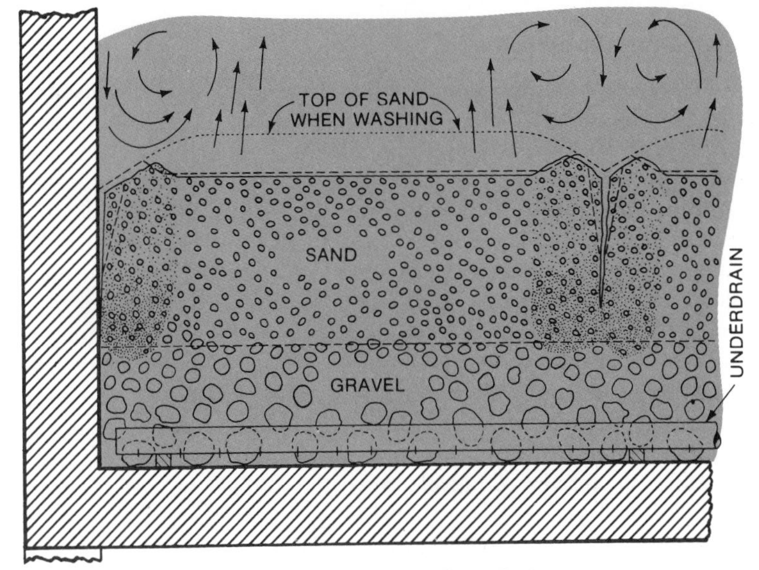

Figure 6-27. Clogged Filter Bed

Figure 6-29. Cracks in the Filter Bed

Figure 6-28. Mudballs on Filter Surface

rod while the filter is out of service. By probing the bed on a grid system and keeping track of the depths at which the gravel is located, it can be determined if serious displacement has occurred. If displacement has occurred, the media must be removed and the gravel regraded. Future displacement can be minimized by placing a 3-in. (80-mm) layer of coarse garnet between the gravel and the media, and by not using excessive backwash rates.

Additional Operating Problems

The following problems are not related to the three major areas just discussed but are quite common.

1. *Air binding.* If a filter is operated so that the pressure in the bed is less than atmospheric (known as NEGATIVE HEAD), the air dissolved in the water will come out of solution and form bubbles within the filter bed. This creates resistance to flow through the filter and leads to very short filter runs. Upon backwashing, the release of the trapped air causes violent agitation, which disturbs the media and can result in the loss of media during backwashing. Negative head typically occurs in filters with less than a 5-ft (1.5-m) water depth above the unexpanded filter bed. If a 5-ft (1.5-m) water depth is not possible, filter runs may have to be terminated at a head loss of about 4.5 ft (1.4 m) to prevent negative head.

Air binding may also occur when cold water supersaturated with air warms up such as in the spring. Unfortunately, not much can be done in this case except keeping water at maximum levels over the filter bed and backwashing frequently.

2. *Media loss.* Some media is always lost during backwashing. This is especially unavoidable if surface washers are used. However, if considerable

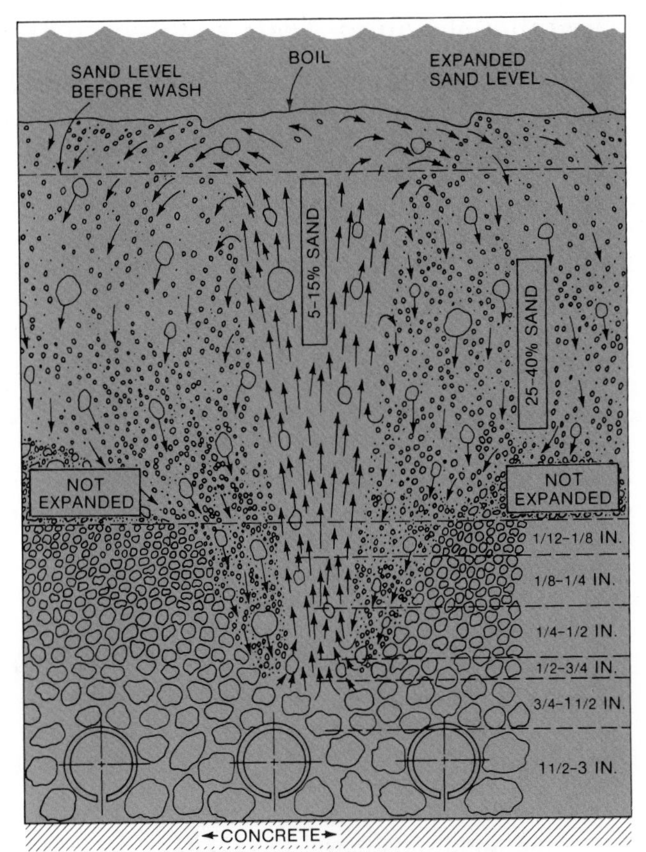

Figure 6-30. A Sand Boil in a Filter Bed

media is being lost, backwashing procedures should be examined. Since the bed is usually completely fluidized at 20 percent expansion, further expansion may not be needed. Turning the surface washers off about 1 to 2 min before the end of the main backwash will help. If serious problems continue, the wash-water troughs may have to be raised.

Monitoring the filtration process is a major responsibility. In addition to checking instruments, the condition of a filter can be checked by observing the backwash and the filter's surface. A bed in good condition with even backwash distribution should appear very uniform with the media moving laterally on the surface. Violent, upswelling boils of water indicate problems. If some areas do not appear to clear up as fast as others, uneven distribution of backwash flow is indicated.

When the filter is drained, its surface should appear smooth. If cracks, mudballs, or ridges appear, problems with backwashing exist. At some point, the filter media will have to be completely replaced or cleaned and regraded. This is a difficult task and expert advice should be sought.

6-5. Operational Control Tests and Record Keeping

Head loss and effluent (filtered-water) turbidity should be monitored and recorded continuously. This provides the operator with the information needed to control the coagulation process, prevent damage to the filter bed, and maintain good-quality filtered water.

A popular method of controlling the direct filtration process is with the pilot filter, as described in Module 3, Coagulation/Flocculation. By filtering a small stream of the chemically treated water through a miniature filter, the operator can be warned of possible problems before the main stream of water reaches the filters. In a conventional plant, a similar warning can be provided by continuously monitoring the turbidity of the settled water.

A NEPHELOMETRIC TURBIDIMETER, sensitive to turbidity levels below 1 NTU, should be used to monitor filtered water. Figure 6-31 shows a typical nephelometric turbidimeter. A sample of filtered water from each filter should be tested at least every three to four hours with a laboratory nephelometer to check the continuous monitors and indicate when they need calibration. Turbidity measurements and the nephelometric principle are discussed further in Volume 4, *Introduction to Water Quality Analyses.*

The condition of the filter beds must be continuously monitored. In addition to watching for filter cracks, media loss, and other noticeable defects, the filter bed should be periodically sampled. Core samples of the beds can be taken, which are examined to determine the volume of mudballs present. Individual media grains can be examined with a microscope to determine if they are clean.

Determining the mudball volume is relatively easy and does not require expensive equipment. The only equipment required is a sampler, a 10-mesh sieve, a bucket, and a 1000-mL graduated cylinder, shown by Figures 6-32A and B. After backwashing, the filter is drained until the water is at least 12 in. (0.3 m) below the sand surface. Five separate samples representative of the filter's

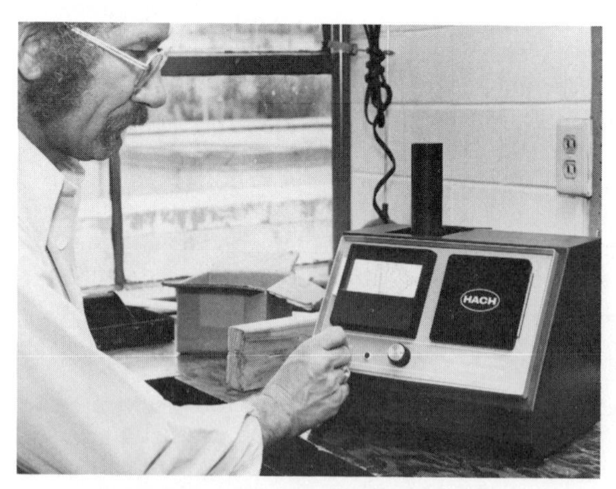

Figure 6-31. Nephelometric Turbidimeter

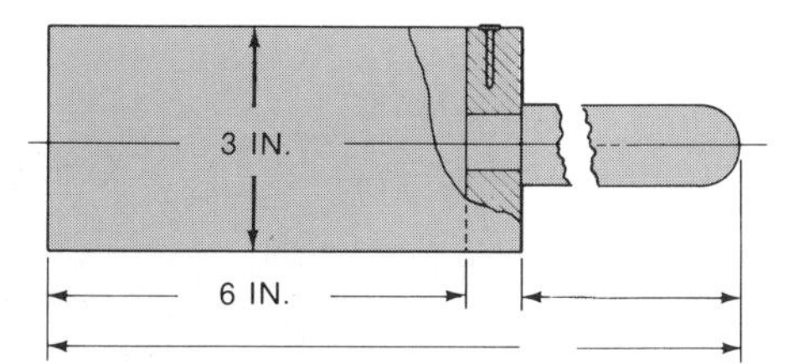

Figure 6-32. A. Mudball Sampler, B. Mudball Volume Measuring Equipment

surface are taken from the top 6 in. (150 mm) of sand with the sampler. The samples are placed on the sieve, which is gently raised and lowered in a bucket of water until the sand washes through. Any mudballs will be left on the sieve as shown in Figure 6-32B.

The graduated cylinder is filled with water to a defined point. When mudballs are added, the increase in volume in the cylinder represents the mudball volume.[4] Values shown in Table 6-2 indicate acceptable mudball volumes. The volume of mudballs should be kept to less than 0.1 percent through proper washing procedures. If a multi-media filter is used, a different sampling procedure is necessary because mudballs will sink through the top coal layer and collect near the sand layer.

Good record keeping for the filtration process can identify problems and indicate proper steps to be taken. The type of filtration records kept will depend on the treatment processes being used. All records should include:

- Rate of flow, million gallons per day or megalitres per day
- Head loss, feet or metres
- Length of filter run, hours
- Backwash water rate, gallons per minute
- Volume of wash water used, gallons
- Volume of water filtered, gallons
- Length of backwash, minutes
- Length of surface wash, minutes
- Filter aid dosage, milligrams per litre.

Figures 6-33 and 6-34 are examples of record-keeping forms.

---

[4] *Basic Science Concepts and Applications*, Mathematics Section, Mudball Calculation.

Table 6-2. Mudball Test Results*

Percentage of Mudballs by Volume	*Condition of Filter Bed*
0.0–0.1	Excellent
0.1–0.2	Very good
0.2–0.5	Good
0.5–1.0	Fair
1.0–2.5	Fairly bad
2.5–5.0	Bad
Over 5	Very bad

*Source: *Water Quality and Treatment*. AWWA. McGraw-Hill Book Company (3rd ed., 1971).

Daily Filter Record
Filter Plant No. 2

FILTER NO. ________ PREV. RUN ________ HOURS DATE: ____/____/ 19____

OPER.	TIME	RATE OF FLOW MGD	LOSS OF HEAD FT	SURFACE WASH MIN	RATE WATER WASH MGD	WATER WASH MIN	POLYMER FEED		WASH WATER USED	WATER FILTERED MIL GAL
							ON	OFF		
	12 MID									
	1 A.M.									
	2									
	3									
	4									
	5									
	6									
	7									
	8									
	9									
	10									
	11									
	NOON									
	1									
	2									
	3									
	4									
	5									
	6									
	7									
	8									
	9									
	10									
	11									
	12 MID									

Figure 6-33. Typical Record-Keeping Form

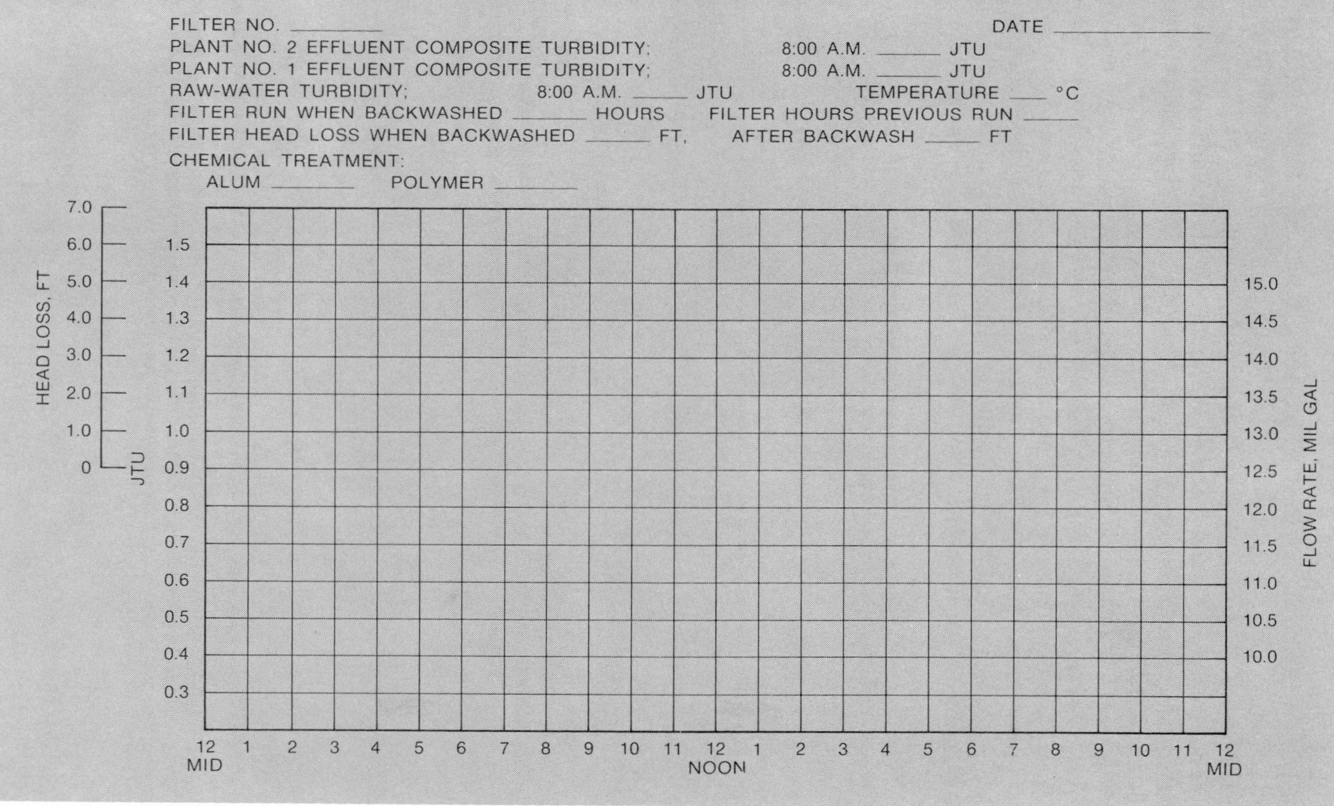

Filter Performance Record

FILTER NO. ________ DATE ________________

PLANT NO. 2 EFFLUENT COMPOSITE TURBIDITY; 8:00 A.M. ______ JTU

PLANT NO. 1 EFFLUENT COMPOSITE TURBIDITY; 8:00 A.M. ______ JTU

RAW-WATER TURBIDITY; 8:00 A.M. _____ JTU TEMPERATURE ___ °C

FILTER RUN WHEN BACKWASHED _______ HOURS FILTER HOURS PREVIOUS RUN _____

FILTER HEAD LOSS WHEN BACKWASHED ______ FT, AFTER BACKWASH _____ FT

CHEMICAL TREATMENT:

ALUM ________ POLYMER ________

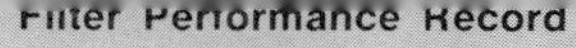

Figure 6-34. Typical Record-Keeping Form

6-6. Safety and Filtration

Guardrails should be installed to prevent falls into the filters. Operators should work in pairs if at all possible and life rings or poles should be located near the filter area for quick rescue in case of an accident.

Pipe galleries, ideal places for accidents, should be well ventilated, well lit, and provided with good drainage to prevent accidents and reduce maintenance problems. If ramps or stairs are in this area, they should be equipped with nonskid treads. Painting the pipes to make them easier to see and prevent possible mistakes when taking samples or performing maintenance is a good practice. Table 6-3 presents a suggested color coding system.

Because polymers used for filter aids are very slippery, spills of polymers should be cleaned up immediately. If the containers or tanks are leaking, they should be repaired or discarded.

Table 6-3. Suggested Piping Color Codes*

Water Lines	
Raw	Olive green
Settled or clarified	Aqua
Finished or potable	Dark blue
Chemical Lines	
Alum	Orange
Ammonia	White
Carbon slurry	Black
Chlorine (gas and solution)	Yellow
Fluoride	Light blue with red band
Lime slurry	Light green
Potassium permanganate	Violet
Sulfur dioxide	Light green with yellow band
Waste Lines	
Backwash waste	Light brown
Sludge	Dark brown
Sewer (sanitary or other)	Dark gray
Other	
Compressed air	Dark green
Gas	Red
Other lines	Light gray

*Source: Recommended Standards for Water Works. Committee of the Great Lakes-Upper Mississippi River Board of State Sanitary Engineers. (1976 ed.) p. 11.

In situations where two colors do not have sufficient contrast to easily differentiate between them, a 6-in. band of contrasting color should be painted on one of the pipes at approximately 30-in. intervals. The name of the liquid or gas should also be painted on the pipe. In some cases it may be advantageous to paint arrows indicating the direction of flow.

Selected Supplementary Readings

Alsaker, Dayton. *Giardia*—'Micro-Monsters' that Contaminate Water. *OpFlow*, 8:3:1 (Mar. 1982).

Arora, M.L. Comparison of Commercial Filter Aids. *Jour. AWWA*, 70:3:167 (Mar. 1978).

Basic Water Treatment Operation. Ministry of the Environment. Toronto, Ontario (1974). Chap. 4.

Benzie, R.E. & Van Norman, Paul. Failure to Inspect and Routinely Maintain Rapid-Sand Filters Results in Extra Labor and Expense for Blissfield, Michigan. *OpFlow*, 4:9:3 (Sept. 1978).

Cleasby, J.L., et. al. Backwashing of Granular Filters. *Jour. AWWA*, 69:2:115 (Feb. 1977).

Cox, Charles R. *Operation and Control of Water Treatment Processes*. World Health Organization. Geneva, Switzerland (1969).

Culp, G.L. & Culp, R.L. *New Concepts in Water Purification*. Van Nostrand Reinhold Company. New York. (1974). Chap. 3.

Curry, M.D. Proper Treatment, Maintenance Aids Filtration Process—Part I. *OpFlow*, 1:9:1 (Sept. 1975).

Curry, M.D. Proper Treatment, Maintenance Aids Filtration Process—Part II. *OpFlow*, 1:10:1 (Oct. 1975).

Curry, Michael. Tips on Proper Care of Your Filters. *OpFlow*, 5:2:3 (Feb. 1979).

Eagleton, Robert. Filter Material Replacement—One Plant's Method. *OpFlow*, 7:5:1 (May 1981).

Finney, J.W. Jr. & Bell, H.K. Polymers Prove Useful in Water Treatment Process. *OpFlow*, 1:10:1 (Oct. 1975).

Hart, D.H. Jr. Inspecting and Maintaining Filtration Systems. *OpFlow*, 8:2:6 (Feb. 1982).

Hart, D.H. Jr. Filtration—The 'Gritty' Process of Removing Impurities. *OpFlow*, 8:6:1 (June 1982).

Hart, D.H. Jr. Backwashing—Flushing the Impurities Out of Filters. *OpFlow*, 8:7:1 (July 1982).

Introduction to Water Quality Analyses. AWWA, Denver, Colo. (1982). Module 5.

Logsdon, G.S. & Lippy, E.C. The Role of Filtration in Preventing Waterborne Disease. *Jour. AWWA*, 74:12:649 (Dec. 1982).

Manual of Instruction for Water Treatment Plant Operators. New York State Dept. of Health, Albany, N.Y. (1975). Chap. 9.

Manual of Water Utility Operations. Texas Water Utilities Assoc., Austin, Texas (6th ed., 1975). Chap. 9.

Operators—Do Utilities Really Need Them? *OpFlow*, 1:6:4 (June 1975).

Propes, Philip. Porosity—the Open Space in a Filter Bed. *OpFlow*, 8:8:3 (Aug. 1982).

Rice, A.H. High-Rate Filtration. *Jour. AWWA* 66:4:258 (Apr. 1974).

Skinner, Jim. Dual Media Filtration in Montana. *OpFlow*, 2:9:1 (Sept. 1976).

Smith, G.K. Bumping Relieves Binding in Filters. *OpFlow*, 2:7:3 (July 1976).

Upgrading Water Treatment Plants to Improve Water Quality. Conf. Seminar Proc. AWWA Denver, Colo. (June 1980).

Water Quality and Treatment. AWWA Handbook. McGraw-Hill Book Company, New York. (3rd ed., 1971). Chapter 7a.

Water Works Operators Manual. Alabama Dept. of Public Health, Montgomery, Ala. (3rd ed., 1972). Chap. XVI.

Glossary Terms Introduced in Module 6

(Terms are defined in the Glossary at the back of the book.)

Adsorption
Anionic
Auxiliary scour
Backwash
Body feed
Breakthrough
Cationic
DE filter
Diatomaceous earth filter
Diatom
Filter agitation
Filter sand
Filter tank
Gravel bed
Head loss
Leopold filter bottom
Mudball
NTU
Negative head
Nephelometric turbidimeter
Nephelometric turbidity unit
Nonionic
On-line turbidimeter
Oxidize
Percolation
Piezometer
Pipe-lateral system
Polyelectrolyte
Polymer
Porous plate
Precipitate
Precoating
Pressure-sand filter
Rate-of-flow controller
Sand boil
Schmutzdecke
Surface washing
Terminal head loss
Underdrain
Wash-water trough
Wheeler bottom

Review Questions

(Answers to Review Questions are given at the back of the book.)

1. What is the major purpose of filtration?

2. What is turbidity?

3. What are three important reasons for removing turbidity?

4. What is the most important mechanism for removing suspended material from water and how does it work?

5. What are three major disadvantages of a slow-sand filter?

6. What functions do the filter underdrains perform?

7. Why are filter surface washers used?

8. What are high-rate filters?

9. What is the major disadvantage of pressure-sand filters?

10. What is diatomaceous earth?

11. Compare the two basic types of filtration processes used in the United States.

12. What are the three basic filtration steps?

13. What is the major disadvantage of a rate-of-flow controller?

14. Generally, when should a filter be backwashed?

15. Why should filtering to waste be practiced?

16. Why are filter aids used?

17. What three major areas result in most of the operating problems with the filtration process?

18. What problems does poor backwashing cause?

19. What are mudballs and how can they be prevented?

20. What parameters should be monitored continuously for good control of filter operation?

Study Problems and Exercises

1. Using the filtration operating records from a local water treatment plant, graph head loss and filtered water turbidity for four filter runs. Answer the following questions using the graph:

   a. What was the head loss and turbidity just before the filters were backwashed?

   b. Had turbidity breakthrough started before the filters were backwashed?

   c. Did turbidity breakthrough occur at the beginning of a filter run?

   d. Do you have any recommendations on how to operate the filter more effectively?

2. Can the filtration process effectively remove viruses and microorganisms, such as *Giardia lamblia*, that form cysts? Discuss any special operational techniques that you feel are necessary.

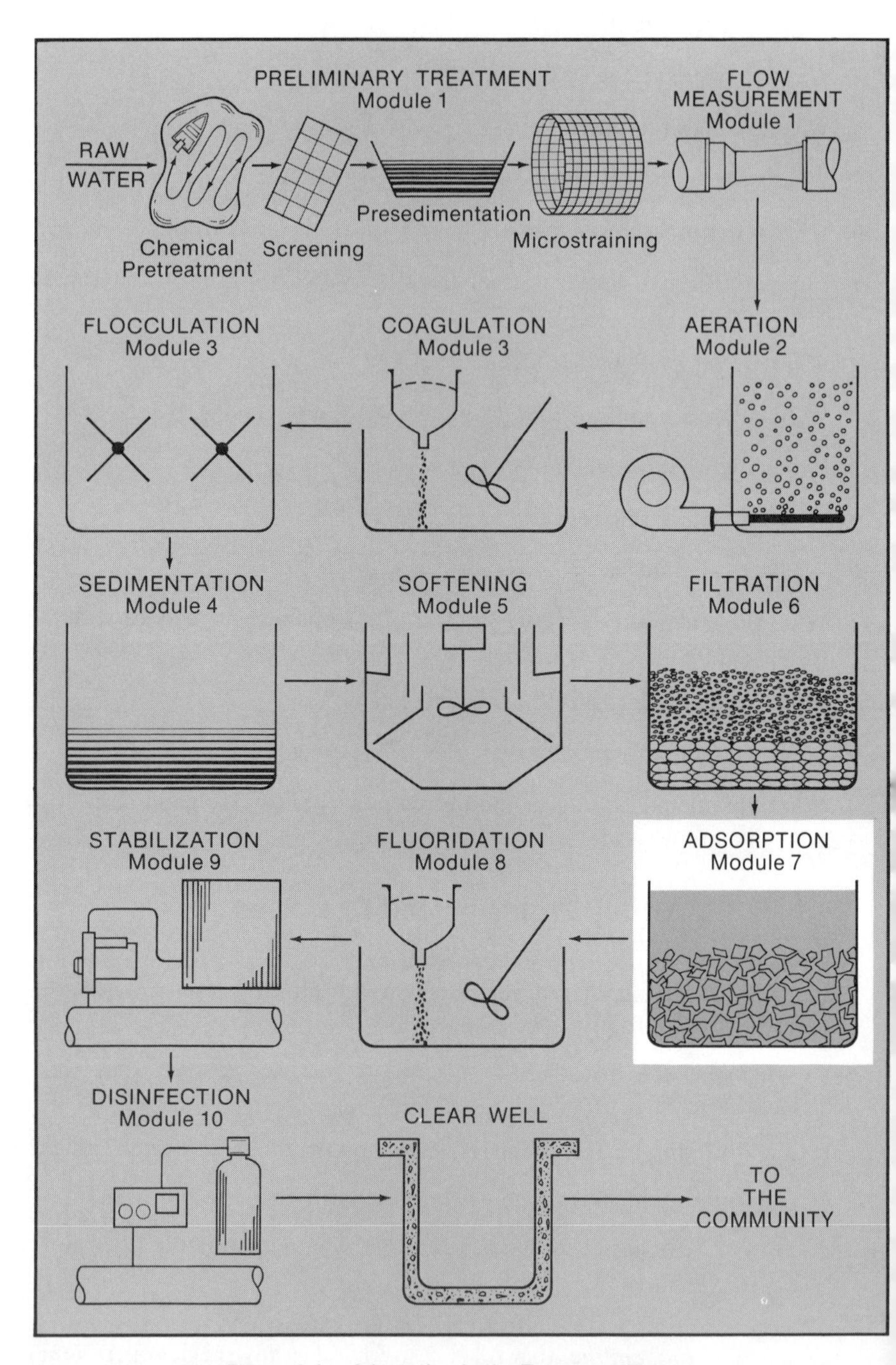

Figure 7-1. Adsorption in the Treatment Process

Module 7

Adsorption

Water contains varying amounts of dissolved ORGANIC SUBSTANCES (carbon-containing compounds, also called ORGANICS). Many organic materials cause color, tastes, and odors in water and may have serious public health effects. For these reasons, maximum contaminant levels for organic compounds such as PESTICIDES and TRIHALOMETHANES are included in the USEPA primary drinking water regulations. Removal of these and other organic compounds from drinking water is important in protecting the public health. It is also a major step in making drinking water more pleasing by reducing objectionable tastes, odors, and color.

The ADSORPTION process is used primarily to remove organic materials. Adsorption can also be used to remove the inorganic ions of fluoride and arsenic. The adsorption process may be one of the later steps in treatment, as shown in Figure 7-1; it may be part of the filtration process; or it may be performed as water first enters the treatment plant.

After completing this module you should be able to

- Describe the purpose of adsorption.
- Explain why organic substances should be removed.
- Identify the types of adsorbents used to remove organics.
- Explain how the adsorption process works.
- List the major operating considerations for using activated carbon.
- Describe common operating problems when using carbon, and recommend appropriate solutions.

- Describe commonly used laboratory control tests to monitor the adsorption process.
- Recommend appropriate safety measures to be considered when using activated carbon.

7-1. Description of Adsorption

A major goal in water treatment is to produce clear, good-tasting water. Adsorption has been used for many years to help achieve this goal by removing organics that cause tastes, odors, and color in drinking water. Recently, it is also becoming an important method for protecting public health by removing toxic organic chemicals from drinking water.

Organics in Raw Water

Raw water contains varying amounts of dissolved organic materials, with surface water usually having greater concentrations than ground water, since surface water receives runoff and a variety of waste discharges. Some of these organic materials, such as HUMIC SUBSTANCES, occur naturally and are found in most surface waters. A multitude of other organics are introduced when domestic, agricultural, and industrial wastes are discharged into surface water or allowed to contaminate ground water. It is becoming evident that a great deal of ground water is being contaminated by organic chemicals, particularly near urban and industrial areas.

In the treatment plant, organics can interfere with treatment processes such as coagulation and flocculation. Organics can also react with chemicals added to water during treatment, primarily chlorine, producing additional undesirable compounds.

Public Health Concern Over Organics

More than 700 different organic chemicals have been discovered in drinking water throughout the United States. In addition to the taste, color, odor, and treatment problems they cause, long-term exposure to organic compounds may pose a serious threat to public health, including a cancer risk.

Two groups of organic compounds are of primary concern. The first group is disinfection by-products, primarily trihalomethanes. Trihalomethanes are produced when naturally occurring organic compounds (mostly humic materials) react with the chlorine used in drinking-water treatment. Trihalomethanes, even in very small concentrations, are suspect CARCINOGENS (cancer-causing agents). Other disinfection by-products form when alternative disinfectants, such as ozone or chlorine dioxide, are used. However, not much is known about the public health impact of these compounds.

The second group of organics are man-made and include a variety of chemicals including solvents, cleaning compounds, INSECTICIDES, and HERBICIDES that enter drinking water sources through waste discharges and runoff. Many of these compounds are known to be toxic to humans. Although federal and state regulations are being revised to control this type of pollution, it will continue to be a major problem because of accidental discharges, spills, and uncontrollable sources, such as surface runoff from agricultural areas sprayed with insecticides and herbicides.

Adsorption Methods

Organic chemicals can be partially removed by using chlorine or potassium permanganate to oxidize the compounds, by aeration (which results in oxidation and air-stripping of the compounds), by coagulation/flocculation/sedimentation, and by filtration. However, these processes cannot remove some organic compounds as efficiently as adsorption.

Adsorption can involve the use of ACTIVATED CARBON or SYNTHETIC RESINS for organics removal. It can also involve the use of ACTIVATED ALUMINA for the removal of the inorganic elements fluoride and arsenic. However, this module will focus on activated carbon since it is the most commonly used material.

Principle of Adsorption

Adsorption works on the principle of ADHESION. In the case of water treatment, organic contaminants are attracted to the adsorbing material (an ADSORBENT such as activated carbon) and adhere (stick) to its surface by a combination of complex physical forces and chemical action.

For adsorption to be effective, the adsorbent must provide an extremely large surface area on which the contaminant chemicals can stick. Also, if the process is to be economical to build and operate, the total surface area needed must be contained in a bed or tank not much larger than that required for a typical rapid-sand or multimedia filter. Porous adsorbing materials help achieve these objectives. Activated carbon is an excellent adsorbent since it has a vast network of pores that vary in size to accept large and small molecules of contaminants. As a result, activated carbon has a very large surface area. For example, 1 lb (0.5 kg) of activated carbon has a total surface area of about 150 acres (0.6 km^2), which can trap and hold (adsorb) over ½ lb (0.25 kg) of carbon tetrachloride. Figure 7-2 is a photograph taken through an electron microscope showing the large

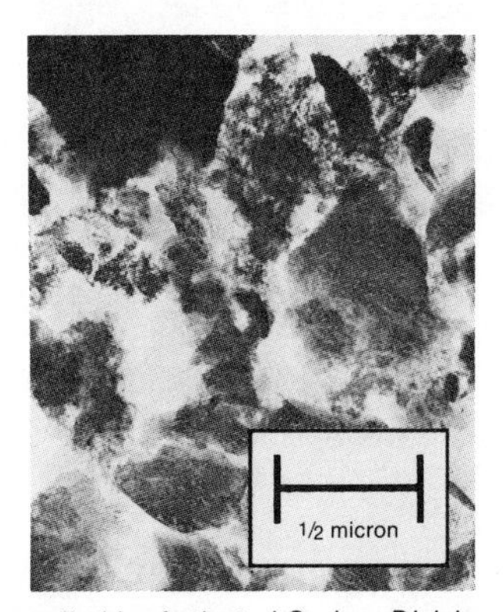

Photograph supplied by Activated Carbon Division, Calgon Corp.

Figure 7-2. Details of the Fine Structure of Activated Carbon

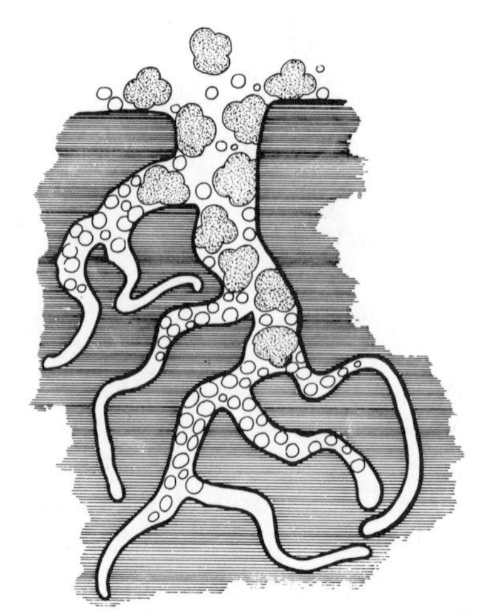

Figure 7-3. Carbon Structure After Activation

number of pores in activated carbon. These pores are created during the manufacturing process by exposing the carbon to very high heat (about 1500° F [800° C]) in the presence of steam. This is known as ACTIVATION of the carbon. Activation oxidizes all particles on the surfaces of the carbon particles, leaving the surfaces free to attract and hold organics. Figure 7-3 shows the structure of the carbon after activation and indicates that different sizes of chemical molecules can be adsorbed within the pores.

Once the surface of the pores is covered, it loses its ability to adsorb. The spent carbon must then be replaced with fresh carbon. The spent carbon can be REACTIVATED by essentially the same process as activation.

7-2. Adsorption Facilities and Equipment

Activated carbon can be made from a variety of materials such as wood, nutshells, coal, peat, and petroleum residues. Activated carbon used in water treatment is usually made from bituminous or lignite coal. The coal is slowly heated in a furnace without oxygen so that it will not burn. This converts the coal to carbon. The carbon is then activated by exposing it to a steam-air mixture. This activated carbon is crushed and screened to obtain the desired particle size.

There are two common forms of activated carbon used in water treatment—POWDERED ACTIVATED CARBON (PAC) and GRANULAR ACTIVATED CARBON (GAC). Table 7-1 compares the general properties of typical types of GAC and PAC. GAC has a larger particle size than PAC and more surface area. Figure 7-4 shows the difference in particle size.

Powdered Activated Carbon

Powdered activated carbon is typically available in 50-lb (23-kg) bags or in bulk form from trucks or railroad cars. It can be fed in dry form or as a slurry

Table 7-1. A Comparison of GAC and PAC Properties

Property	*Types of Carbon* GAC	PAC
Density, lb/ft^3 (g/cm^3)	26–30 (0.42–0.48)	20–45 (0.32–0.72)
Surface area, m^2/g	650–1150	500–600
Mean particle diameter, mm	1.2–1.6	less than 0.1

with water. Small plants and plants that use PAC only for periodic taste and odor problems typically use the dry-feed method. The PAC can be stored in bags or in bulk form in steel tanks. The tanks should be located so they can feed directly into the hoppers of the dry feeders by gravity.

If carbon is used in dry form, chemical feeders specifically designed to handle the carbon should be used. Such feeders can quickly, easily, and accurately feed over a wide range, including high dosages for emergency treatments. They rapidly wet the carbon and discharge the resulting thin suspension (SLURRY) to the point of application. The hopper above the feeder should have walls slanting at a 60-degree angle and a hopper agitator mechanism to keep the PAC from ARCHING. This will ensure free gravity flow. Although tight seals around the feeder confine black carbon dust, unusual circumstances may call for additional dust-collection apparatus. To keep dust to a minimum, dry carbon feeders should be located in a confined enclosure or area of the plant, as shown in Figure 7-5. Because of the large amount of fine, combustible dust created by dry PAC, special handling and storage practices must be followed, as discussed in the safety section of this module.

Because of the handling problems encountered with dry PAC, many plants that consistently use PAC use a slurry feed system of the type shown in Figure 7-6. If delivered in bulk form, the PAC is removed from the tank car or truck using an EDUCTOR, which works on the principle of the Venturi tube, and a carbon slurry pump. The slurry formed in the eductor, which contains about 1 lb of PAC per 1 gal of water (0.1 kg/L), is transferred to a storage tank.

Figure 7-4. Difference in Particle Size Between GAC and PAC

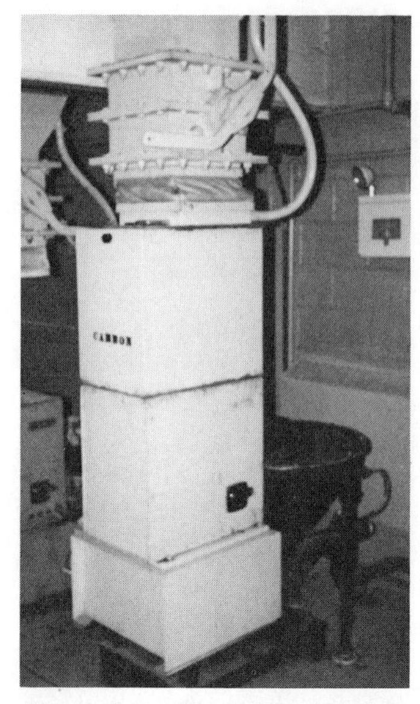

Figure 7-5. Dry Carbon Feeder

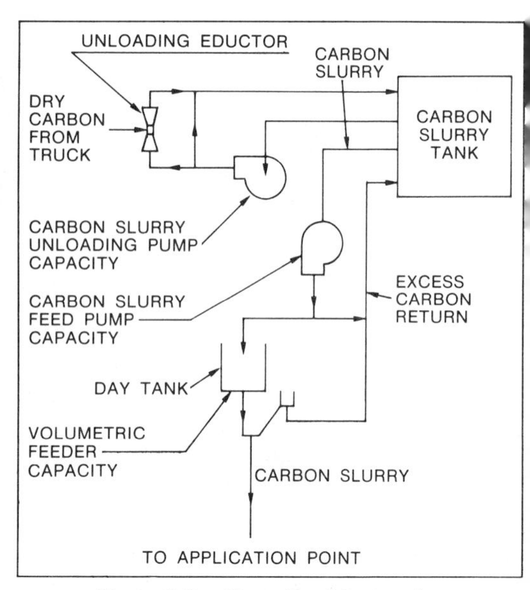

Figure 7-6. Slurry Feed System for Handling PAC

A plant should have at least two slurry tanks so that a shipment can be placed into one tank before the other is empty. The tanks should be square and made of steel or concrete with a capacity about 20 percent greater than the maximum carbon load delivered by railroad car or truck. The tanks should have an epoxy or bitumastic lining to protect them from corrosion. Mechanical agitators should be provided in the storage tank to keep the PAC slurry in suspension. If delivered in bags, the carbon is added directly to the storage tank.

The slurry is usually pumped to a day tank, which holds the volume of slurry that will be fed over the next few hours. This tank should be plastic or steel with a corrosion-resistant lining. To keep the slurry in suspension the tank should be equipped with a mixer. From the day tank the slurry can flow by gravity to a volumetric feeder, which should be easy to clean and maintain. One type of feeder for this purpose, which can handle a wide range of dosages, is shown in Figure 7-7. This feeder is easily washed down when not in use, since it consists of a dipper wheel rather than a pump.

An eductor can be used to move the slurry from the feeder to the application point. The piping should slope downgrade to the application point, with provisions for flushing any carbon that may settle out and clog the pipe. The piping carrying the slurry should be of corrosion- and erosion-resistant material such as rubber, plastic, or stainless steel. The pump impellers and the mixing blades in the slurry tank and day tank should be of stainless steel to resist corrosion and erosion.

PAC is usually added to the water ahead of the normal coagulation/flocculation, sedimentation, and filtration steps because it must be removed from

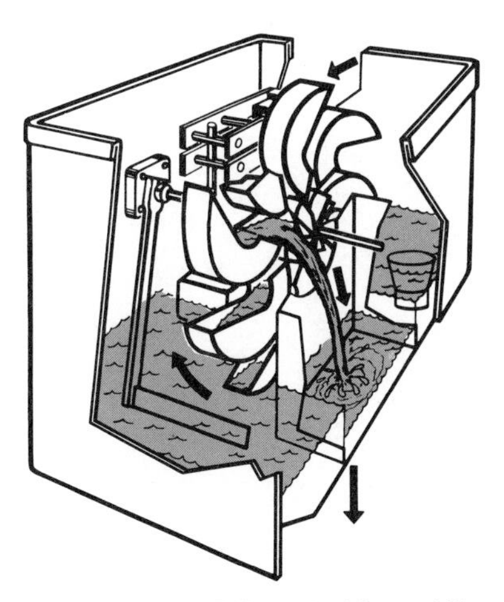

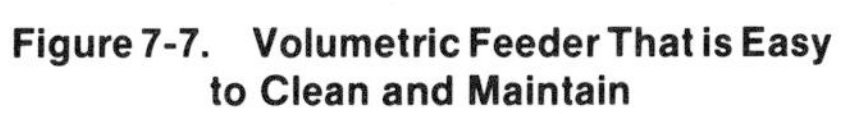
Courtesy of BIF, a unit of General Signal

Figure 7-7. Volumetric Feeder That is Easy to Clean and Maintain

Courtesy of the Cincinnati Water Works

Figure 7-8. GAC Contactor

the water to avoid interfering with the disinfection process or possibly moving through the plant and into the distribution system. Since the PAC is discharged as part of the sludge in the sedimentation basin and the backwash water from the filters, it is not practical to recover and reuse it.

Granular Activated Carbon

Granular activated carbon is typically used when carbon is needed continuously to remove organics. GAC is used like filter media. It can be used as a replacement for existing media in a conventional filter, or it can be installed in separate, closed pressure columns known as CONTACTORS (Figure 7-8). The decision of which method to use must be based on the types of organic compounds to be removed and the plant layout. For most situations, using GAC as a partial or complete replacement for the regular filter media (Figures 7-9 and 7-10) is the most economical approach. A minimum of 24 in. (610 mm) of GAC is suggested to provide effective removal of turbidity and most organic compounds causing tastes and odors. However, deeper depths of GAC are recommended for improved organics removal and to prolong BED LIFE of the carbon (the length of time the carbon is effective in removing organics). GAC has proven to be as good a filtering media as sand or anthracite coal, and it is available in the same particle-size ranges. The depth of carbon possible in a particular plant will depend on how much room the filters have below the wash-water troughs to allow for bed expansion during backwash. In some cases, the troughs may have

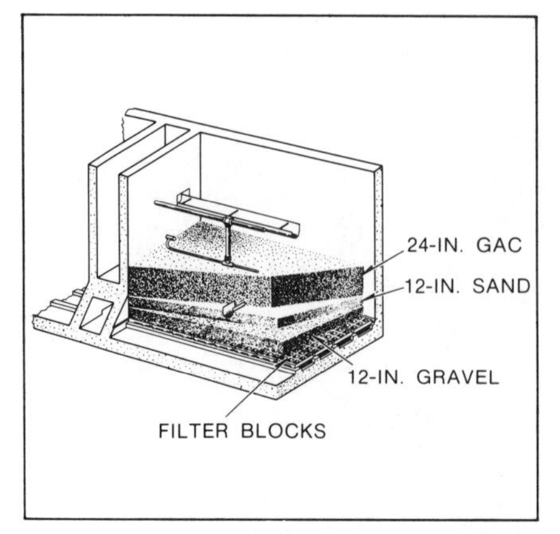

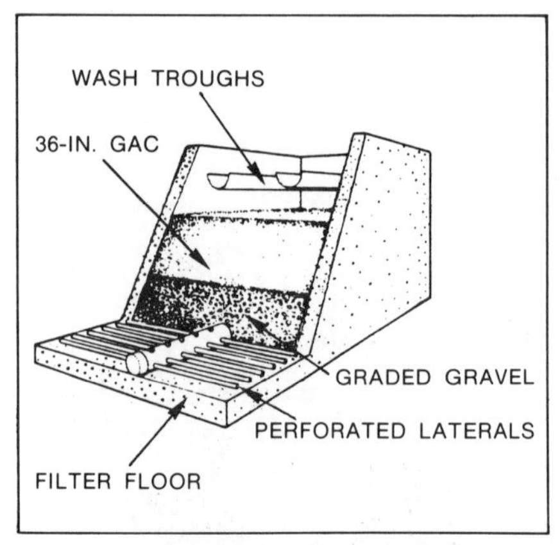

Artwork supplied by Activated Carbon Division, Calgon Corp.

Figure 7-9. Partial Replacement with GAC **Figure 7-10. Complete Replacement with GAC**

to be elevated to allow the recommended minimum of 24 in. (610 mm) of GAC to be placed in the filters and to allow for proper bed expansion to take place.

GAC contactors are used when a greater depth of carbon (usually more than 36 in. [0.9 m]) is needed to provide the contact time necessary to remove certain organic compounds. Contactors may also be used if the life of the carbon before reactivation is expected to be only a few months, since the removal and replacement of GAC is easier with contactors than with filters. When treating surface waters, the contactors should be placed after the regular filters in order to use the contactors primarily for adsorption and not filtration. This will extend the life of the carbon because it will not become covered by floc, which reduces the adsorptive capacity of the GAC. For ground waters, extensive treatment before the contactors is usually not necessary.

After a period of a few months to 3–4 years, depending on the type and concentration of organic compounds in the water being treated, the GAC loses its ability to adsorb. The old GAC must then be removed and replaced with fresh carbon. The used GAC can be reactivated and reused. Reactivation (Figure 7-11) can be done on-site, but this is usually only practical when the carbon needing reactivation (spent carbon) is greater than 1500 lb (680 kg) per day. If the spent carbon is less than this amount, then it is usually more cost effective to dispose of it in a landfill or by some other means. Reactivation usually consists of passing the spent carbon through a regeneration furnace (usually the multihearth type) where it is heated to 1500–1700° F (820–930° C) in a controlled atmosphere that oxidizes the adsorbed impurities. Because about 5 percent of the carbon is lost during this process, additional carbon must be added to the contactors or filters if on-site reactivation is practiced.

GAC is available in 60-lb (27-kg) bags or in bulk form delivered either by truck or rail. GAC is usually placed in the filters or contactors in a slurry form using an eductor system to facilitate handling and reduce dust.

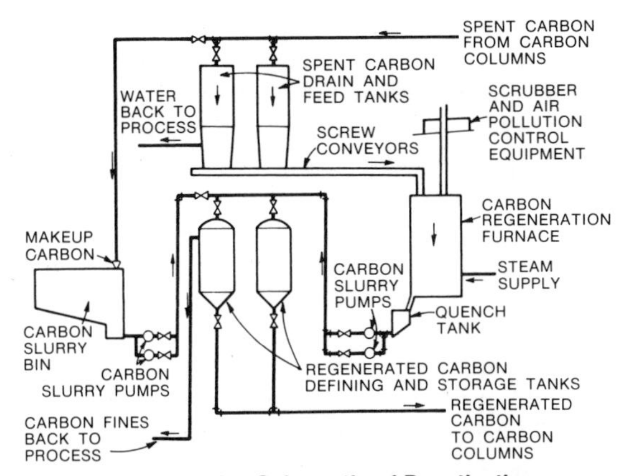

Figure 7-11. Schematic of Reactivation

7-3. Operation of the Adsorption Process

The operating procedures for adsorption differ depending on whether PAC or GAC is used.

PAC Adsorption

PAC is used primarily to help control those organic compounds responsible for tastes and odors. However, it is also used to remove compounds that will form trihalomethanes if allowed to react with chlorine. In addition, the carbon particles often aid coagulation by providing NUCLEI for proper floc formation, which is an additional benefit particularly where the natural turbidity is low.

PAC can be applied in the treatment plant at almost any point before filtration. However, the following items should be considered when selecting an application point:

- The contact time between the PAC and the organics is important and depends on the ability of the carbon to remain in suspension; at least 15 min of contact time should be provided.
- The surfaces of the PAC particles lose their capacity to adsorb if coated with coagulants or other water treatment chemicals.
- PAC will adsorb chlorine.

The above factors indicate that the raw-water intake line is the most advantageous point for PAC addition, if adequate mixing facilities are available. Using a raw-water application point allows a relatively long contact time and avoids the problems that can be caused if PAC is added after prechlorination. The chemical changes that occur when chlorine reacts with organic compounds can produce other compounds, such as trihalomethanes, which are difficult to adsorb. Also, the carbon will adsorb the chlorine and result in increased dosages of both if carbon is added after prechlorination.

If addition to the raw water is not possible, PAC can be added at other points before filtration, but dosages usually have to be higher to account for shorter contact times and the interference by other chemicals, such as coagulants and chlorine. If the PAC is added to the effluent from the sedimentation basins or as the water enters the filters, particular care must be taken in filter operation. The finely divided carbon can pass through the filter and cause "black water" complaints from customers. An efficient way to apply PAC is to use two or more application points with part of the carbon being added to the raw water and smaller doses being added later (before filtration) to remove remaining taste- and odor-causing compounds.

The PAC dosage used will depend on the type and concentration of organic compounds that are present. Common dosages range from 2 to 20 mg/L, but doses can go as high as 100 mg/L to handle a severe taste-and-odor problem or a spill of an organic chemical such as carbon tetrachloride. The dosage needs to be established based on the actual water quality and treatment conditions.

GAC Adsorption

GAC adsorption is used when the removal of organic compounds is necessary on a continuous basis and when long contact times are needed for effective removal. The most common procedure is to replace some of the media in a conventional filter with a minimum of 24 in. (610 mm) of GAC. The media is removed using an eductor like the one shown in Figure 7-12. The GAC is placed in the filter as a slurry (Figure 7-13) for ease of handling. Once the carbon is

Figure 7-12. Filter Media Removal with an Eductor

Photograph supplied by Activated Carbon Division, Calgon Corp.

Figure 7-13. Placement of GAC Slurry

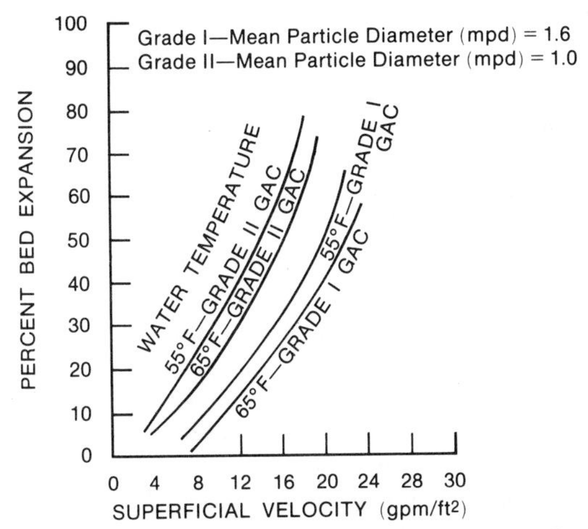

Figure 7-14. A Curve for the Selection of Proper Backwash Rate (Example)

placed, the filters should be backwashed to remove entrapped air and carbon fines. The filters are operated in the same manner as regular rapid-sand filters with a filtration rate of about 2 gpm/sq ft (1.4 mm/s). This provides an EMPTY BED CONTACT TIME (EBCT) of 7.5–9 min, which is sufficient to remove many organic compounds. EBCT is a term used for GAC adsorption beds and is equal to the volume of the empty bed (or space) into which the GAC will be placed divided by the flow rate through the bed. Therefore, EBCT is calculated much the same way as detention time.[1]

The filter bed is backwashed using the same general procedure as for conventional filters. The backwash rate will depend on the grade of GAC (based on particle diameter and density), water temperature, and the amount of bed expansion desired. For proper cleaning of the bed, a 50-percent expansion of the GAC is recommended. Curves are available from GAC manufacturers to help select the proper backwash rate. Figure 7-14 is an example of such a curve. Use of surface washers is also recommended to ensure adequate cleaning of the carbon layer so mudballs cannot form within the bed.

Good control of backwashing is very important to achieve adequate cleaning of the bed by preventing the GAC from being washed away. If the GAC has replaced sand, additional care is particularly needed since the particle density of GAC (about 1.4 g/cm^3) is much less than that of sand (2.65 g/cm^3). If the GAC has replaced anthracite coal, the backwash rate can remain about the same since the particle densities are almost equal.

The two major causes of carbon loss are entrapment within the bed and excessive backwash rates. Prior to backwashing, the filter should be drained to a few inches (millimetres) below the backwash troughs. Surface washers should be turned on for about 2 min to break up any mat that may have formed on the

[1] *Basic Science Concepts and Applications*, Mathematics Section, Detention Time.

surface, to break up mudballs within the bed, and to clean floc particles from the carbon. This is particularly important if polymers are used as a coagulant or filter aid, which will cause the floc to be sticky and hard to remove. The agitation created by the surface washers also helps remove any air that may be entrapped within the bed. Backwashing should begin at 2–3 gpm/sq ft (1.4–2.0 mm/s) to allow any remaining air to be released slowly so it does not cause carbon to be lost into the wash-water troughs. This should be increased to 5–6 gpm/sq ft (136–163 mm/s) for about 1 min and then gradually to the rate necessary to achieve 50-percent bed expansion. Backwashing should continue until the flow into the backwash troughs is clear. The time required for backwashing and the frequency of backwashing depends on the amount of floc and other suspended matter the filter has to remove, as discussed in Module 6, Filtration.

Some loss of carbon will occur in the backwashing process. It is important to keep track of the bed depth so that the lost carbon can be replaced and so that operating procedures can be changed, if necessary, to prevent serious losses from occurring. Bed depth can be monitored either by taking routine measurements of the distance from the top of the backwash troughs to the top of the carbon bed or by placing a permanent reference mark (such as a stainless-steel plate) in each filter, to indicate where the top of the carbon should be. It is not unusual for approximately 1 in. (25 mm) of carbon per year to be lost from the filters.

The actual life of the GAC bed depends on the concentration and type of organic compounds being removed. For typical taste- and odor-causing compounds, bed life may be as long as three years. However, for trihalomethane compounds such as chloroform, bed life can be as short as one month. Generally speaking, as the influent concentration increases the bed life decreases. In addition, some organic compounds are not as strongly adsorbed as others, depending on their molecular weight and other physical-chemical properties. For example, pesticides, phenols, and taste- and odor-causing compounds are strongly adsorbed, while alcohols and trihalomethanes are weakly adsorbed. The more weakly adsorbed a compound is, the shorter the bed life will be.

A major advantage GAC beds have is that they are not all used up at once; therefore, breakthrough of the contaminants takes place over a long period of time rather than suddenly. Figure 7-15 indicates a typical breakthrough pattern for a 30-in. (762-mm) GAC bed. This pattern is important since it allows the replacement of the GAC to be phased so all of the filters do not have to be out of service at the same time. The spent GAC is removed with an eductor to a truck for hauling to a disposal or reactivation site. Methods to help determine when to replace the GAC are discussed under Operational Control Tests and Record Keeping.

GAC can also be placed in a closed container (Figure 7-8) much like a pressure filter. These containers are known as contactors or adsorbers. They are used primarily to remove contaminants such as trichloroethylene (TCE) from ground water. In these cases the water is pumped from the wells and through the contactors.

An advantage of using contactors is that they can be manufactured to the size necessary to provide the desired EBCT. As a result, contactors might also be

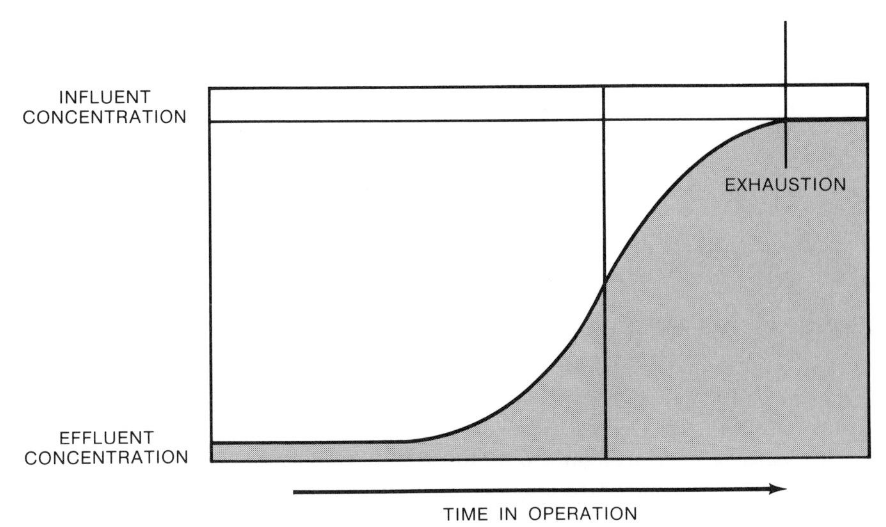

Figure 7-15. Breakthrough Pattern for 30-in. GAC Bed

very cost effective in surface-water systems that have high concentrations of organic compounds, particularly those that are weakly adsorbed. Under these conditions, a longer EBCT is needed than the 2 ft (0.6 m) of GAC in a gravity filter can provide while still maintaining a reasonable bed life. Table 7-2 shows the effect of these factors in a particular water treatment situation. The increase in bed depth and EBCT greatly increase the bed life.

The decision to use contactors and the contactor design should be based on pilot tests at the treatment plant where they will be installed. If used in a conventional surface-water treatment plant, the contactors should be placed after the filtration process. This will allow the suspended matter to be removed and the GAC to be used primarily for adsorption. The life of the GAC will be extended because it will not become coated with floc particles and will not have to be backwashed as often.

Because not many contactors are in full-time use, there is little operational experience available. Usually, two or more contactors are operated in parallel downflow operation, like gravity filters. However, the start-up of the units is staggered so that the beds will not all be exhausted at the same time. The beds

Table 7-2. Removal of Chloroform by GAC Adsorption (Example)

Bed Depth		*EBCT*	*Average Influent Chloroform Concentration*	*Time to Bed Exhaustion**
ft	m	min	mg/L	Weeks
2.5	0.8	6.2	67	3.4
5.0	1.5	12	67	7.0
7.5	2.3	19	67	10.9
10.0	3.0	25	67	14.0

*Bed is exhausted when effluent concentration is equal to influent concentration.

must be backwashed periodically to remove suspended material and carbon fines. Usually, the frequency of backwashing will be determined by the increase in pressure drop as the adsorbers are operated. When monitoring indicates a bed is exhausted, the GAC must be replaced with fresh carbon.

7-4. Common Operating Problems

The operating problems encountered with adsorption treatment process depend on the type of material used—PAC or GAC.

Powdered Activated Carbon

The most common operating problem with PAC is handling of the chemical. Since it is a fine powder, dust can be a major problem, particularly if a dry feed system is used. If PAC is used continuously or if large quantities are used periodically, a slurry system should be considered. Dust can still be a problem, but it can be minimized by good handling practices, discussed in the safety section of this module.

PAC passing through the filters and entering the distribution system can result in "black water" complaints from consumers. Black water is usually caused by inadequate coagulation/sedimentation or high doses of PAC being added just before the filters. To solve the latter problem, the application point can be moved so the PAC is fed into the raw-water intake or into the rapid-mix basin. If continuous high doses are necessary to control the organic compounds, it may be more cost effective to use GAC filtration.

Another problem with PAC is that regardless of the PAC dose used, taste and odor problems remain or often seem to get worse. This problem usually results because the PAC and chlorine are being added too close to one another. The chlorine reacts with the organic compounds producing additional compounds that are more difficult to adsorb. One solution to this problem is to add the PAC so that it has at least 15 min contact time before the chlorine is added. Another solution is to stop continuous prechlorination, which will also reduce the amount of additional compounds that might be formed.

Granular Activated Carbon

The operating problems when using GAC as a filter media and adsorbent in a conventional gravity filter are the same as with any rapid-sand or multimedia filter. Most of the problems, as discussed in Module 6, Filtration, concern chemical pretreatment before the filter, control of the filter flow rate, and backwashing the filter.

It is very important that the coagulation, flocculation, and sedimentation processes be operated to reduce the amount of suspended material going into the filters to the greatest extent possible. Although GAC is an effective filter media, adsorption capacity is lost as the surfaces of the carbon particles become coated with floc. This, in turn, results in shorter bed life and higher operating costs. Proper monitoring and control of the processes ahead of the filters can extend the bed life from weeks to months, allowing a considerable savings of time and money.

Filter flow rates should be controlled so they do not fluctuate rapidly. Such fluctuations can wash previously deposited material through the filter and can form channels through the bed. The channels will reduce the filtering capacity of the bed and greatly reduce the contact time between the organic compounds and the GAC. Therefore, flow rates should be changed gradually and the control equipment should be carefully maintained so malfunctions do not cause extreme variations in flow.

Proper backwashing is essential for effective filtration and adsorption. The backwash rate must be sufficient to achieve the bed expansion (50 percent is recommended for GAC filters) necessary for proper cleaning of the bed. If the bed is not adequately cleaned, filtration and adsorption capacity are lost. The accumulation of floc material on the carbon particles will cause mudballs to form, which can result in bed cracking and shrinkage. This is a particular problem where high polymer doses are being used as coagulant and filter aids. The polymer causes the floc to be very sticky, and as a result, it will strongly adhere to the carbon. Therefore, surface-wash equipment is highly recommended for GAC filters.

Since GAC is less dense (lighter) than other filter media, it can easily be washed away during backwash if care is not taken. Backwash rates may have to be lowered to prevent loss. In addition, manufacturers recommendations for backwash rates relating to the particle size (grade) of GAC being used and the water temperature should be followed closely. Figure 7-14 illustrates these relationships. The water pressure used for the surface washers may also have to be decreased somewhat to reduce carbon loss. In addition, the washers should be turned off before full bed expansion. If carbon loss is more than about 2 in. (50 mm) per year, the backwashing procedures should be reviewed to see if changes can be made to reduce the loss. The height of the wash-water troughs above the filter bed should also be examined. It may be cost effective to raise the troughs to prevent further carbon loss.

A special problem with GAC filters is the rapid growth of bacteria within the bed. Since carbon removes organic compounds that the bacteria use as food, the filter provides an excellent environment in which the bacteria thrive. The carbon also adsorbs chlorine, making any chlorine added ahead of the filters to control bacteria ineffective. Chlorination ahead of the filters could also produce new organic compounds that the carbon may not be able to adsorb. As a result of the bacterial growth on the carbon, the filtered water may have bacterial concentrations thousands of times higher than the water going onto the filters. Research has indicated that the high bacterial populations in the bed actually enhance organics removal because the bacteria break down complex compounds to simpler products that are more easily adsorbed by the GAC. However, it is not good practice to allow high concentrations of bacteria to pass into the distribution system and on to the consumer. Therefore, close monitoring of the bacterial content of the finished water and control of the final chlorination process is necessary. Some treatment plants backwash with chlorinated water to help control the bacterial levels.

If GAC contactors are used after conventional filters, it is very important that the suspended material coming through the filters be kept to a minimum so the GAC surfaces will not become coated and the pore spaces will not fill with floc. If the length of time between backwashing starts getting shorter or if bed life is much shorter than it should be, floc carryover from the filters may be a problem. Continuous turbidity monitoring of the water from the filters is essential to identify the need for operational changes before the contactors are seriously affected. Bacterial growth in the contactors can also be a problem, but can be handled in the same manner as discussed for GAC in conventional filters.

7-5. Control Tests and Record Keeping

Operational control tests for adsorption range from jar tests to sophisticated water quality analyses using a gas chromatograph, as described in Volume 4, *Introduction to Water Quality Analyses*, and in *Standard Methods for the Examination of Water and Wastewater*. Although analyses for specific organic compounds or general organic parameters such as TOTAL ORGANIC CARBON (TOC) are highly recommended, many treatment plants do not have the analytical equipment or laboratory personnel to perform these tests. This section will discuss operational tests that can be performed by the operator.

Powdered Activated Carbon

The major operational control tests for a PAC system are (1) determination of the proper dosage and (2) monitoring of the treated water to determine if organic compounds are being removed.

Dosage can be determined with a modification of the jar test used for coagulants. The stirring apparatus as well as all critical glassware should be cleaned with a nonscented detergent and rinsed thoroughly with odor-free water (see *Standard Methods for the Examination of Water and Wastewater* for the method to produce odor-free water). One-litre samples of the raw water are then dosed with varying amounts of a well-shaken stock PAC suspension, such as 5, 10, 20, and 40 mg/L. The stock solution is prepared by adding 1 g of PAC to 1 L of odor-free water. Each millilitre of this solution when added to a 1-L sample of raw water is equal to a dosage of 1 mg/L. The four dosed samples and a fifth sample to which no PAC is added are stirred for a time period that approximates the contact time the PAC will have with the water in the plant. At the end of that time, each sample is filtered through glass wool or filter paper to remove the PAC. The first 200 mL of sample through the filter is discarded and the remainder subjected to the threshold odor test to arrive at a threshold odor number (TON) for each sample. (The threshold odor test is described in Volume 4, *Introduction to Water Quality Analyses*.) The results can then be plotted (see the example in Figure 7-16) to obtain an optimum PAC dosage. In the example, the dosage needed to reduce the TON to 3 (recommended by the Secondary

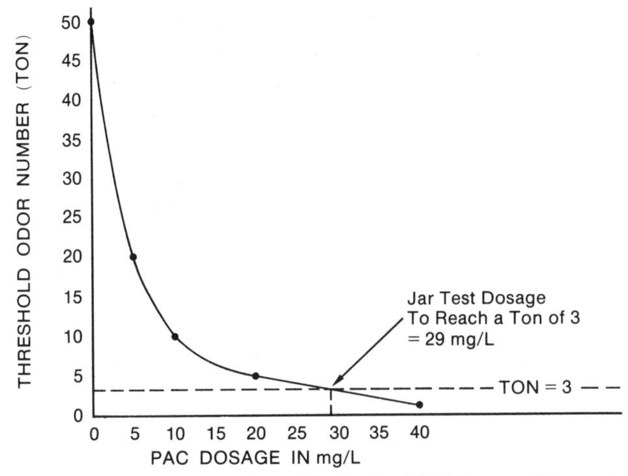

Figure 7-16. Determination of Optimum PAC Dosage (Example)

Drinking Water Regulations) would be 29 mg/L. Experience has indicated that plant-scale application is more efficient than that obtained by the jar test. However, the plant should begin with the dosage indicated by the jar-test result and then gradually reduce it. The threshold odor test on the raw and treated water should be conducted daily while the PAC is being fed, to determine if changes in the dosage are required.

If PAC is used to remove compounds other than those causing tastes and odors, different procedures are needed to determine proper dosage. For example, if the goal is to remove trihalomethane PRECURSOR COMPOUNDS (humic substances that react with chlorine to produce THMs), then pilot testing should be done to determine the removal of those specific compounds.

Good record keeping can help prevent taste- and odor-causing compounds from reaching the consumer and prevent the wasting of PAC. The records should include the dates on which PAC was being added, the calculated dosage (from the jar test), the actual plant dosage that was effective, and the TON values for the raw and treated water. The amount of PAC fed should be calculated and recorded each day to help determine the time to order more chemical. It is also helpful to record the type of odor or taste in the water—for example, musty, septic, or rotten-egg odors and sweet or bitter tastes. This might give a clue as to the source of the problems so that preventative action can be taken at the source. All of these records will give the operator a history of the problems that have occurred each year and allow planning so proper addition of PAC can be accomplished when it is needed.

GAC Used in Conventional Gravity Filters

Since use of GAC in a conventional gravity filter is a combination of filtration and adsorption, most of the control tests are those used for filtration, as described in Module 6, Filtration. Head loss and turbidity of the filtered water should be monitored and recorded continuously to ensure proper backwashing frequency and good quality water. The condition of the filter bed should be checked frequently, particularly during and after backwashing, to determine if there are problems with cracks, shrinkage, channeling, or mudballs.

In addition, certain additional tests should be conducted to monitor the adsorption process. The distance between the top of the wash-water trough (or some other predetermined reference point) and the top of the carbon should be measured and recorded at least every three months to determine the rate of carbon loss. Samples of backwash water can also be tested to see if excessive carbon is being lost. Excessive loss may show a need for changes to the backwashing procedures. Core samples of the carbon bed should be taken at the time of installation and at least every six months to determine the amount of bed life remaining. The sample should represent the carbon from top to bottom of the bed. GAC manufacturers can provide the necessary information on how to sample the bed, what tests can be used, and what the tests mean.

If the filter is removing taste- and odor-causing compounds, the threshold odor test should be conducted routinely on the raw and finished water as a check on the effectiveness of the filter. If the filter is intended to remove other types of organic compounds, analyses should be conducted for these compounds.

Since bacteria can thrive in GAC filters, standard-plate-count analyses should be done on the filtered water and the water after final chlorination each day. This will help determine the final chlorination dose and will indicate whether chlorination of the backwash water is needed. The plate count should be kept well below 500 organisms/mL.

In addition to the records recommended for the filtration process (Module 6, Filtration), records should be kept on the amount of carbon lost, the bed life of the GAC, and the results of the threshold odor test and other analyses performed.

GAC Contactors

Since GAC contactors are used for organic compounds other than those causing tastes and odors, more sophisticated control tests are usually needed. Generally, a specific compound or a group of compounds (such as trihalomethanes) is removed; therefore, the effectiveness of the contactors must be judged by testing for those compounds. This requires special laboratory equipment and trained personnel. In addition to these chemical analyses, monitoring of other factors is required. Head loss through the contactors should be monitored and recorded continuously, so that backwashing is performed at the proper time. Core samples of the carbon bed should be taken at least every three months so that approximate remaining bed life can be determined. Turbidity of the contactors' effluents should be monitored and recorded continuously to determine if carbon fines or other suspended matter are passing through the GAC bed. Standard plate count analyses of the water from the conventional filters, the contactors, and after final chlorination should be conducted daily to help determine what chlorine dosages are needed in the plant to control bacterial growth.

In addition to the above, records should be kept of the amount of water treated, the bed life of the GAC, the time when a contactor is backwashed, and the amount of backwash water used. When fresh GAC is placed in a contactor the amount added should be recorded so losses can be calculated.

7-6. Safety and Adsorption

Since PAC and GAC are fine, dusty materials, care must be taken in handling and storing them. If either PAC or GAC are being used in bags, the bags should be stored in a clean, dry place on pallets so air can circulate underneath. The bags should be stacked in single or double rows with access aisles around every stack to allow easy handling and fire inspections. Carbon bags should never be stored in stacks higher than 6 ft (2 m). Although activated carbon, even as a dust, is not considered explosive, it will burn. Like charcoal used in barbecues, it burns without producing smoke or flame and glows with the release of intense heat. Such fires are difficult to detect and when found are hard to handle. This is a major reason for using slurry storage if large quantities of carbon are being used.

The storage area should be fireproof, with self-closing fire doors separating the carbon storage area from other areas. Storage bins for dry bulk carbon should be fireproof and equipped for fire control.

Burning carbon should never be doused with a large stream of water since the fire will be spread quickly as the burning carbon particles fly in all directions. A fine mist or spray from a hose, or chemical foam from a fire extinguisher are effective in controlling carbon fires. Also, smoking should be prohibited at all times during the handling and unloading of carbon and in the storage and feeding areas.

Activated carbon should never be stored along with gasoline, mineral oils, or vegetable oils. Such materials, when mixed with carbon, will slowly oxidize until ignition temperature (600–800° F [316–427° C]) is reached. Never mix or store carbon with chlorine compounds or potassium permanganate, since spontaneous combustion can occur. Carbon should not be stored where overhead electrical equipment could start a fire.

Because activated carbon is an electrical conductor, dust must be minimized. The heat from a motor may ignite the accumulated dust, which could short-circuit the motor and cause a fire. Therefore, all tanks receiving dry carbon should be vented and provided with dust-control equipment, such as bag-type dust collectors. Even slurry storage tanks should be so equipped. If a PAC dry-feed system is used, the hoppers and feeders should be enclosed in a separate room so dust will not spread throughout the plant. As an additional precaution, the electrical equipment should be cleaned frequently. Explosion-proof electrical wiring and apparatus should be used for any equipment operated near carbon storage areas, especially while feeding is being performed.

Since oxygen is removed from air in the presence of wet activated carbon, slurry tanks or other enclosed spaces containing carbon may have seriously reduced oxygen levels. Therefore, care should be taken when entering such spaces. Devices to indicate the amount of oxygen present should be used before

Figure 7-17. Safety Clothing to Be Worn When Handling Carbon

entering a closed space. Anyone entering a tank or other enclosure should have a line attached to them and to someone standing close by so they can be pulled from danger if necessary.

Dust masks, face shields, gauntlets, and aprons (Figure 7-17) should be worn by personnel when carbon in bag or bulk form is being unloaded or handled. In addition, adequate shower facilities should be provided if personnel handle carbon routinely.

Selected Supplementary Readings

Barnett, R.H. & Trussell, A.R. Controlling Organics: The Casitas Municipal Water District Experience. *Jour. AWWA*, 70:11:660 (Nov. 1978).

Carns, K.E. & Stinson, K.B. Controlling Organics: The East Bay Municipal Utility District Experience. *Jour. AWWA*, 70:11:637 (Nov. 1978).

Cohen, R.S., et al. Controlling Organics: The Metropolitan Water District of Southern California Experience. *Jour. AWWA*, 70:11:647 (Nov. 1978).

Controlling Organics in Drinking Water. AWWA Conf. Seminar Proc. AWWA, Denver, Colo. (June 1979).

Culp, G.L. & Culp, R.L. *New Concepts in Water Purification.* Van Nostrand Reinhold Company, New York. (1974). Chap. 8.

Ferrara, A.P. Addition of GAC to Filter Beds. *OpFlow*, 3:5:3 (May 1977).

Ferrara, A.P. Techniques of Media Removal for Replacement With GAC. *OpFlow*, 2:10:3 (Oct. 1976).

Handbook of Taste and Odor Control Experiences in the US and Canada. AWWA, Denver, Colo. (1976).
Hansen, R.E. Granular Carbon Filters for Taste and Odor Removal. *Jour. AWWA*, 64:3:176 (Mar. 1972).
Hyndshaw, A.Y. Activated Carbon to Remove Organic Contaminants From Water. *Jour. AWWA*, 64:5:309 (May 1972).
Hyndshaw, A.Y. Treatment Application Points for Activated Carbon. *Jour. AWWA*, 54:1:91 (Jan. 1962).
Introduction to Water Quality Analyses. AWWA, Denver, Colo. (1982). Modules 1 and 5.
Lange, A.L. & Kawczynski, Elizabeth. Controlling Organics: The Contra Costa County Water District Experience. *Jour. AWWA*, 70:11:653 (Nov. 1978).
McBride, D.G. Controlling Organics: The Los Angeles Department of Water and Power Experience. *Jour. AWWA*, 70:11:644 (Nov. 1978).
Safe Handling of Chemicals: Activated Carbon. *OpFlow*, 6:1:3 (Jan. 1980).
Safety Practice For Water Utilities. AWWA Manual M3. AWWA, Denver, Colo. (1983).
Treatment Techniques for Controlling Trihalomethanes in Drinking Water. AWWA, Denver, Colo. (1982).
Water Treatment Plant Design. ASCE, AWWA, and CSSE. AWWA, Denver, Colo. (1969). Chap. 10 and 14.

Glossary Terms Introduced in Module 7

(Terms are defined in the Glossary at the back of the book.)

Activated alumina
Activated carbon
Activation
Adhesion
Adsorbent
Adsorption
Arching
Bed life
Carcinogen
Contactor
Eductor
Empty bed contact time (EBCT)
Granular activated carbon (GAC)
Herbicide
Humic substance
Insecticide
Nuclei
Organic substance (organic)
Pesticide
Powdered activated carbon (PAC)
Precursor compound
Reactivated
Slurry
Synthetic resin
Total organic carbon (TOC)
Trihalomethane

Review Questions

(Answers to Review Questions are given at the back of the book.)

1. What is the purpose of adsorption?
2. Why should treatment plant operators be concerned with removing organic materials from drinking water?
3. Some organic materials found in water come from natural sources. Why are these of concern?
4. Why are man-made organics of concern?
5. List at least four methods of removing organics.
6. What is activated carbon and how is it made?
7. What is adsorption?
8. Name three types of adsorbents.
9. What are the two basic types of activated carbon?
10. What are the major differences between PAC and GAC?
11. Why should PAC dry feeders be kept in a separate enclosure?
12. What is carbon slurry and why is it used?
13. What type of material should be used for piping that carries carbon slurry?
14. How is GAC used?
15. When is GAC used rather than PAC?
16. What minimum depth of GAC is recommended in filters?
17. What is reactivation?
18. What three points should be considered when selecting a PAC application point?
19. What is EBCT?

20. What are two major causes of carbon loss in a GAC filter?

21. What does the life of GAC beds depend on?

22. What is an advantage of contactors?

23. What is the most common operating problem with PAC?

24. What unique problem is associated with GAC beds?

25. What is an effective method of determining PAC dosage?

26. Why should carbon not be stored with chlorine or potassium permanganate?

27. How should a carbon fire be extinguished?

Study Problems and Exercises

1. You are the chief operator of a water treatment plant that treats surface water. The water is highly colored much of the year and has a terrible taste and odor each spring. Your THM analyses for the past year shows an average concentration of 350 mg/L. Treatment processes consist of prechlorination, alum coagulation, flocculation, sedimentation, rapid-sand filtration, and final chlorination. What corrective action would you recommend to the town council to solve these water quality problems?

2. Using the jar test data given below, calculate the optimum PAC dosage to achieve a TON of 3.

PAC Dosage *mg/L*	TON
0	80
5	30
10	20
20	7
40	2

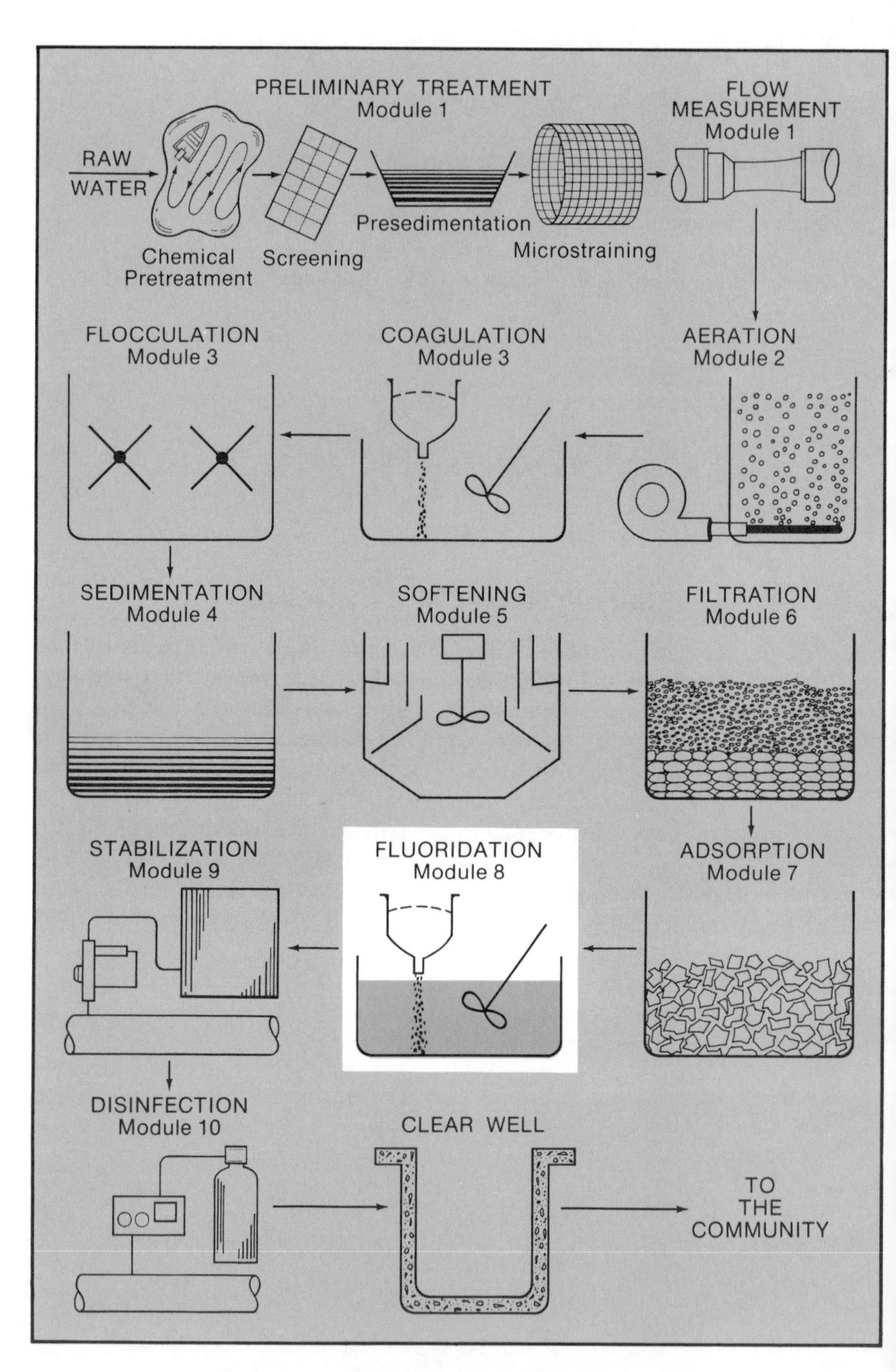

Figure 8-1. Fluoridation in the Treatment Process

Module 8

Fluoridation

Fluoridation is used to maintain fluoride concentrations in drinking water at levels known to reduce tooth decay in children. The benefits of fluoride were discovered by comparing tooth decay rates of children from areas having high and low concentrations of natural fluoride in their drinking water. Fluoridation is a safe, effective, and economical process endorsed by the American Water Works Association and public health groups worldwide. As shown in Figure 8-1, fluoride is added to water after filtration and just before disinfection in the typical treatment plant.

Fluoride is an ION originating from the element fluorine. It is a constituent of the earth's crust and consequently is found naturally to some degree in all drinking water sources. Fluoride is essential for proper tooth and bone formation. However, in excessive amounts it can cause FLUOROSIS, a discoloration (mottling) of teeth, or more serious damage such as pitting.

After completing this module you should be able to

- Explain the benefits of fluoridation.
- Describe typical fluoridation processes.
- Describe the facilities used for fluoridation.
- Describe common operating problems and explain how they can be corrected, avoided, or controlled.
- Discuss how fluoridation is controlled.
- Describe the control tests and record keeping used to monitor fluoridation.
- Describe the safety considerations involved in the fluoridation process.

8-1. Description of Fluoridation

Research over the past 80 years shows that fluoride is essential to man's normal growth and development, and that fluoridating drinking water provides the optimum level of fluoride children need to develop teeth resistant to decay. At optimum levels, fluoride can reduce the incidence of tooth decay among children by 65 percent. Figure 8-2 shows that the tooth decay rate decreases with increasing fluoride concentration, particularly up to 1 mg/L.

The effects of fluoride are based on the total amount consumed each day. In addition to the concentration of fluoride in the drinking water supply, the amount consumed depends on how much water a person drinks. For example, a person who drinks 1 gal (3.8 L) of water containing 0.6 mg/L fluoride every day receives the same total amount of fluoride as a person who drinks 0.5 gal (1.9 L) of water containing 1.2 mg/L fluoride every day.

Water consumption depends on the temperature in a region. People in warm climates drink much more water than people in cold climates. Consequently, the optimum fluoride concentration that should be in water varies from place to place, depending on the average air temperature. The fluoride levels in Table 8-1 are for annual averages of maximum daily air temperatures, which must be determined for a 5-year period. Each state health department uses this information to establish optimum fluoride levels for their state.

The optimum fluoride concentrations listed in Table 8-1 are based on an average consumption of ½ gal (2 L) of fluoridated water each day. However, in some communities only drinking water in the schools is fluoridated. Since this water is only available to the children when they are in school (about six hours a day), the available fluoride concentration is increased to four times the community's optimum level. This assures that each child will receive an effective daily intake of fluoride. Fluorosis will not result since much less than ½ gal (2 L) of this water is consumed each day.

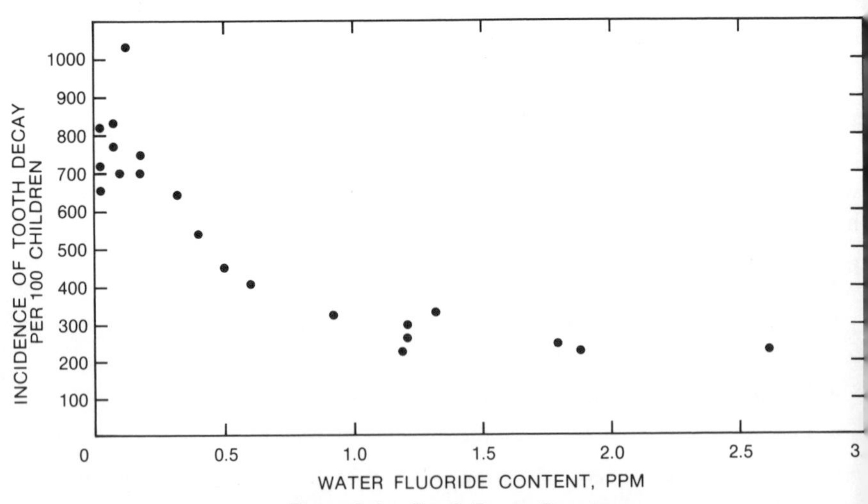

Figure 8-2. Tooth Decay Decrease

Table 8-1. Optimum Fluoride Concentrations and Fluoride MCLs*

nnual Average of Maximum Daily Air Temperature F	C	Recommended Control Limits of Fluoride Concentration mg/L Lower	Optimum	Upper	Maximum Contaminant Level mg/L
3.7 and below	12.0 and below	0.9	1.2	1.7	2.4
53.8–58.3	12.1–14.6	0.8	1.1	1.5	2.2
58.4–63.8	14.7–17.6	0.8	1.0	1.3	2.0
63.9–70.6	17.7–21.4	0.7	0.9	1.2	1.8
70.7–79.2	21.5–26.2	0.7	0.8	1.0	1.6
79.3–90.5	26.3–32.5	0.6	0.7	0.8	1.4

*Based on temperature data for a minimum of 5 years.

To achieve the maximum benefits of fluoridation, the optimum fluoride concentration in the public water supply must be maintained.[1] A drop of only 0.3 mg/L below the optimum concentration can reduce fluoride's benefits by as much as two-thirds. As shown in Figure 8-2, concentrations above 1.5 mg/L do not significantly reduce tooth decay any further, indicating that overfeeding should be avoided. Occasional overfeeding, however, will not result in fluorosis.

Fluorosis results from excessive amounts of fluoride in drinking water. In its mildest form, fluorosis appears as very slight, opaque, whitish areas (MOTTLING) on the tooth surface. More severe fluorosis usually causes teeth to darken, turning from shades of gray to black. When the fluoride concentration is over 4 mg/L, pitting of the teeth can occur. Teeth are then more susceptible to cavities and wear.

Studies have shown that fluorosis starts to occur when children less than eight years old regularly drink water containing twice the optimum fluoride level for three months or longer. For this reason, the USEPA drinking water regulations include maximum contaminant levels (MCLs) for fluoride that are twice the optimum levels (Table 8-1). Since such high levels will rarely, if ever, occur in a properly operated treatment plant, the MCLs apply mainly to localities where the natural fluoride concentrations are high.

8-2. Fluoridation Facilities

The facilities used for fluoridation are similar to those used for feeding other water treatment chemicals. The type of equipment used depends primarily on how much water is treated and the type of fluoride chemical chosen.

Fluoride Chemicals

The three common chemical compounds used for fluoridation are:

- Sodium fluoride
- Hydrofluosilicic acid
- Sodium silicofluoride.

[1] *Basic Science Concepts and Applications*, Chemistry Section, Dosage Problems (Fluoridation Calculations).

The characteristics of these compounds are summarized in Table 8-2. On
chemicals meeting applicable AWWA standards should be used.

Sodium fluoride. SODIUM FLUORIDE, (NaF), the first compound to be used i
controlled fluoridation, is still widely used today. It is a white, odorless materi
that comes in powder or crystalline form. The crystalline form is preferred whe
manually handling chemicals because less dust results. The solubility of sodiu
fluoride is almost constant—4 g NaF/100 mL of water—within the commo
range of water temperatures, 32 to 77° F (0 to 25° C). Sodium fluoride
available in 100-lb (45-kg) multi-ply paper bags, fiber drums that hold up to 40
lb (180 kg), and in bulk.

Hydrofluosilicic acid. HYDROFLUOSILICIC ACID, (H_2SiF_6), sometim
referred to as fluosilicic acid or "silly acid," is a clear, colorless, fuming, corrosi
liquid. It has a pungent odor and can cause skin irritation. All commerci
solutions of hydrofluosilicic acid have a low pH, ranging from 1.0 to 1.5. In high
alkaline waters, the addition of hydrofluosilicic acid will usually not affect pH
However, in low alkaline (poorly buffered) waters the addition of hydrofluosilic
acid can reduce pH, so it is not recommended.

Table 8-2. Characteristics of Fluoride Compounds

Item	*Sodium Fluoride NaF*	*Sodium Silico-fluoride Na_2SiF_6*	*Hydro-fluosilicic Acic H_2SiF_6*
Form	Powder or crystal	Powder or very fine crystal	Liquid
Molecular weight	42.00	188.05	144.08
Commercial purity, %	90–98	98–99	22–30
Fluoride ion, % (100% pure material)	42.25	60.7	79.2
Pounds required per mil gal for 1.0 ppm F at indicated purity	18.8 (98%)	14.0 (98.5%)	35.2 (30%)
pH of saturated solution	7.6	3.5	1.2 (1% solutio
Sodium ion contributed at 1.0 ppm F, ppm	1.17	0.40	0.00
F ion storage space, cu ft/100 lb	22–34	23–30	54–73
Solubility at 77° F (25° C), g/100 mL water	4.05	0.762	Infinite
Weight, lb/cu ft	65–90	55–72	10.5 lb/gal (30%)
Shipping containers	100-lb bags, 125–400-lb fiber drums, bulk	100-lb bags, 125–400-lb fiber drums, bulk	13-gal carboys 55-gal drums, b

Hydrofluosilicic acid is available in bulk (tank car or tank trucks) for large users, and in 13-gal (50-L) polyethylene carboys or 55-gal (210-L) rubber-lined drums for small users. Because it contains a high proportion of water (about 70 percent), hydrofluosilicic acid is relatively costly to ship compared to dry chemicals used for fluoridation.

Sodium silicofluoride Hydrofluosilicic acid is easily converted into various salts. One of them, SODIUM SILICOFLUORIDE, is the most inexpensive chemical available for fluoridation. Sodium silicofluoride (Na_2SiF_6) is a white, odorless, crystalline or granular material in purities of 98 percent or greater. Unlike sodium fluoride, its solubility decreases as water temperature decreases. At 60 to 70° F (16 to 21° C), it takes 60 gal (230 L) of water to dissolve 1 lb (0.45 kg) of Na_2SiF_6. The pH of the saturated solution is quite low, usually between 3.0 and 4.0, but this is not a problem since the solution is diluted once it is added to the drinking water. Sodium silicofluoride is available in the same-sized containers as sodium fluoride.

Chemical Feeders

Fluoride is fed into the water system by either a dry feed system or a solution feed system. Either system will feed a specific quantity of chemical into the water at a preset rate. Table 8-3 lists the required equipment and other characteristics of several methods of feeding fluoride.

Dry feeders. A dry feeder meters a dry powdered or crystalline chemical at a given feed rate. It is generally used to add fluoride to systems that produce 1 mgd or more. The two basic types of dry feeders are volumetric dry feeders and gravimetric dry feeders.

VOLUMETRIC DRY FEEDERS (Figures 8-3 and 8-4) are simpler to operate and less expensive to purchase and maintain than gravimetric dry feeders; however, they

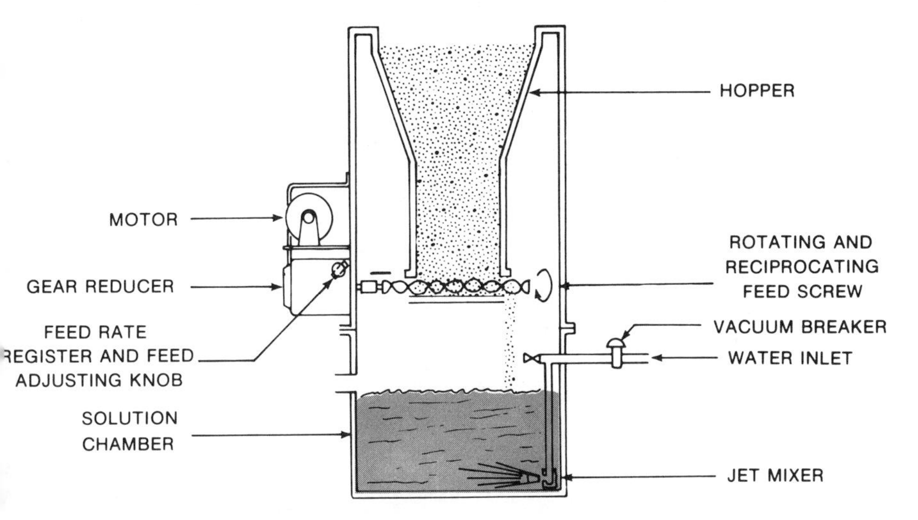

Figure 8-3. Screw-Type Volumetric Dry Feeder

Table 8-3. Fluoridation Check List

Operating Parameters	*Sodium Fluoride Manual Solution Preparation*	*Sodium Fluoride Automatic Solution Preparation*	*Hydrofluosilicic Acid Diluted*	*Hydrofluosilicic Acid 23-30%*	*Sodium Silicofluoride Dry Feed*	*Sodium Flouride Dry Feed*
Water flow rate	Less than 500 gpm	Less than 2000 gpm	Less than 500 gpm	More than 500 gpm	More than 100 gpm	More than 2 mgd
Population served by system or each well of multiple-well system	Less than 5000	Less than 10,000	Less than 10,000	More than 10,000	More than 10,000	More than 50,000
Equipment required	Solution feeder, mixing tank, scales, mixer	Solution feeder, saturator, water meter	Solution feeder, scales, measuring container, mixing tank, mixer	Solution feeder, day tank, scales, transfer pump	Volumetric dry feeder, scales, hopper, dissolving chamber	Gravimetric dry feeder, hopper, dissolving chamber
Feed accuracy	Depends on solution preparation and feeder	Depends on feeder	Depends on solution preparation and feeder	Depends on feeder	Usually within 3%	Usually with 1%
Chemical specifications and availability	Crystalline NaF, dust-free, in bags or drums; generally available	Downflow—coarse crystalline NaF in bags or drums, may be scarce; upflow—fine crystalline NaF, generally available	Low-silica or fortified acid in drums or carboys; generally available	Bulk acid in tank cars or trucks; available on contract	Powder in bags, drums, or bulk; generally available	
Handling requirements	Weighing, mixing, measuring	Dumping whole bags only	Pouring or siphoning, measuring, mixing, weighing	All handling by pump	Bag loaders or bulk-handling equipment required	
Feeding point	Injection into filter effluent line or main	Injection into filter effluent line or main	Injection into filter effluent line or main	Injection into filter effluent line or main	Gravity feed from dissolving chamber into open flume or clear well, pressure feed into filter effluent line or main	
Other requirements	Solution water may require softening	Solution water may require softening	Dilution water may require softening	Acid-proof storage tank, piping, etc.	Dry storage area, dust collectors, dissolving-chamber mixers, hopper agitators, eductors, etc.	
Hazards	Dust, spillage, solution preparation error	Dust, spillage	Corrosion, fumes, spillage, solution preparation error	Corrosion, fumes, leakage	Dust, spillage, arching and flooding in feeder and hopper	

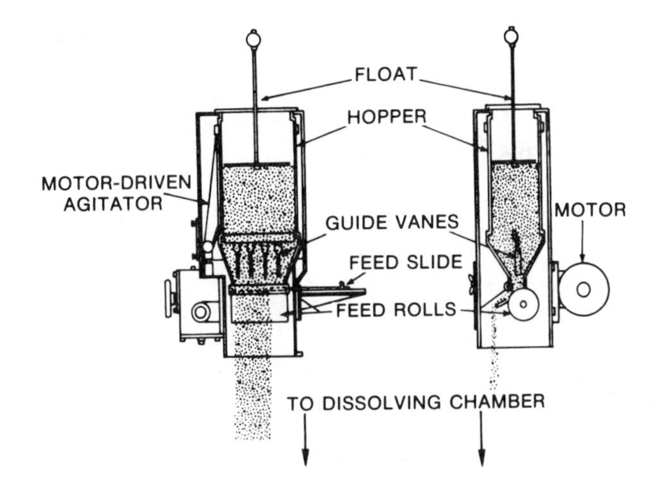

Figure 8-4. Roll-Type Volumetric Dry Feeder

are less accurate and generally lower in capacity. The feed mechanism (feed rolls or feed screw) delivers exactly the same volume of dry chemical to the dissolving tank with each complete revolution. Varying the speed of rotation varies the feed rate.

GRAVIMETRIC DRY FEEDERS deliver large quantities of chemicals and are extremely accurate, but they are expensive. They can be readily adapted to automatic control and recording. The belt-type feeder (Figure 8-5) delivers a certain weight of material with each revolution of the conveyor belt. The feed rate is varied by varying the speed of the belt. The loss-in-weight type feeder (Figure 8-6) matches the weight lost by the feed hopper to the preset weight of the required dosage. Because gravimetric feeders control the weight of material, not the volume, variations in density have no effect on feed rate. This accounts for the extreme accuracy of this type feeder.

The steady stream of dry material discharged from a dry feeder falls into a solution chamber or tank where it is dissolved into water. The tank is usually equipped with a mixer to encourage dissolving of the chemical. The resulting fluoride solution either flows by gravity into the clear well or is pumped into a pressure pipeline. Either sodium fluoride or sodium silicofluoride can be used in a dry feeder. Since sodium silicofluoride is less expensive, it is most often used with this system.

Solution feeders. Solution feeders are the most economical way for smaller water systems to fluoridate water (Table 8-3). The feeders are small pumps that feed a sodium fluoride or hydrofluosilicic acid solution from a tank or saturator into the water system at a preset rate.

Chemical feed pumps used for fluoridating must be accurate and capable of delivering a constant chemical dosage. They must also be able to deliver against pressure. For this reason, POSITIVE-DISPLACEMENT PUMPS (diaphragm pumps and piston pumps) generally are used.

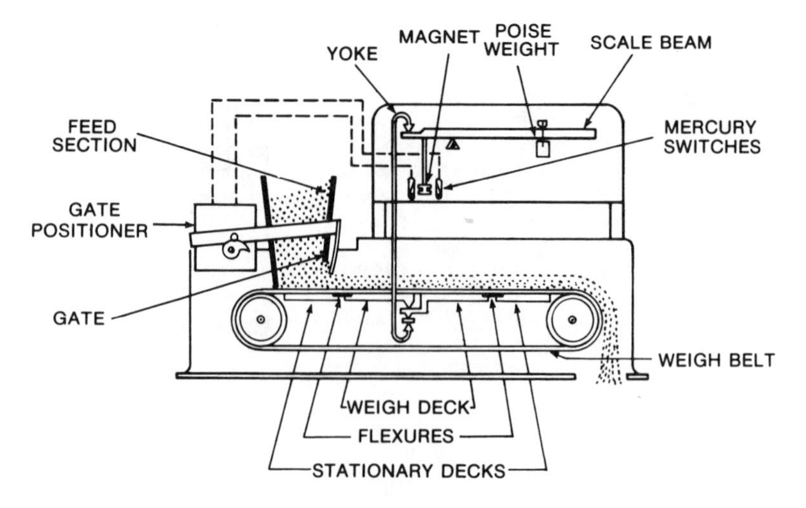

Figure 8-5. Belt-Type Gravimetric Dry Feeder

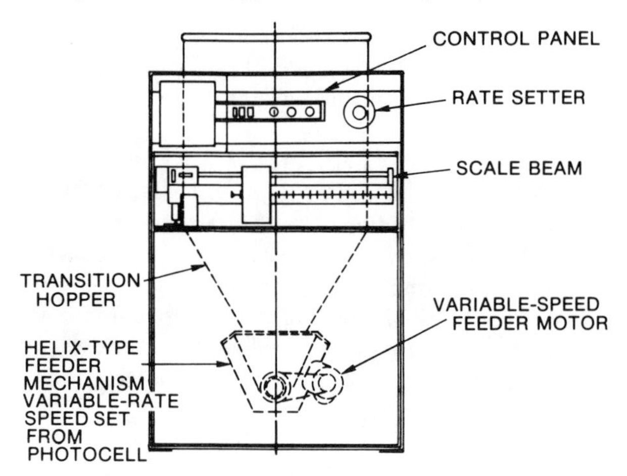

Figure 8-6. Loss-in-Weight-Type Gravimetric Dry Feeder

The most widely used is the DIAPHRAGM PUMP, which is also a common feeder for hypochlorination systems (Figure 8-7). It consists of a flexible pumping diaphragm made of rubber, plastic, or thin metal. The diaphragm is actuated by a reciprocating pump shaft driven by a cam. To vary the feed rate, the cam drive speed or the pump shaft stroke is adjusted. These pumps are often electronically controlled based on flow rate.

The PISTON PUMP is similar to the diaphragm pump. Instead of the flexible diaphragm, a rigid piston moves back and forth within a cylinder to serve as the metering chamber for chemicals being pumped. By varying the stroke length, the feed rate can be varied. Some piston pumps are capable of delivering chemicals at very low rates and against high pressures.

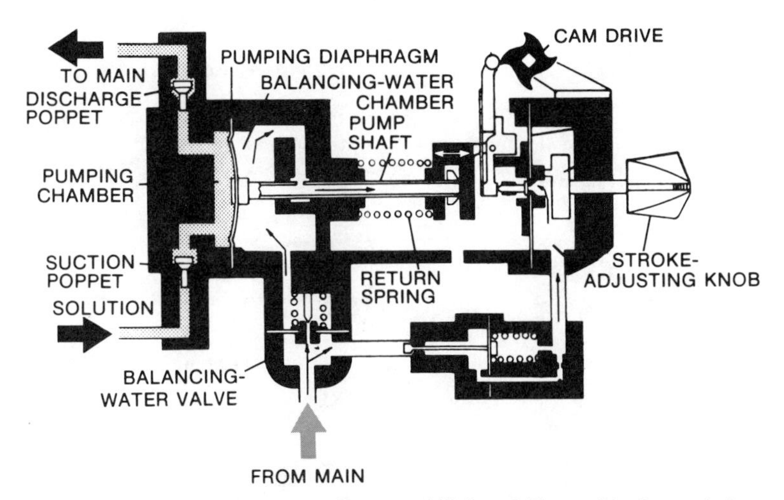

Courtesy of Wallace & Tiernan Div.; Pennwalt Corp.

Figure 8-7. Diaphragm Chemical Feed Pump

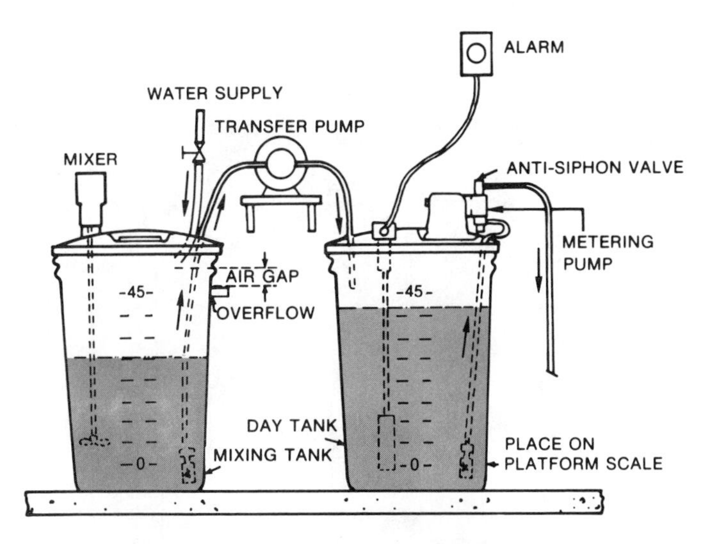

Figure 8-8. Manual Solution Feed Installation

The most common installations using chemical feed pumps are the manual-solution feed, saturator, and acid-feed systems. The installation chosen will depend on water production, chemical cost, chemical availability, and available space.

Manual-solution feed. The MANUAL SOLUTION FEED method, as shown in Figure 8-8, is used with sodium fluoride for systems treating less than 500 gpm (32 L/s). The chemical is added to the mixing tank for dissolving. The concentrated solution is transferred to the day tank and fed with a metering

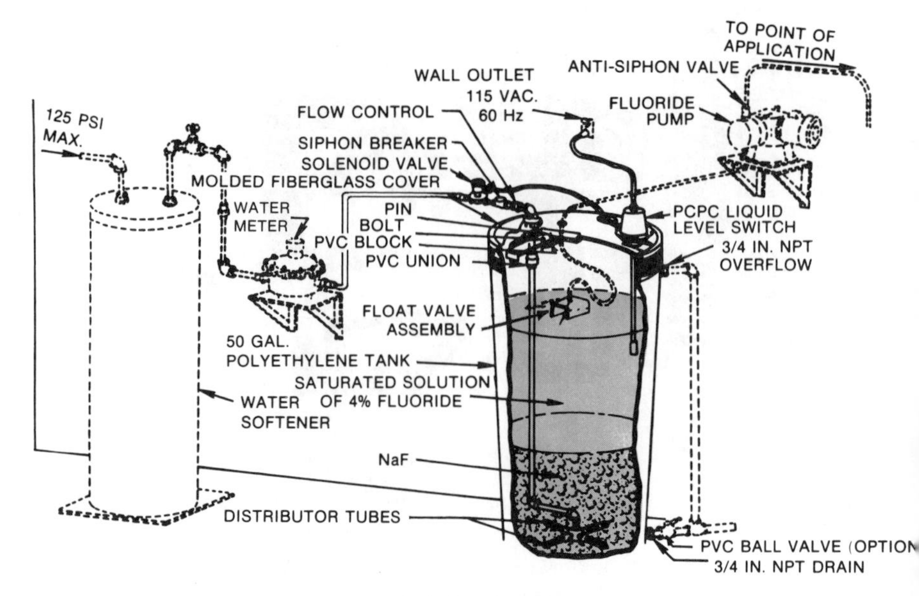

A. Upflow Saturator

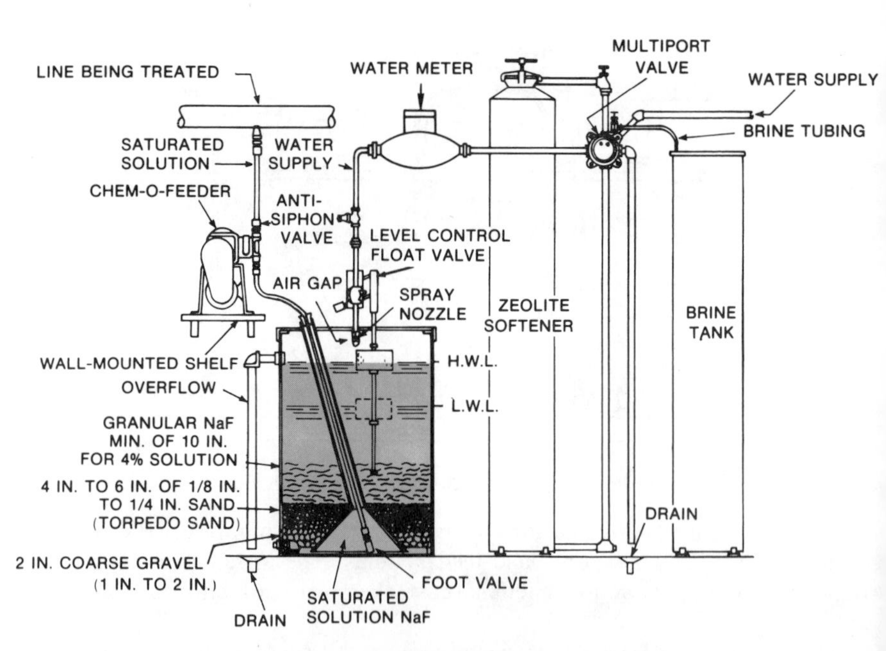

B. Downflow Saturator

Figure 8-9. Two Types of Saturators

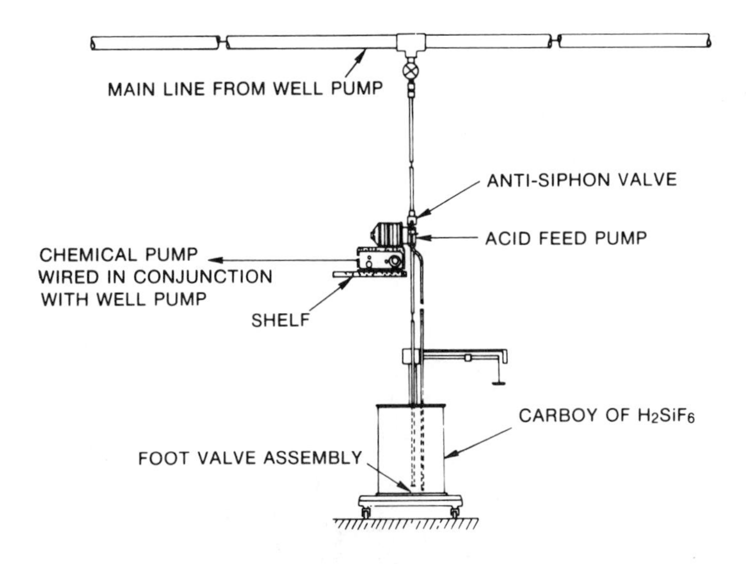

Figure 8-10. Acid Feed Installation

pump. The day tank is placed on a platform scale, which allows determination of how much solution is fed so that the pump can be properly set. The major disadvantage of this installation is that the sodium fluoride must be manually weighed and added to the mixing tank.

Saturator. A SATURATOR retains the simplicity of the manual-feed system but eliminates the disadvantage of constant chemical handling. The principle of the saturator is that a saturated solution of 40 g/L sodium fluoride (18 g/L fluoride ion, F^-) will result if water at any temperature between 32° F and 77° F (0° C and 25° C) is allowed to trickle through a bed of sodium fluoride crystals. This 4-percent NaF solution is then pumped into the water supply at a controlled rate with a metering pump. A saturator is ideal for small systems because of its low maintenance requirements and ease of operation.

There are two types of saturators: upflow and downflow. With upflow saturators (Figure 8-9A), water is introduced at the bottom of a 55-gal (210-L) polyethylene tank, beneath crystals of sodium fluoride. The water moves slowly upward, forming the saturated fluoride solution, which is then pumped to the point of application.

With downflow saturators (Figure 8-9B) the sodium fluoride rests on a bed of sand and gravel, which in turn rests on a collection system such as a perforated pipe manifold or funnel. Water enters the top of the tank, moves slowly through a layer of sodium fluoride (at least 6 in. [0.15 m] thick), through the sand and gravel, and then into the collection system. It is then pumped into the water system.

Acid feed systems. An acid feed system, one of the simplest installations, feeds hydrofluosilicic acid directly from the shipping container into the water supply, as shown in Figure 8-10. The shipping container rests on a scale, which allows determination of the amount of acid used. However, for water systems treating less than 500 gpm (32 L/s), it is usually impractical to feed the acid directly since the metering pump cannot be turned down far enough to pump the small amounts of acid required. In such cases, the dilution system shown in Figure 8-11 can be used.

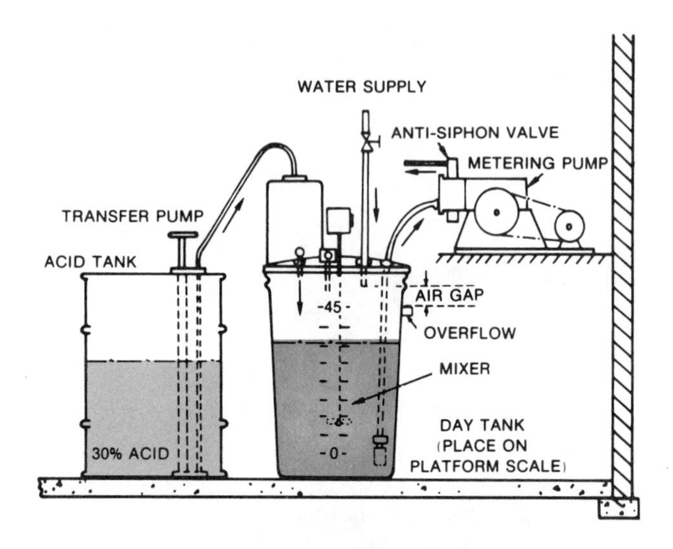

Figure 8-11. Diluted Acid Feed System

Figure 8-12. Platform Scale Installation

Auxiliary Equipment

Following are descriptions of the more common types of auxiliary equipment used with the feed systems described.

Scales. Scales are needed for determining the quantity of solution fed and for determining the quantity of dry fluoride compound or hydrofluosilicic acid delivered by the appropriate feeders. They are necessary in all systems except those using saturators. The most common is the platform scale, on which a solution tank, carboy of acid, or an entire volumetric dry feeder is placed. There are scales designed for this purpose, which are supplied by manufacturers with volumetric dry feeders. However, an ordinary platform scale (like those commonly used in hardware stores) is satisfactory if it has sufficient capacity and sensitivity. Some minor modifications, such as removing the wheels or rotating the beam, may be necessary. The scale must be able to weigh the tank and its contents when full or the volumetric feeder and its hopper when full, with measurements to the nearest pound (half kilogram) or better. An example of a platform scale installation is shown in Figure 8-12.

Softeners. When using sodium fluoride solutions with water having more than 75 mg/L hardness as calcium carbonate ($CaCO_3$), insoluble compounds of calcium and magnesium fluoride can form. These compounds can clog the feeder, the feeder suction line, the gravel bed in a downflow saturator, and other equipment. For this reason, water used with the sodium fluoride solution should be softened if the hardness exceeds 75 mg/L as $CaCO_3$. The entire water supply need not be softened—only the make-up water used for solution preparation. Since the volume of water to be softened is usually quite small, a household-type ION-EXCHANGE WATER SOFTENER is usually adequate. It can be installed directly in the pipeline used for solution make-up (Figure 8-9A).

Dissolving tanks. The dry material discharged from a volumetric or gravimetric feeder is continuously dissolved in a chamber beneath the feeder. This chamber, referred to as the solution pot, dissolving tank, solution tank, or dissolving chamber, may be a part of the feeder or it may be separate. Only after the chemical passes through the dissolving tank can it be pumped into the water to be treated. Slurry feed cannot be used because the buildup of undissolved chemical can cause inaccurate feed rates and may clog feeders and create deposits in the tanks or water system.

Mixers. Whenever solutions are prepared (manual preparation of sodium fluoride, dilution of hydrofluosilicic acid, or solution of dry materials) it is important that the solution be completely and evenly mixed. A fractional horsepower mixer with a stainless-steel shaft and propeller is satisfactory for preparing sodium fluoride. A similar mixer with a corrosion-resistant alloy or plastic-coated shaft and propeller is satisfactory for hydrofluosilicic acid. A jet mixer can dissolve sodium silicofluoride in the solution tank of a dry feeder (Figure 8-3), but a mechanical mixer is preferred. Because of sodium silicofluoride's low solubility, particularly in cold water, and the limited detention time available for dissolving, violent agitation is necessary to prevent discharge of a slurry. Mechanical mixers should never be used on saturators.

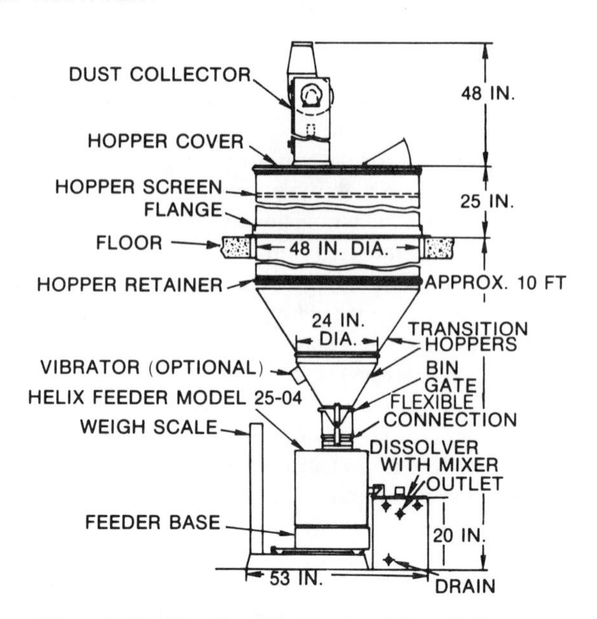

Figure 8-13. Dry Feed Hoppers and Dust Collectors

Saturators are designed to produce a thoroughly mixed solution of sodium fluoride.

Water meters. The flow through the treatment plant must be metered in order to calculate fluoride and other chemical feed rates. In addition, the solution make-up water for a saturator must be metered in order to accurately determine the fluoride feed rates (Figure 8-9). Since the make-up water flow is small, a ½-in. (12-mm) meter is usually sufficient.

Flow meters. A flow meter measures the rate of flow. The flow meter serves two important functions for effective fluoridation, especially when flow rate in a plant varies. First, the meter indicates the flow rates the feeder must accommodate. Second, it can be designed to provide a signal (hydraulic, pneumatic, or electric) that will allow automatic adjustment (known as pacing) of the feed rates.

Day storage tank. A day storage tank is a plastic tank that holds about one day's supply of fluoride solution. Day tanks are essential for large systems that feed hydrofluosilicic acid from bulk storage. Enough acid to treat one day's or one shift's water production is siphoned into a small tank mounted on a platform scale. This acid is then fed into the water main. As shown in Figures 8-8 and 8-11, day tanks are also used with manual-solution and diluted-acid feed systems.

Hoppers. Most dry feeders come equipped with a small hopper. In large installations, an additional or extension hopper is provided above the main hopper to provide the needed total chemical storage capacity. As shown in Figure 8-13, this extension hopper is located one floor above the feeder. Chemicals, if stored on the second floor, are conveniently loaded into the hopper.

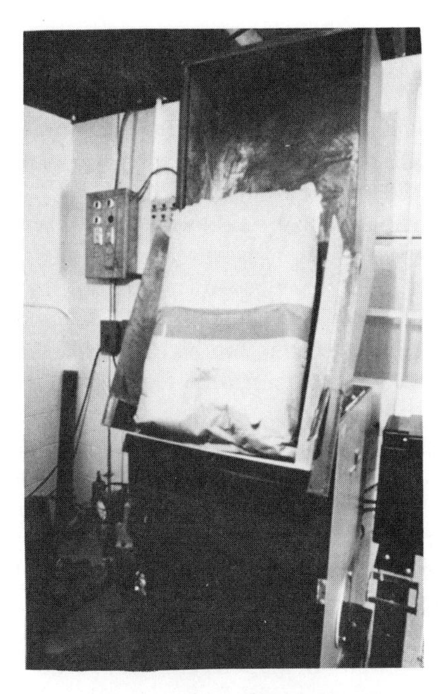

Figure 8-14. Bag Loader

In small plants, it is best to have the chemical hopper large enough to hold slightly more than the entire bag or drum of chemical. The hopper will not be completely empty before there is enough room in it for the contents of a fresh bag or drum. By loading an entire container this way, handling of chemical is minimized, which also minimizes dust and spillage. An agitator should be installed (Figure 8-13) to gently vibrate the hopper and keep the chemical flowing smoothly.

Bag loaders. When the hopper of a dry feeder is directly above the feeder and the operator must lift the bag of chemical a considerable height to fill the hopper, a bag loader is more a necessity than a convenience. A bag loader is essentially a hopper extension large enough to hold a single 100-lb (45-kg) bag of chemical (Figure 8-14). The front of the loader is hinged so that it will swing down to a more accessible height. The operator fastens the bag by running an attached rod through the bottom of the bag. He then opens the bag, and then swings the loader back into position. This device makes emptying the bag easier and minimizes dust. For safety reasons, an operator should wear a respirator during this operation.

Dust collectors and wet scrubbers. Handling powdered dry chemicals always generates dust. When small quantities of fluoride are being handled, ordinary care will minimize dust, and good housekeeping plus an exhaust fan will keep the storage and loading area relatively dust-free. However, when larger quantities (more than one bag at a time) are handled, dust prevention and collection facilities should be provided.

A dust canopy, completely enclosing the hopper-filling area and equipped with an exhaust fan, prevents dust from spreading throughout the loading area. To prevent dust from escaping into the atmosphere and into the area surrounding the water plant, dust filters are incorporated into the exhaust system. Dust collectors and exhaust fans are sometimes incorporated into the hoppers of larger dry feeders (Figure 8-13).

Wet scrubbers remove dust from exhausted air. The air flows through a chamber in which there is a continuous water spray. The air is thus "scrubbed" clean and the dust particles are carried down the drain by water.

Weight recorders. If a platform scale is used to weigh the dry chemicals or solutions that are added, a recorder can be attached to keep a record of the weight of chemical fed. Many volumetric dry feeders have recorders available as an accessory (along with the scales).

Alarms. To prevent underfeeding or even loss of feed, alarm systems can be included in either solution or dry feed systems. The alarm alerts the operator when the solution level in the day tank is low or when a new bag of dry chemical should be put into the hopper. An alarm can also signal that the water supply to a saturator or dissolving tank has either stopped or diminished. The alarms are triggered by level switches, flow switches, or pressure switches.

Vacuum breakers. Any time there is a water connection to a chemical solution, there is the possibility of a CROSS CONNECTION and BACKFLOW. This can result in a concentrated solution being drawn into the treated-water system. A cross connection can occur in the supply line to a dissolving tank or in the discharge of a solution feeder.

The simplest way to prevent backflow is to provide an AIR GAP in the line—a vertical separation between the water line and tank or device receiving the water (Figures 8-8 and 8-11). When pressure must be maintained in the system, a VACUUM BREAKER must be used (pressure is lost in an air-gap system). A vacuum breaker, also known as an ANTI-SIPHON DEVICE, is a valve that is kept closed by water pressure. When the water pressure fails, the valve of the vacuum breaker opens to the atmosphere, allowing air rather than potentially hazardous solutions to be drawn into the system.

The vacuum breaker should be installed as close to the chemical solution as possible. Vacuum breakers must be installed on water make-up lines to upflow saturators and dry feeder solution tanks. In some states, mechanical vacuum breakers are not permitted and air gaps must be used. In such cases, repumping may be necessary to recover lost pressure.

8-3. Operation of the Fluoridation Process

To insure uninterrupted and unvarying feeding of fluoride chemicals, proper operation and maintenance of the equipment is required. The following paragraphs give general guidelines for operation and maintenance. For details of an individual system, the manufacturer's manual should be consulted.

Dry feeders. In addition to making sure the hoppers have sufficient chemicals, the operator must inspect and clean dry feeders routinely to prevent costly breakdowns. The belts, rolls, and discs or screws (depending on the type of feeder) must be regularly inspected for signs of wear. Replacements should be made before the equipment fails. A lubrication schedule should be established and followed based on the manufacturer's recommendations. The dissolving tanks should be inspected and cleaned when precipitate buildup is evident. If the deposits are a result of hardness, softening the make-up water should be considered. If softening is not practical, frequent cleaning is essential.

The calibration of dry feeders should be checked occasionally, particularly if maintaining a constant fluoride concentration is a problem. To do this, a small set of scales and a stop watch or a watch with a sweep second hand are needed. To begin the calibration procedure, a shallow pan or sheet of cardboard is inserted between the feeder's measuring mechanism and dissolving chamber to collect the chemical being fed. The chemical is then collected at short intervals, such as five periods of 5 min each. Each of the amounts collected is weighed and totaled. The individual samples indicate the uniformity of feed, which should show less than a 10 percent variation. The total is used to check the measured feed rate against the setting on the feeder.

Solution feeders. Saturators are the most common solution feed system. In addition to maintaining the pump, the operator must make sure that at least 6 in. (0.15 m) of chemical are in the saturator to insure that a constant 4 percent solution is maintained. If more than 1000 gpm (3780 L/m) of water are being treated, the minimum depth of chemical is 10 in. (0.25 m). Lines drawn on the outside of the translucent containers can help determine when to add chemical.

When using a downflow saturator, only crystalline grade sodium fluoride should be used—powdered sodium fluoride will quickly clog the gravel bed. Any form of sodium fluoride can be used with upflow saturators, but crystalline grade is often used since less dust is produced. Sodium silicofluoride should never be used with either type of saturator since it will not dissolve to produce the constant 4 percent solution.

The saturator tank (and the gravel bed in a downflow saturator) should be cleaned from one to three times a year depending on the scale buildup due to water hardness.

The pump is the key component of a solution feed system. Diaphragm pumps are most often used. They have proven to be quite reliable, but routine maintenance according to manufacturer's recommendations is essential. The pump head should be dismantled for cleaning and inspecting the check valves and the diaphragm. If the valves are worn, they should be replaced. If the diaphragm is cracked, it should be replaced.

Fluoride injection point. Regardless of the feed system used, the injection point can cause operational problems. The injection point should be located after the filtration process to reduce any fluoride loss. If possible, a point at which all water passes at a fixed rate should be selected. When adding water treatment chemicals containing calcium (lime or calcium hypochlorite), the fluoride injection point should be as far away as possible to prevent precipitation. The injector must be cleaned one to three times a year to prevent blockage caused by scale formation.

Chemical storage. Chemical storage areas must be kept clean and orderly. Poor storage, shown in Figure 8-15, can cause safety hazards and loss of chemicals. The storage area for fluoride chemicals should be isolated from areas used to store other chemicals to prevent mix-ups. Bags of dry chemicals should be piled neatly on pallets as close to the feeding equipment as possible (Figure 8-16). Whenever possible, whole bags should be emptied into hoppers, since partially empty bags are difficult to store without spillage and dust. Spills should be cleaned immediately. Metal and fiber drums should also be stored on pallets and kept tightly closed. Empty chemical containers should never be used for any purpose. They should be disposed of in a manner acceptable to local or state health authorities.

8-4. Common Operating Problems

Some of the problems commonly encountered in fluoridation include varying fluoride readings, low fluoride readings, and high fluoride readings.

Variable fluoride readings. Fluoride levels in the distribution system will vary considerably just after the start of a fluoridation program. This is caused primarily by the dilution of fluoridated water with the unfluoridated water held

Figure 8-15. Poor Storage of Chemicals

Figure 8-16. Proper Chemical Storage

in storage tanks prior to the start of fluoridation. The fluoride levels will drop whenever water is drawn from storage and return to normal whenever storage is being refilled. Depending on the amount of water in the storage tank or reservoir and on the amount actually used each day, the dilution effect can last from several weeks to several years. Usually this problem disappears as the tanks eventually become filled with fluoridated water. Varying readings may indicate that a dry feeder needs recalibration. A variation of 0.2 to 0.3 mg/L over a two- to three-day period is not in itself a serious problem nor does it create a health hazard. However, an operator should investigate the variation because it may indicate a more serious problem.

Low fluoride readings. When the fluoride concentration measured by laboratory testing is consistently lower than the calculated concentration, a number of problems may be indicated.

Assuming the calculations and the weight and flow data that they are based on are correct, the problem may be an interference in the laboratory test procedure. For example, aluminum, introduced by alum coagulation, interferes noticeably with the ALIZARIN-VISUAL TEST method and causes erroneously low readings. Checking the results using the electrode method or comparing results with the state or local health department will determine if this is the cause of the problem.

The most common cause of low readings is underdosing due to inadequate chemical depth in a saturator or incomplete mixing in a dissolving tank. Deposits of undissolved chemical in the dissolving tank of a dry feeder indicate incomplete mixing. This can be due to inadequate baffling or inadequate make-up water flow rate. As the fluoride is dissolved, a high reading may result.

If the fluoride level is low in a sample from the distribution system, unfluoridated water may be mixing with the treated water. In some systems, ground-water sources are used periodically to supplement normal flows from the treatment plant. Distant wells may be used periodically to supply water to meet peak needs. The water produced from these wells is often disinfected at the well head and pumped directly into the distribution system. If the ground water is low in natural fluoride, it will lower the treated water's fluoride concentration and cause a low reading. If fluoride is added before filtration, significant losses will occur that will result in low readings. Either the addition point should be moved downstream of the filters or additional fluoride should be added to make up for the loss.

High fluoride readings. If testing indicates a fluoride concentration consistently higher than the calculated concentration, several problems may be indicated.

Polyphosphates (used for water stabilization) can cause high readings when using the SPADNS METHOD. This can be checked by using the electrode method or comparing results with the local or state health department.

Sometimes high readings occur because the natural fluoride in the water has not been measured or considered in the dosing calculations. Consequently, more fluoride than needed may be added. The natural fluoride level in surface-water supplies can vary greatly and should be measured daily so that the correct dosage can be calculated. The natural fluoride level in most ground-water supplies varies only slightly from month to month. However, a high reading could be caused if water from a well with high natural fluoride ground water is combined with fluoridated water.

8-5. Operational Control Tests and Record Keeping

Successful fluoridation requires (1) measuring the fluoride level before fluoridation so that the operator knows how much fluoride to add and (2) measuring the level after fluoridation to confirm that the correct amount of chemical is being added.

Daily control testing. The fluoride concentration of treated water should be measured daily. Samples should be taken at proper locations in the plant and at representative points in the distribution system. The operator can then make necessary adjustments to feeders for maintaining the proper fluoride concentration. If the water source is a surface supply, daily raw-water samples should also be tested since the natural fluoride concentration may vary. This usually is not necessary with a ground-water source since natural concentrations remain fairly constant.

Continuous control testing. Continuous control testing, performed by automatic monitors, provides the operator with an uninterrupted record of fluoride levels. Most continuous control testing is done with continuous or semicontinuous analyzers, which automatically perform the colorimetric or electrometric analyses. These devices analyze the fluoride level of a flowing stream of water. The results are often displayed on an indicating-type meter and simultaneously recorded on circular or strip-chart recorders.

Some water plants monitor both raw water and treated water. The raw-water monitor alerts the plant operator to changes in the raw-water fluoride level, which necessitate changes in the fluoride solution feed rate.

Monitors equipped with special controller equipment and circuitry can be used to automatically control fluoride solution feed rates. The monitor measures the fluoride ion concentration in the treated water. This concentration is subtracted electronically from a desired concentration preset into the equipment. If the comparison shows the fluoride level to be too high, the controller automatically decreases the solution feed rate. If the fluoride level is too low, the controller increases the solution feed rate. With any monitor, even one used as a controller, water plant personnel still need to manually sample water and perform fluoride analyses at routine intervals. These routine spot checks serve to verify the proper functioning and accuracy of the monitor.

Besides the results of daily or continuous control testing, daily records of the weight of fluoride chemicals fed and the volume of water treated should be kept. From these records, the theoretical concentration can be calculated and compared to the actual concentration on a daily basis.[2] This provides an additional check on the adequacy of the treatment process.

---

[1] *Basic Science Concepts and Applications*, Chemistry Section, Dosage Problems (Fluoridation Calculations).

8-6. Safety and Fluoridation

Fluoride chemicals at the water treatment plant present a potential health hazard to water plant operators. Operators are exposed to high fluoride levels since they handle the concentrated dry chemicals and solutions. Overexposure to these chemicals can result from ingestion, inhalation, or bodily contact. However, properly handling fluoride chemicals and using adequate safety equipment will minimize health hazards. These same practices are recommended for all water treatment chemicals.

Ingestion. The most common way to ingest fluoride chemicals is through contaminated food or drink. Fluoride chemical may be mistaken for sugar or salt if meals are eaten in areas where fluorides are stored or applied.

To prevent fluoride ingestion, no one should eat, drink, or smoke where fluorides are stored, handled, or applied. Personnel handling fluoride chemicals should not put their hands to their faces unless they have washed thoroughly and are in an area designated specifically for eating, drinking, or smoking.

Inhalation. Accidental inhalation of dust is the most common type of overexposure. Masks and other protective devices are strongly recommended and the causes of dust should be minimized. Bags should not be dropped, an even slit should be made at the top of the bag to avoid tearing down the side, and the bag's contents should be poured gently into hoppers. In addition, crystalline chemicals and bag-loading hoppers can be used to reduce dust in the air. Good ventilation is absolutely necessary in work areas, even if there is no visible dust and masks are worn.

Bodily contact. The operator should receive detailed safety instructions before handling fluoride compounds. To avoid bodily contact with fluorides the following should be worn when handling the chemicals:

- Chemical goggles
- Respirator or mask approved by the National Institute of Occupational Safety and Health
- Rubber gloves with long gauntlets, rubber apron, and rubber boots.

In addition, the operator should wear clothing that covers the skin as completely as possible. Any protective covering needed to cover open cuts or sores should also be worn.

Acid Handling

Hydrofluosilicic acid requires special handling. Spilling acid on the skin, splashing it into the eyes, and inhaling its vapors are all serious hazards. The absorption of fluoride through the skin is not a hazard, but hydrofluosilicic acid is corrosive and constitutes a hazard in skin contact. Careful handling, quick rinsing of spills or splashes, use of respirators, and adequate ventilation are the best safety measures. To protect personnel from acid spray, acid pumps should be furnished with clear plastic shields around packing glands and revolving parts.

Selected Supplementary Readings

About Fluoridation—The Community Way to Reduce Tooth Decay. Channing L. Beter Co., Inc., Greenfield, Mass. (1970).

Fluoridation Facts—Answers to Questions About Fluoridation. Amer. Dental Assoc., Chicago, Ill. (1974).

Hansen, Ed. Fluoride Compounds Used to Prevent Tooth Decay. *OpFlow*, 9:5:6 (May 1983).

Introduction to Water Quality Analyses. AWWA, Denver, Colo. (1982). Module 5.

Maintenance of Sodium Fluoride Saturator Installations. *Jour. AWWA*, 65:9:609 (Sept. 1973).

Manual of Instruction for Water Treatment Plant Operators. New York State Dept. of Health, Albany, N.Y. (1975). Chap. 16.

Safety Practices for Water Utilities. AWWA Manual M3, AWWA, Denver, Colo. (1983).

Saferian, Sam. Fluoride—Handling Safety. *Jour. AWWA*, 64:9:604 (Sept. 1972).

Water Fluoridation Principles and Practices. AWWA Manual M4. AWWA, Denver, Colo. (1977).

Water Quality and Treatment. AWWA Handbook. McGraw-Hill Book Company, New York (3rd ed., 1971). Chap. 12.

Water Works Operators Manual. Alabama Dept. of Public Health, Montgomery, Ala. (3rd ed., 1972). Chap. 22.

Glossary Terms Introduced in Module 8

(Terms are defined in the Glossary at the back of the book.)

Air gap
Alizarin-visual test
Anti-siphon device
Backflow
Cross connection
Diaphragm pump
Fluorosis
Gravimetric dry feeder
Hydrofluosilicic acid
Ion
Ion-exchange water softener
Manual solution feed
Mottling
Piston pump
Positive-displacement pump
SPADNS method
Saturator
Sodium fluoride
Sodium silicofluoride
Vacuum breaker
Volumetric dry feeder

Review Questions

(Answers to Review Questions are given at the back of the book.)

1. What is the purpose of fluoridation?

2. What variable determines the optimum fluoride ion concentration in drinking water for a given community?

3. How does the factor in Question 2 relate to the amount of fluoride consumed each day?

4. What is the difference between a natural fluoride ion and that added at the treatment plant?

5. What is dental fluorosis?

6. What is the relationship between the optimum fluoride levels and the fluoride MCLs set by the USEPA drinking water regulations?

7. What are the three commonly used fluoride compounds?

8. What are the two major types of feeding systems?

9. Compare the two types of dry feeders.

10. Sodium fluoride has an unusual solubility characteristic. What is it and why is it important?

11. What is the weight of fluoride ions contained in 50 lb of NaF, 95 percent pure?

12. What is the percentage by weight of fluoride ions contained in Na_2SiF_6, 98.5 percent pure?

13. A particular treatment plant is fluoridating at a rate of 24.0 lb NaF per day. The NaF is put into solution using a simple saturator-type dissolver. What is the required solution feed rate in gallons per hour?

14. Discuss important considerations in locating the fluoride injection point.

15. What records should be kept for the fluoridation process? Why?

16. How can an operator reduce the potential for accidental ingestion of fluoride?

17. What personal protection equipment should an operator wear when handling fluoride compounds?

Study Problems and Exercises

1. Prepare a complete listing of fluoridation equipment and components in use at a local water treatment plant (include equipment name and manufacturer's name). Include a sketch showing the major pieces of equipment involved. Identify chemical storage areas.

2. Based on the type and purity of the fluoride compound delivered to the local treatment plant, calculate the following:
   - Percent of F^- ion contained by weight
   - Pounds of F^- ion per gallon of solution
   - Amount of solution needed to raise level of F^- from the natural level of 0.3 mg/L to a treated level of 1.1 mg/L.

3. The feed rate of a gravimetric dry feeder was checked using the method described in the module. The following results were obtained. The feeder was set to feed 0.925 lb/hour. The weights of sodium silicofluoride collected in 5 min periods were:
   37 g
   42 g
   48 g
   31 g
   42 g
   What does this tell you about the uniformity of feed and the feed rate of the feeder?

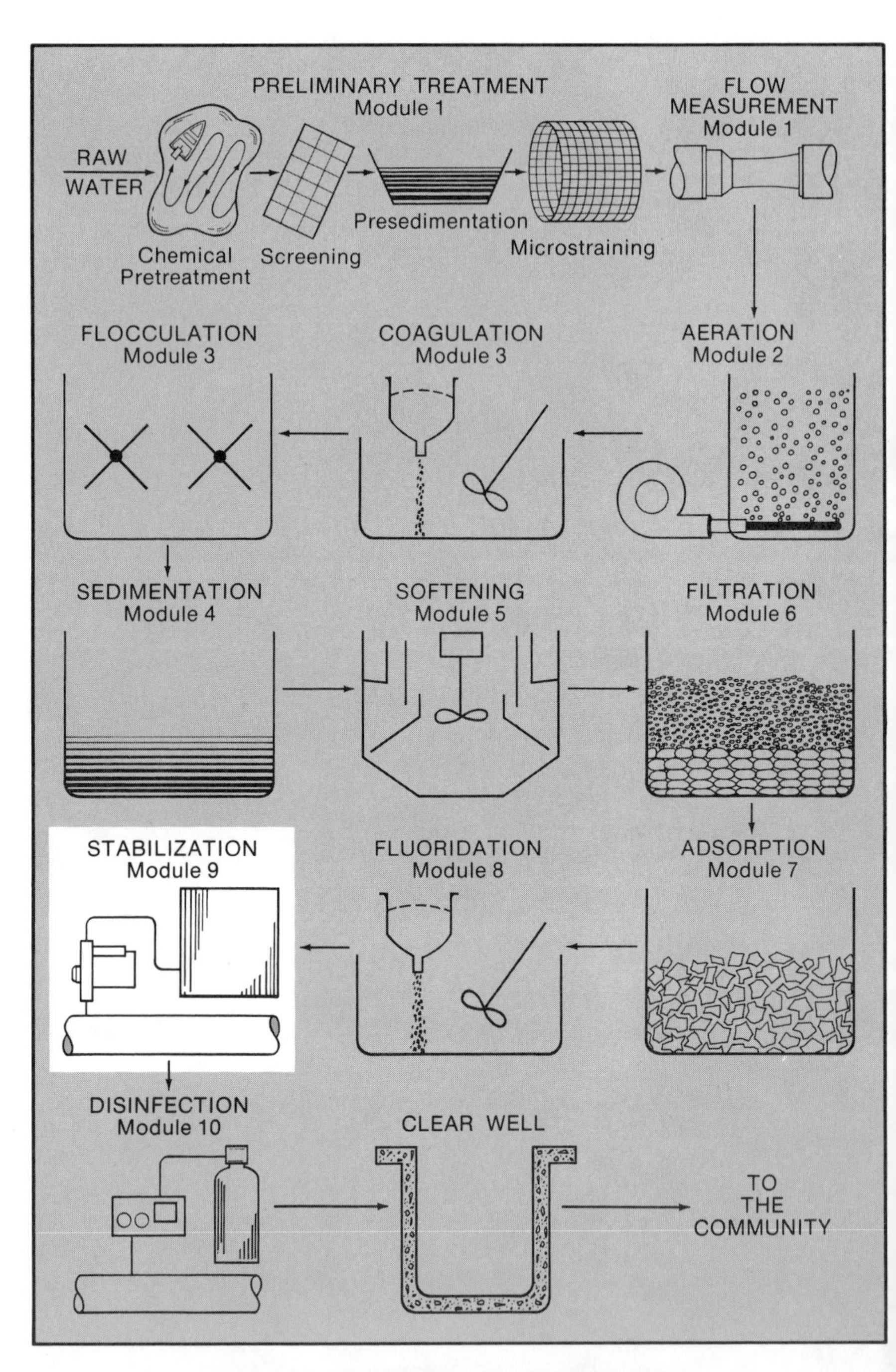

Figure 9-1. Stabilization in the Treatment Process

Module 9

Stabilization

Many waters either cause corrosion or deposit scale on pipelines and plumbing fixtures. Such waters are said to be UNSTABLE. The treatment process used to reduce or eliminate the problems of corrosion and scaling is STABILIZATION. As shown in Figure 9-1, stabilization is normally one of the last processes in the treatment plant. Water stabilization requires a basic understanding of the chemistry of the treatment processes used in the plant and an ability to monitor and control treated water chemistry within relatively narrow limits. Monitoring and control at the plant and in the distribution system are essential to the success of stabilization.

After completing this module you should be able to

- Describe the purpose of stabilization.
- Identify the primary problems caused by unstable waters.
- Identify the different operating methods practiced in controlling unstable water.
- Discuss major operating considerations associated with stabilization processes.
- Identify typical operating problems and solutions.
- Describe and perform the common operational control tests recommended for monitoring stabilization processes.
- Recommend appropriate record-keeping procedures essential to process control.
- Explain the basic safety considerations related to the stabilization processes.

9-1. Description of Stabilization

Purpose of Stabilization

The purpose of stabilization is to control both corrosive and scale-forming tendencies of drinking water before the water enters the distribution system. Distributing unstable water can cause problems related to public health, aesthetics, and economics.

Public health problems. Corrosive water can dissolve toxic metals into the drinking water from distribution and household plumbing systems. Lead and cadmium are the metals most likely to be a problem, since they are commonly present in household plumbing. Lead service lines are common in older water systems and lead solder is used to join copper pipes in household plumbing. Cadmium is a constituent of galvanized piping. Corrosive water in contact with either lead or cadmium can result in concentrations of metals that pose a threat to public health.

Corrosion of cast-iron mains in the distribution system can cause the formation of iron deposits, called TUBERCULES, in the mains. These deposits protect bacteria and other microorganisms from residual chlorine, allowing large populations of microorganisms to thrive. Pressure or velocity changes of the water in the main can cause the microorganisms to be released into the water, creating a potential for disease outbreaks. Some types of bacteria can also accelerate the corrosion process.

Aesthetic problems. Color, taste, and odor problems can all result from corrosive water attacking metal pipes in the distribution and household plumbing systems. RED WATER problems can occur when iron is dissolved from cast-iron mains by corrosive water. The dissolved iron stains plumbing fixtures and laundry and makes water unappealing. The dissolved iron also acts as a food source for a group of microorganisms, called IRON BACTERIA, which can cause serious taste and odor problems. Corrosion of copper pipe can result in water with a metallic taste and can cause blue-green stains on plumbing fixtures and laundry.

Economic problems. Unstable water results in significant costs to water systems and customers each year. Aggressive water can significantly reduce the life of valves and unprotected metal or asbestos–cement pipe in the utility distribution system, as well as affecting the service life and performance of consumer's plumbing systems and hot-water heaters. Buildup of corrosion products (tuberculation) or uncontrolled scale deposits can seriously reduce pipeline capacity and increase resistance to flow. This, in turn, reduces distribution-system efficiency and substantially increases pumping costs. If scale deposits or tuberculation go unchecked, pipes may become completely plugged, requiring expensive repair or replacement. Scaling can also cause operation of hot-water heaters and boilers to become more expensive due to reduced volume and heating capacity (Table 9-1).

In many systems, monitoring and control of water stability has been neglected because personnel are unaware of the extensive damage that can result from

Table 9-1. Estimated Effect of Scale on Boiler Fuel Consumption

Scale Thickness in.	*Fuel Consumption* Increase
1/50	7%
1/16	18%
1/8	39%

unstable water. In most cases, the costs of monitoring and stabilizing the water are outweighed by the savings associated with more efficient system operation and reduced maintenance.

The US Environmental Protection Agency has established a primary drinking water regulation that requires monitoring to determine water stability. If the water is found to be corrosive, the regulations suggest that a control plan be developed. In addition, the Secondary Regulations of the Safe Drinking Water Act indicate that drinking water should be noncorrosive.

Chemistry of Water Stability

The stability of treated water depends on several chemical characteristics. To understand how these characteristics can be adjusted to control stability, it is necessary to understand the basic principles of corrosion and scale formation.

Corrosion. CORROSION can be broadly defined as the wearing away or deterioration of a material due to chemical reactions with its environment. The most familiar example of this is the formation of rust—oxidized iron—when an iron or steel surface is exposed to moisture. Corrosion is usually distinguished from EROSION, the wearing away of material due to physical causes such as abrasion. Water that promotes corrosion is known as CORROSIVE or AGGRESSIVE water.

In water treatment operations, corrosion can occur to some extent with almost any metal that is exposed to water. Concrete and asbestos–cement can also be affected by aggressive water, which will dissolve the cement. However, this process involves leaching rather than the electrochemical corrosion reactions that take place with metal pipe. Whether corrosion of a material will be so extensive as to cause problems depends on several interrelated factors, including the material involved, the chemical and biological characteristics of the water, and the electrical characteristics of the material and its environment. The interrelation of these factors, as well as the process of corrosion itself, is quite complex. As a result, it is difficult to make general statements about what combinations of water and equipment will or will not have corrosion problems. The discussions in this module cover only basic principles; the operator faced with persistent corrosion in a given installation may require the assistance of corrosion-control specialists.

Chemistry of corrosion. The chemical reactions that occur in the corrosion of metals are similar to those that occur in an automobile battery. In fact, corrosion generates an electric current, which flows through the metal being corroded. The chemical and electrical reactions that occur during CONCENTRATION CELL CORROSION of iron pipes are illustrated in Figure 9-2.

In Figure 9-2A, minor impurities and variations in the metal (found in all metal pipe) have caused one spot on the pipe to act as an electrical ANODE in relation to another spot that is acting as an electrical CATHODE. At the anode, atoms of iron (Fe^{+2}) are breaking away from the pipe and going into solution in the water. As each atom breaks away, it ionizes by losing two electrons, which travel through the pipe to the cathode.

In Figure 9-2B, chemical reactions within the water balance the electrical and chemical reactions at the anode and cathode. Many of the water molecules (H_2O) have dissociated into H^+ ions and OH^- radicals—this is a normal condition, even with totally pure water. The Fe^{+2} released at the anode combines with two OH^- radicals from dissociated water molecules to form $Fe(OH)_2$, ferrous hydroxide. Similarly, two H^+ ions from the dissociation of the water molecules near the cathode pick up the two electrons originally lost by the iron atom, then bond together as H_2, hydrogen gas.

The formation of $Fe(OH)_2$ leaves an excess of H^+ near the anode, and the formation of the H_2 leaves an excess of OH^- near the cathode. This change in the normal distribution of H^+ and OH^- accelerates the rate of corrosion and causes increased pitting in the anode area (the "concentration cell"), as shown in Figure 9-2C.

If the water contains dissolved oxygen (O_2), which most surface water does, then $Fe(OH)_3$, ferric hydroxide, will form (Figure 9-2D). Ferric hydroxide is common iron rust. The rust precipitates, forming deposits called tubercules (Figure 9-2E). The existence of tubercules further concentrates the corrosion, increasing both pitting at the anode and growth of the tubercule. Tubercules can grow into large nodules (Figure 9-3), significantly reducing the carrying capacity of the pipe. During rapid pressure or velocity changes, some of the $Fe(OH)_3$ can be carried away, causing "red water."

Factors affecting corrosion. The rate of corrosion depends on many site-specific conditions, such as the characteristics of the water and pipe material. Therefore, there are no established guidelines that determine the rate at which a pipe will be corroded.

Chemical reactions play a critical role in determining the rate of corrosion at both the cathode and anode. Any factor that influences these reactions will also influence the corrosion rate. Factors that affect the stability of water include:

- Dissolved oxygen
- Total dissolved solids (TDS)
- pH
- Alkalinity
- Temperature
- Type of metal used for pipes and appurtenances

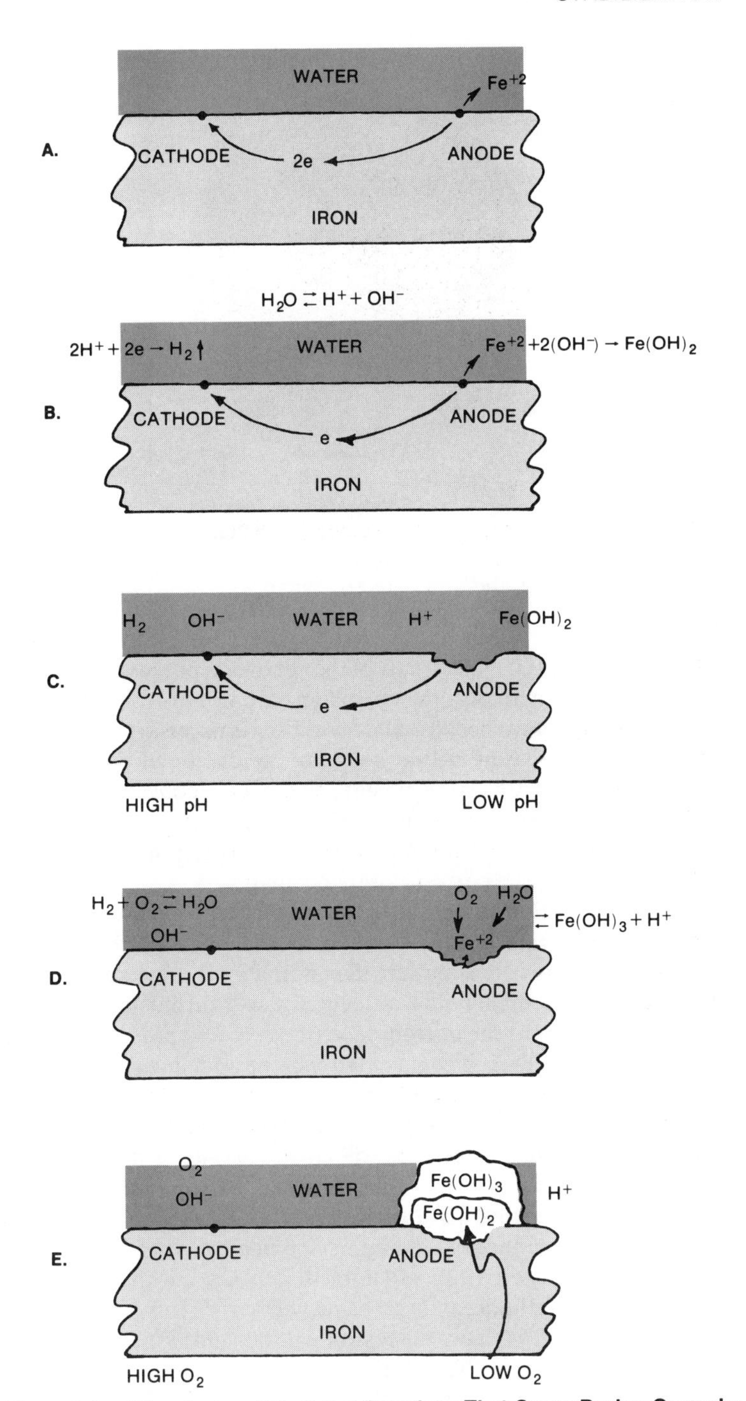

Figure 9-2. Chemical and Electrical Reactions That Occur During Corrosion

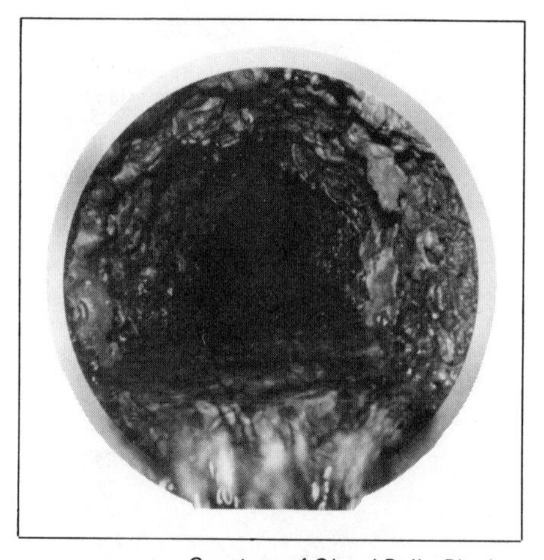

Courtesy of Girard Polly-Pig, Inc.

Figure 9-3. Tuberculated Pipe

- Electrical insulation and stray electric currents
- Bacteria.

Dissolved oxygen (DO) is a key part of the corrosion process; therefore, as the concentration of DO increases; the corrosion rate will also increase. The TDS concentration is important because electrical flow is necessary for the corrosion of metal pipe to occur. Pure water is a poor conductor of electricity since it contains few ions to carry electric current. As the TDS concentration increases, water becomes a better conductor thereby increasing the corrosion rate. Alkalinity and pH also affect the rate of chemical reactions. Generally, as pH and alkalinity increase, the corrosion rate decreases.

Since chemical reactions occur faster at higher temperatures, an increase in water temperature generally increases the corrosion rate. Flow velocity can increase or decrease the corrosion rate depending on the nature of the water. If the water is corrosive, higher flow velocities cause turbulent conditions that bring dissolved oxygen to the corroding surface more rapidly, which increases the corrosion rate. However, if chemicals are being added to stabilize the water, the higher velocities will decrease the corrosion rate by allowing the chemicals (such as $CaCO_3$) to deposit on the pipe walls more quickly.

Metals that easily give up electrons will corrode easily. Table 9-2 lists metals commonly used in water systems, with those at the top being most likely to corrode. This listing is called the GALVANIC SERIES of metals. Where dissimilar metals are electrically connected and immersed in a common flow of water, the metal highest in the galvanic series will immediately become the anode, the other metal will become the cathode, and corrosion will occur. This is termed GALVANIC CORROSION. The rate of galvanic corrosion will depend largely on how widely separated the metals are in the galvanic series—widely separated metals will exhibit extremely rapid corrosion of the anode metal (the metal highest in the series); the cathode metal will be protected from corrosion. A common example

Table 9-2. Galvanic Series For Metals Used in Water Systems

Corroded End (Anode)	MOST ACTIVE
Magnesium	+
Magnesium alloys	
Zinc	Corrosion Potential
Aluminum	
Cadmium	
Mild steel	
Wrought (black) iron	
Cast iron	
Lead-tin solders	
Lead	
Tin	
Brass	
Copper	
Stainless steel	−
Protected End (Cathode)	**LEAST ACTIVE**

of galvanic corrosion occurs when a brass corporation cock is tapped into a cast-iron main and attached to a copper service line. The copper will be protected at the expense of the brass and cast iron.

The electric current generated by corrosion is very slight, and insulation that interrupts the electrical current between dissimilar metals will prevent galvanic corrosion. However, when a strong current is passed through any corrodible metal, usually due to stray currents from improperly grounded household electrical systems, rapid corrosion will result.

Certain types of bacteria can accelerate the corrosion process because they produce carbon dioxide (CO_2) and hydrogen sulfide (H_2S) during their life cycles, which can increase the corrosion rate. They can also produce slime, which will entrap precipitating iron compounds and increase the amount of tuberculation and red-water problems. Two groups of bacteria cause the most problems. Iron bacteria, such as *Gallionella*, and *Crenothrix*, can form considerable amounts of slime on pipe walls particularly if the water has enough dissolved iron present (from the source water or from corrosion of the pipe material) to allow them to thrive. Beneath this layer, CO_2 production by the bacteria can significantly lower the pH, which will speed up the corrosion rate. The periodic sloughing of these slime accumulations can cause other major problems such as tastes and odors. The slimes can also prevent the effective deposition of a protective $CaCO_3$ lining by enmeshing it within the slime layer. As this layer sloughs away it carries away the $CaCO_3$ and leaves the pipe surface bare. Sulfate-reducing bacteria, such as *Desulfovibrio desulfuricans*, can accelerate corrosion when sulfate (SO_4) is present in the water. They reduce the SO_4 under anaerobic conditions (which occur under the slime layer where the oxygen is depleted by other bacteria) to form H_2S and iron sulfide (Fe_2S_2) causing obnoxious odors and a black-colored water. Carbon dioxide production can also lower the pH.

All of these factors interact with each other and the pipe material. This combination of factors and their influence on the corrosion reactions will help determine how quickly the corrosion proceeds. As the factors change, the corrosion rate will also change. The only factors over which the operator has significant control are pH, alkalinity, and the bacteriological content of the water. The techniques used to control these parameters are discussed later in this module.

Types of corrosion. Corrosion in water systems can be divided into two broad classes—localized and uniform. LOCALIZED CORROSION, the most common type in water systems, attacks metal surfaces unevenly, as shown in Figures 9-2 and 9-4. Localized corrosion is usually a more serious problem than uniform corrosion since it leads to a more rapid failure of the pipe. Two common types of corrosion that produce pitting are galvanic corrosion and concentration-cell corrosion, discussed in the preceding paragraphs. UNIFORM CORROSION takes place at an equal rate over the entire surface. It usually occurs where waters having very low pH and low alkalinity act on unprotected surfaces.

Scale formation. Scale-forming water can protect pipes from corrosion by depositing a layer of scale that separates the corrodible pipe material from the water. However, uncontrolled scale deposits can significantly reduce the carrying capacity of the distribution system (Figure 9-5).

Courtesy of Johnson Controls, Inc.

Figure 9-4. Pitting of Pipe

Courtesy of Johnson Controls, Inc.

Figure 9-5. Scaling of Pipe

Chemistry of scale formation. Scale is formed when the divalent metallic cations[1] associated with hardness, primarily magnesium and calcium, combine with other minerals dissolved in the water and precipitate to coat pipe walls. (See Module 5, Softening, for further discussion of hardness.) The most common form of scale is calcium carbonate ($CaCO_3$). Other scale-forming compounds include magnesium carbonate ($MgCO_3$), calcium sulfate ($CaSO_4$), and magnesium chloride ($MgCl_2$).

Factors affecting scale formation. Water can hold only so much of any given chemical in solution—if more of the chemical is added, it will precipitate instead of dissolve. The point at which no more of the chemical can be dissolved is called the SATURATION POINT. The saturation point varies with other chemical characteristics of the water, including pH, temperature, and total dissolved solids (TDS).

The saturation point of calcium carbonate depends primarily on the pH of the water. For example, if a water with a certain temperature and TDS concentration can maintain 500 mg/L of $CaCO_3$ in solution at pH 7, then the same water will hold only 14 mg/L of $CaCO_3$ in solution if the pH is raised to 9.4. Temperature also affects the saturation point, although not as dramatically as pH. The solubility of $CaCO_3$ in water decreases as temperature increases. The increased temperature in hot-water heaters and boilers causes major problems with scale, which significantly reduces their energy efficiency. Since the presence of other minerals in the water affects the solubility of $CaCO_3$, the TDS concentration must be known in order to determine the $CaCO_3$ saturation point. As the TDS concentration increases, the solubility of $CaCO_3$ increases.

Stabilization Methods

The basic methods used for stabilizing water or protecting the distribution system against unstable water are:

- Adjustment of pH and alkalinity
- Use of protective coatings
- Use of corrosion inhibitors and sequestering agents.

Selection of the method or methods to be used in a given application depends on the chemical characteristics of the raw water and on the effects of other treatment processes used. Discussion of the factors to be considered in selecting a method of stabilization is found in the Operation of the Stabilization Process section of this module.

Adjustment of pH and alkalinity. An increase in pH and alkalinity levels can prevent corrosion; a decrease can prevent scale formation. Lime is generally used to increase both pH and alkalinity since it is less expensive than other chemicals having the same effect. Soda ash (sodium carbonate) may be added along with the lime to further increase the alkalinity. Sodium bicarbonate is often used instead of sodium carbonate since it will also increase alkalinity

---

[1] *Basic Science Concepts and Applications*, Chemistry Section, The Structure of Matter—Atomic Structure.

without as much increase in pH. The increased alkalinity buffers the water against pH changes in the distribution system. This has proven particularly effective in controlling corrosion of lead and copper service pipes. Instead of lime, caustic soda with soda ash or sodium bicarbonate can also be used to increase pH and alkalinity.

As discussed in Module 5, Softening, lime-softened water can cause severe scale problems if it is not stabilized. Stabilization after softening is accomplished by adding carbon dioxide (recarbonation) or sulfuric acid. Both chemicals lower the pH so calcium carbonate will not precipitate in the distribution system.

Use of protective coatings. Since corrosion attacks the surface of the pipe, a protective coating on the pipe surface can inhibit corrosion. A lining of cement or asphaltic material is commonly applied to cast-iron and other distribution pipes during the manufacturing process. Although these coatings form an effective barrier against corrosion, aggressive water can attack the coatings or reach the pipe where gaps in the coating occur. As a result, many systems apply an additional protective coating by controlling the chemistry of the water flowing through the pipes.

A common protective-coating technique is to adjust the pH of the water to a level just above the saturation point of calcium carbonate. When this level is maintained, calcium carbonate will precipitate and form a protective layer on the pipe walls. This process must be closely controlled—a pH that is too low may result in corrosion, and a pH that is too high may result in excessive precipitation, causing clogging of service pipes and increased flow restriction in the distribution system. Lime, soda ash (sodium carbonate), sodium bicarbonate, or sodium hydroxide can be used to raise the pH level. Lime is often used since it also adds needed calcium (hardness) and alkalinity.

Use of corrosion inhibitors and sequestering agents. If waters do not contain enough calcium or alkalinity to make the formation of calcium carbonate coatings economical (water obtained primarily from snowmelt is an example, since it can have alkalinity and calcium concentrations as low as 2 mg/L as $CaCO_3$), then other chemical compounds can be used to form protective coatings. The most common compounds are polyphosphates and silicates. The chemical reactions by which these compounds combine with corrosion products to form a protective layer are not completely understood; however, the chemicals have proven successful in many systems.

Some polyphosphates can also be used as SEQUESTERING AGENTS to prevent scale formation. These compounds sequester, or chemically tie up, the scale-forming ions (primarily calcium and magnesium) so they cannot react to form scale. Since these compounds remain in solution they are eventually ingested by the consumers. Therefore, any sequestering agent selected must be suitable for use in drinking water.

Polyphosphates also sequester iron, whether it is dissolved in water from the source or from corrosion of the system. This prevents the precipitation of the iron compounds so red water will not result. However, this effect does not prevent corrosion—it merely prevents the corrosion by-products (iron in this case) from causing additional problems.

9-2. Stabilization Chemicals and Facilities

Whatever method or methods are used for stabilization, the facilities required for the process consist primarily of selected chemicals and the equipment required to store, handle, and feed these chemicals. A wide range of chemicals is available for stabilization. Some of these are primarily intended for industrial or other nonpotable applications. Therefore, all chemicals used to stabilize drinking water must be suitable for such use. Chemicals should also comply with AWWA standards to ensure quality. Table 9-3 lists the chemicals commonly used for stabilization of potable water.

Chemical Storage and Handling

Chemicals should be purchased in quantities that will assure on-site availability of a minimum 30-day supply. This practice will guard against interruption of service due to temporary shortages and other unforeseen events. In determining the quantity of chemical to be stored at any one time, the operator should consider storage space, discounts for purchasing large quantities, and manufacturer's shelf-life rating for the chemical.

Lime. Lime is available either in unslaked or slaked form. Unslaked lime, also called quicklime or calcium oxide (CaO), is available in a variety of sizes

Table 9-3. Chemicals Commonly Used For Stabilization of Potable Water

Treatment Method	*Chemical Name*	*Chemical Formula*
Increase pH and alkalinity	Lime	
	—Unslaked lime (quicklime)	CaO
	—Slaked lime (hydrated lime)	$Ca(OH)_2$
	—Sodium bicarbonate	$NaHCO_3$
	—Sodium carbonate (soda ash)	Na_2CO_3
	—Sodium hydroxide (caustic soda)	$NaOH$
Decrease pH and alkalinity	Carbon dioxide	CO_2
	Sulfuric acid	H_2SO_4
Formation of protective coatings	Sodium silicate (water glass)	$Na_2O(SiO_2)_n$*
	Sodium hexametaphosphate (sodium polyphosphate, glassy)	$(NaPO_3)_n \cdot Na_2O$†
	Sodium zinc phosphate	$(MPO_3)_n \cdot M_2O$‡
	Zinc orthophosphate	$Zn_3(PO_4)_2$
Sequestering agents	Sodium hexametaphosphate (sodium polyphosphate, glassy)	$(NaPO_3)_n \cdot Na_2O$†
	Tetrasodium pyrophosphate	$Na_4P_2O_7 \cdot 10H_2O$

*Typically n = 3.
†Typically n = 14.
‡M = Na and/or 1/2 Zn, typically n = 5.

from a powder to a pebble form. Although available in bags, it is usually delivered in bulk form because it is more economical and easier to handle. As a result, quicklime is generally only used by treatment plants that use considerable quantities, especially since it should not be stored more than three months due to deterioration. Quicklime is noncorrosive in the dry form and can be stored in steel or concrete bins. A minimum of two bins should be available to provide for maintenance and ease in unloading. The bulk deliveries are transferred from the rail cars or trucks to the bins by mechanical or pneumatic conveyors.

For treatment plants that do not use large quantities of lime, hydrated or slaked lime (calcium hydroxide—$Ca(OH)_2$) is more cost effective. Hydrated lime is a finely divided powder available in 50- or 100-lb (22.5- or 43-kg) bags and in bulk form. The bulk form is unloaded and stored like quicklime and should be used within three months. The bagged hydrated lime can be stored for up to one year without serious deterioration. Both types should be stored in a dry, well-ventilated area. This is particularly important for quicklime since water will start the slaking process, generating a tremendous amount of heat, which could start a fire if combustible materials are near.

Soda ash (sodium carbonate, Na_2CO_3). Soda ash—a white alkaline chemical—is available as granules and powder in bulk or in 100-lb (43-kg) bags. Since soda ash absorbs moisture and will readily cake, it should be stored in a dry, well-ventilated area. Dry soda ash is not corrosive, so it can be stored in steel or concrete bins.

Sodium bicarbonate ($NaHCO_3$). Sodium bicarbonate, more commonly known as baking powder, is an alkaline white chemical in a powdered or granular form. It is available in 100-lb (43-kg) bags or barrels up to 400 lb (181 kg). Sodium bicarbonate should also be stored in a cool, dry, well-ventilated area, since it decomposes rapidly as the temperature nears 100° F (38° C). Neither sodium bicarbonate nor sodium carbonate should be stored near acid-containing products. If acid comes into contact with either chemical, large quantities of carbon dioxide can be released creating a safety problem. Corrosion-resistant materials should be used for storing and transporting any sodium bicarbonate solution.

Sodium hydroxide (NaOH). Sodium hydroxide, or caustic soda, is available as a liquid or in dry form as flakes or lumps. Regardless of the form, sodium hydroxide must receive careful handling because of the many hazards involved. As a liquid, caustic soda is available in solution concentrations of 50-percent NaOH and 73-percent NaOH. The 50-percent solution will begin to crystallize if its temperature drops below 54° F (12° C) and the 73-percent solution will crystallize as its temperature drops below 145° F (63° C). Therefore, special storage procedures may be necessary to keep the temperature high enough to prevent the solution from crystallizing. This can be accomplished by further dilution or by installing heaters in the storage tanks. The manufacturer's recommendations should be followed closely to prevent problems. The storage tanks should be constructed of a nickel-clad or nickel-alloy steel, lined with caustic-resistant material such as rubber or PVC (polyvinyl-chloride plastic). Tanks made of PVC or fiberglass can also be used. If the dry form of caustic soda is used, it is usually dissolved immediately on delivery in a storage tank to a concentration where crystallization will not be a problem at local temperatures.

With either the dry or liquid forms, the addition of water for dilution generates a considerable amount of heat. Therefore, the rate of dilution must be carefully controlled so boiling or splattering of the solution does not occur. A source of flushing water should be readily available in case of spills. Additional safety precautions are discussed in the Safety and Stabilization section.

Sulfuric acid (H_2SO_4). Sulfuric acid is a corrosive, dense, oily liquid available in strengths containing 62-percent, 78-percent, or 93-percent H_2SO_4. It can be purchased in 55- or 110-gal (210- or 420-L) barrels or in bulk (tank car or truck). Sulfuric acid must be stored in corrosion-resistant tanks (such as PVC). If diluted, the acid must always be added to the water rather than the water to the acid to prevent severe splattering caused by the rapid generation of heat.

Sodium silicate (water glass). Sodium silicate is an opaque, syrupy, alkaline liquid available in 55-gal (210-L) barrels and in bulk. The barrels can be stored or the liquid can be added immediately to storage tanks. Bulk deliveries are unloaded directly into storage tanks. The tanks can be constructed of steel or plastic, since the chemical is noncorrosive.

Carbon dioxide (CO_2). Carbon dioxide is a colorless, odorless gas, which can be generated on-site or purchased in bulk in a liquified form. On-site generation involves producing the carbon dioxide in a furnace or in an underwater burner.

The storage units for liquified CO_2 are insulated, refrigerated pressure tanks. The tanks can range in capacity from 6000 to 100 000 lb (2700 to 45 000 kg) depending on the plant's requirements. The carbon dioxide is kept at about 0° F (−18° C) and 300 psig (2070 kPa, gauge) within the tank so it will remain in liquid form. Carbon-dioxide vapor forms above the liquid surface and is withdrawn and piped to the application point. The tank has a built-in vaporizer, which helps maintain a constant vapor supply. The storage tank should be located as close as possible to the application point, in a place convenient to delivery trucks.

Phosphate compounds. There are a number of phosphate compounds that can be used for stabilization. Three of the most common are sodium hexametaphosphate (sodium polyphosphate, glassy), sodium zinc phosphate, and zinc orthophosphate. Sodium hexametaphosphate is available in 50- or 100-lb (22.5- or 45-kg) bags or in 60- to 350-lb (27- to 159-kg) drums in a form that looks like broken glass. The other two compounds are generally available in dry form (50-lb [22.5-kg] bags) and in liquid form (55-lb [210-L] drums or in tank trucks). The concentrated liquid solutions are slightly acidic, so they should be stored in corrosion-resistant materials such as fiberglass, PVC, or stainless steel.

Chemical Feed Equipment

Lime. The type of equipment needed for lime feeding depends on the type of lime being used. If slaked lime is being used, a gravimetric or volumetric dry feeder adds the lime to a solution chamber where water is added to form a slurry

(Figure 9-6). Where unslaked lime is used, the dry feeder adds lime to a SLAKER, where the lime and water are mixed together to slake the lime ($CaO + H_2O \rightarrow Ca(OH)_2$) and form a slurry (MILK OF LIME). Module 5, Softening, describes slakers in more detail. The hoppers or bins supplying the dry feeders must be equipped with agitators, since both types of lime will pack during storage and not flow freely.

Regardless of the type of lime used, the slurry must be added to the water. It is best to minimize the distance from the feeders to the application point since lime slurry will cake on any surface. Open troughs or flexible lines should be used to carry the slurry to the application point, since they can be easily cleaned and flushed. Any solution feeders used with the slurry should be of the type that can be easily cleaned.

Soda ash and sodium bicarbonate. Soda ash and sodium bicarbonate can be fed with gravimetric or volumetric dry feeders. Agitators should be provided on the hoppers to prevent arching and interrupting the flow of the dry chemical into the feeder. Soda ash solutions can be fed using conventional solution feeders and lines. Since soda ash goes into solution slowly, larger dissolving chambers than usual, equipped with good mixers, are needed to provide proper detention time and mixing. Sodium bicarbonate solutions are caustic. Therefore, metering pumps, dissolving chambers, and pipes should be constructed or lined with caustic-resistant materials, such as stainless steel or PVC.

Sodium hydroxide. Sodium hydroxide solutions can be fed by metering pumps designed for handling corrosive and caustic solutions. All piping should also be of caustic-resistant materials such as PVC. No copper, brass, bronze, or aluminum in valves or fittings should be allowed to come into contact with caustic soda solutions, since these materials will deteriorate rapidly.

Sulfuric acid. Sulfuric acid can be fed directly from the shipping container with a corrosion-resistant metering pump and piping. When making up dilutions, always add the acid to the water in the dilution tank; violent splattering of the solution will occur if water is added to acid. The diluted acid

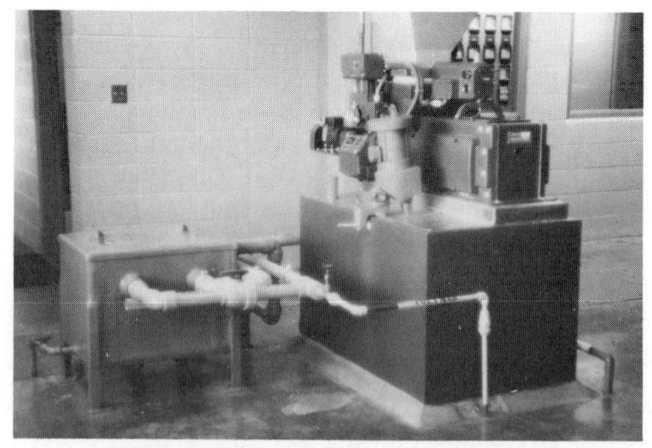

Figure 9-6. Lime-Feeding Equipment

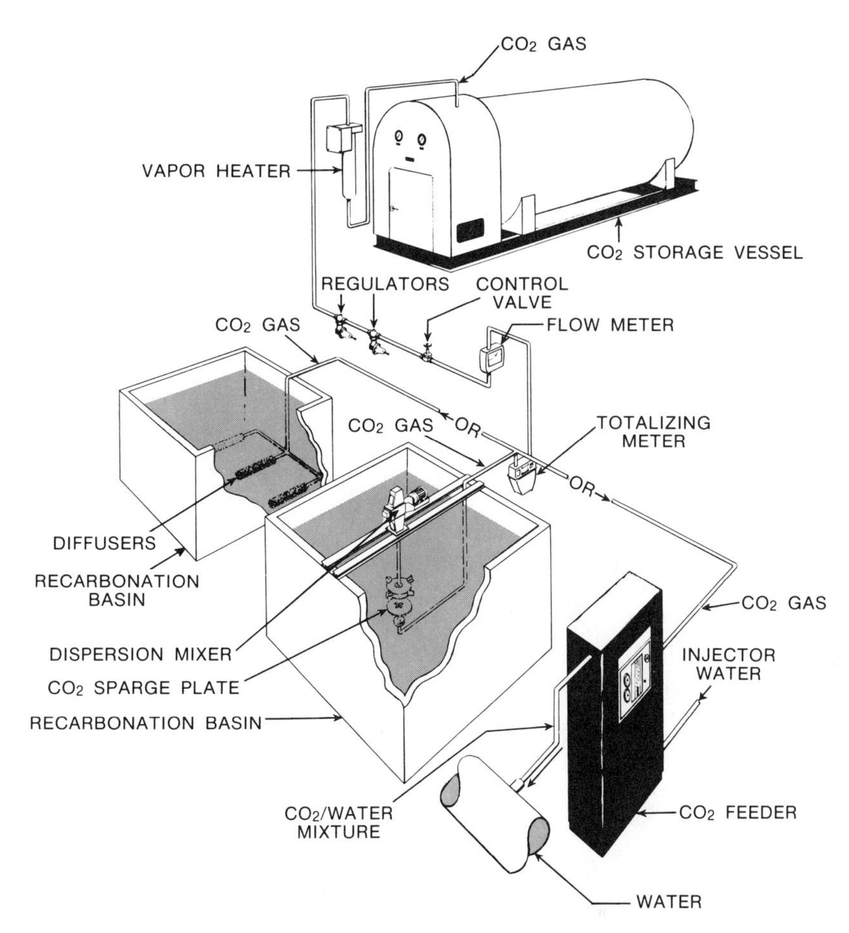

Courtesy of TOMCO₂ Equipment Co.

Figure 9-7. Liquid CO_2 Recarbonation System

solution is even more corrosive than the pure acid and must be handled carefully. Corrosion-resistant metering pumps, dilution tanks, and piping must be used.

Sodium silicate. Sodium silicate can be fed directly from the 55-gal (210-L) barrels or from storage tanks with a conventional metering pump and piping.

Carbon dioxide. If carbon dioxide is generated on-site, the feeding system consists of the generation equipment and related piping. The most typical generation system consists of a combustion unit located beneath the surface of the water in a reaction basin. The CO_2 is produced and mixed by the same unit. The submerged combustion generator is disscussed in Module 5, Softening. Another common system uses CO_2 gas drawn from a tank of liquid CO_2. The equipment necessary is shown in Figure 9-7. The CO_2 can be fed to a recarbonation basin or directly to a pipeline for pH control.

Phosphate compounds. Phosphate compounds are fed directly from storage tanks. If dry compounds are used they are usually dissolved in the storage tanks by placing the chemical in stainless steel baskets and suspending the baskets in the tanks. PVC or stainless steel piping, valves, and fittings should be used. Chemical metering pumps designed for corrosive materials are required.

9-3. Operation of The Stabilization Process

The day-to-day operation of the stabilization process consists mainly of operating and maintaining chemical feed facilities. However, for utilities not currently practicing stabilization, or where the results of existing stabilization procedures have been unsatisfactory, the first step is the selection of a stabilization process.

Selecting a Stabilization Process

The selection of a stabilization process should be based on a thorough review of information concerning the water system. The information considered should include water quality data, distribution system condition, the condition of home plumbing systems, and customer complaints. Careful consideration of this data will ensure an intelligent selection of stabilization methods.

Water quality data. Water quality data for the source water and for several points within the distribution system should be examined. This data should cover the following:

- pH
- alkalinity
- conductivity
- temperature
- iron
- hardness
- chloride
- sulfate
- fluoride
- color
- zinc
- copper
- lead
- cadmium
- dissolved oxygen
- carbon dioxide
- total dissolved solids

Data for the raw water should be compared with data for samples taken from points within the system. If water that has passed through the system shows a significant increase in constituents such as iron, copper, zinc, lead, or cadmium, then the water is probably causing corrosion and should be stabilized. Water quality data can be used to calculate stability indices or indicators. The most common of these is the LANGELIER INDEX (LI).[2] The LI provides an indication of whether the water is likely to form or dissolve calcium carbonate; it is not an exact measure of corrosion or scaling potential. The LI is calculated by the following equation:

$$LI = pH\ (actual) - pH_s$$

[2] *Basic Science Concepts and Applications*, Chemistry Section, Chemistry of Treatment Processes, Chemical Methods for Scale and Corrosion Control.

The pH (actual) is the measured pH of the water, and the pH_s is the theoretical pH at which the water will be saturated with calcium carbonate. The pH_s is calculated using a formula that takes into account calcium ion concentration, alkalinity, pH, temperature, and total dissolved solids. If the LI equals zero, the water is considered stable. If the LI is a positive value, calcium carbonate will precipitate and the water has scale-forming tendencies. If the LI is a negative value, calcium carbonate will be dissolved and the water has corrosive tendencies. (Additional information on LI calculation is given in *Introduction to Water Quality Analyses*, Physical/Chemical Tests Module, Calcium Carbonate Stability.)

Other stability indices can be calculated to indicate the corrosion potential of water. Table 9-4 shows how two of these indices compare to the Langelier Index. However, none of the indices is an exact measure of corrosion or scaling. Additional data from the system must be evaluated to determine the actual effect of the water.

Distribution system conditions. Ultimately, the purpose of stabilization is to prevent corrosion and scaling in the distribution system. Therefore, the physical condition of the distribution system gives the most accurate indication of the need for stabilization. The system's condition can be evaluated to some extent by examining records of main breaks and leaks. Patterns of deterioration or comments on the records may indicate problems with corrosion or scaling. An actual, physical inspection of the system's condition at various points should be performed before making any major changes to the stabilization process. Where convenient, the operator can examine sections of pipe or valves removed during repair, as well as pieces of pipe (coupons) removed during tapping.

Home plumbing and customer complaints. Local plumbers can give information on the condition of home plumbing systems. Since home plumbing consists of pipe with much smaller diameter and lesser wall thickness than the pipe in the distribution system, problems will show up faster. A review of customer complaints can also provide a good indication of stability problems. Complaints of red water, dirty water, and leaks can indicate corrosion. Loss-of-pressure complaints warn of excessive tuberculation or buildup of scale.

Evaluating system needs. An evaluation of the information detailed in the preceding paragraphs will indicate whether corrosion or scaling problems exist in the distribution system. The selection of the stabilization process should be based primarily on the problem identified. Other considerations should include

Table 9-4. Comparison of Common Stability Indices*

	Stability Index		
Stability Characteristics	*Langelier Index (LI)*	*Aggressive Index (A.I.)*	*Ryznar Index (R.I.)*
Highly aggressive	<−2.0	<10.0	>10.0
Moderately aggressive	−2.0 to <0.0	10.0 to <12.0	6.0 to <10.0
Nonaggressive	>0.0	>12.0	<6.0

*LI = $pH - pH_s$; A.I. = $pH + \log_{10}(A) + \log_{10}(Ca)$; R.I. = $2pH_s - pH = pH - 2(LI)$; pH = pH actual; pH_s = saturation pH; A = total alkalinity in mg/L $CaCO_3$; Ca = calcium hardness as mg/L $CaCO_3$.

costs, availability of the chemicals, ease of handling of the chemicals, and operator safety. The existing treatment processes used in the plant should be considered to ensure that the stabilization chemicals do not interfere with them. Pilot plant studies can be useful in selecting the most cost-effective approach.

Table 9-5 lists some treatment methods available for stabilization. In addition to treatment, corrosion control measures considered should include the use of corrosion-resistant materials, such as cement-lined ductile-iron pipe in the distribution system or copper and plastic piping in home and building plumbing systems.

Table 9-5. Summary of Treatment Techniques for Water Stabilization

Treatment	*Application*	*Effectiveness*	*Comments/Problems*
Prevent Corrosion			
Lime and sodium bicarbonate or lime	Increase pH Increase hardness Increase alkalinity	Most effective in water with low pH and hardness. Excellent protection for copper, lead, and asbestos-cement pipe in stabilized waters. Good protection for galvanized and steel pipe.	May be best overall treatment approach. Over-saturation may cause calcium deposits.
Sodium hydroxide	Increase pH	Most effective in waters with sufficient hardness and alkalinity to stabilize water. May provide adequate protection against lead corrosion in low-alkalinity, soft waters.	Should not be used to stabilize waters without the presence of adequate alkalinity and hardness. May cause tuberculation in iron pipes at pH 7.5–9.0.
Sodium hydroxide and sodium bicarbonate	Increase pH Adjust alkalinity	Most effective in waters with low pH and sufficient hardness. Excellent protection for lead corrosion in soft waters at pH 8.3.	Combination of high alkalinity and hardness with low pH is more effective than combination of high pH with low hardness and alkalinity.
Inhibition with phosphates (primarily sodium zinc phosphate and zinc orthophosphate)	Formation of protective film on pipe surfaces	Effective at pH levels above 7.0. Good protection for asbestos-cement pipe. Addition of lime may increase effectiveness of treatment for copper, steel, lead, and asbestos.	May cause leaching of lead in stagnant waters. May encourage the growth of algae and microorganisms. May cause red water if extensive tuberculation is present. May not be effective at low pH levels.

Table 9-5. Summary of Treatment Techniques for Water Stabilization *(continued)*

Treatment	*Application*	*Effectiveness*	*Comments/Problems*
Prevent Corrosion			
Inhibition with silicates	Formation of protective film on pipe surfaces	Most effective in waters having low hardness and pH below 8.4. Good protection for copper galvanized and steel pipe.	May increase the potential of pitting in copper and steel pipes. May not be compatible with some industrial processes.
Prevent Scale Formation			
Carbon dioxide or sulfuric acid	Decrease pH Decrease alkalinity	Effective with high-pH, high-alkalinity water such as lime-softened water.	Overfeeding can cause low pH and corrosion.
Sequestering with phosphates (primarily sodium hexametaphosphate and tetrasodium polyphosphate.	Sequester scale-forming ions.	Effective in controlling scale formation from lime-softened waters and iron in the source water.	Can loosen existing deposits and cause red-water complaints. Compounds lose sequestering ability in hot water heaters causing precipitation of $CaCO_3$ or iron.

Table 9-6. Alkalinity Adjustment by Chemical Treatment

Chemical 1 mg/L	*Alkalinity Change* as $CaCO_3$
Hydrated lime	Increase 1.35 mg/L
Soda ash	Increase 0.94 mg/L
Caustic soda	Increase 1.23 mg/L

Operation of Chemical Facilities for Adjustment of pH and Alkalinity

Table 9-6 shows the alkalinity adjustments possible using common pH/alkalinity-adjustment chemicals. The dosage in a given system depends on the stabilization goals identified as appropriate for that system's water.

Lime. The most common chemical used for pH and alkalinity adjustment is lime, because it is readily available, inexpensive compared to other chemicals, and relatively easy to feed. However, lime systems can create many maintenance problems if not operated properly. The problems generally relate to the scaling caused by the lime slurry. Because of the high pH of the slurry and the low solubility of lime in water, $CaCO_3$ will precipitate on anything the slurry touches. Solution chambers, piping, and pumps must be cleaned frequently to prevent clogging and equipment damage.

The lime slurry is usually added to the filtered-water conduits or into the clear well. It is best to minimize the length of lime slurry-feeding lines by having the feeder as close as possible to the application point. Pumps can be eliminated if the slurry can flow by gravity into a reaction chamber. However, for low rate of lime application, a suitable metering pump should be used. The parts of the pump that contact the slurry must be cleaned routinely to maintain accurate feed rates.

Caustic soda. Caustic soda can be used to increase both pH and alkalinity. The NaOH solution is fed by chemical metering pumps into a filtered-water conduit or the clear well. Since NaOH is a strong caustic chemical, the pumps, tanks, and feed lines should be routinely inspected for leaks to prevent safety hazards and damage to equipment. Any leaks should be repaired immediately and spills cleaned up promptly.

Other chemicals. Soda ash or sodium bicarbonate is often fed in combination with lime for additional alkalinity. Neither chemical poses difficult operation or maintenance problems. Carbon dioxide is used primarily for recarbonation of lime-softened water. The operation of this process is discussed in Module 5, Softening. Sulfuric acid, used to lower pH, requires the same precautions as caustic soda.

Operation of Processes Designed to Create Protective Coatings

The addition of lime (with soda ash or sodium bicarbonate in some cases) can be used to precipitate a $CaCO_3$ scale on the pipe walls in the distribution system. This coating of scale will protect the pipe from corrosion. If the process is not properly controlled, excessive scale can result, which will cause increased head loss and clogging of home plumbing systems.

For best results, the protective coating should be dense and provide uniform coverage. An acceptable coating can be produced by treated water that meets all of the following characteristics:

- Alkalinity and calcium concentrations are both maintained at 40 mg/L as $CaCO_3$ (minimum).
- The water is slightly oversaturated with $CaCO_3$ (4–10 mg/L over the saturation concentration).
- pH is held within the 6.8–7.3 range.

The stabilization process should never be operated with a widely fluctuating pH in an effort to dissolve previously deposited $CaCO_3$ or to try and maintain a certain thickness. Such attempts will only result in increased corrosion or scaling and operational problems.

Polyphosphates can also be used to form a protective coating on the pipe walls. Sodium zinc phosphate and zinc orthophosphate are typically used for this purpose. The dosage will depend on local conditions but will generally range between 0.5 and 3 mg/L. If sodium hexametaphosphate is used, the dosage may range up to 10 mg/L. With all of the compounds, a somewhat higher dose is applied for the first few weeks to help spread the chemical throughout the system.

The polyphosphate solution is usually transferred from the bulk storage or mixing tank to a day tank sized to store about one day's feeding requirement. Phosphate is a bacterial nutrient, and bacterial growth in the storage tanks can be a problem. A small amount of chlorine added to the storage tanks will inhibit bacterial growth. The solution is pumped from the day tank and added to the water just after filtration or in the clear well through a diffuser.

Sodium silicate can also be used to form a protective film and is fed in much the same way as the polyphosphate compounds. The initial dose used is generally 15–30 mg/L, which is then reduced to 5–10 mg/L as a continuous dose.

Sequestering Agents and Corrosion Inhibitors

Sodium hexametaphosphate can also be used to sequester scale-forming compounds such as $CaCO_3$ and $Fe(OH)_3$. The operating procedures are the same as those used to form coatings. Dosages range from 0.5 to 10 mg/L.

9-4. Common Operating Problems

Scaling or corrosion of plant equipment, discussed in the preceding section, is the major in-plant operating problem. Problems within the distribution system include excessive scaling and persistence of red-water problems, even after stabilization.

Excessive Scaling

The chief cause of excessive scaling is poor control of the stabilization process, especially when lime (and in some cases soda ash or sodium bicarbonate) is being added to form a $CaCO_3$ coating. If the pH is not kept in a narrow range near the saturation pH, excessive calcium carbonate can precipitate and cause head loss and clogging of household plumbing. The same problems can occur if the pH of lime-softened water is not adjusted down to near the saturation pH, either by recarbonation or addition of sulfuric acid. Proper monitoring of the treated water and necessary adjustments to the chemical feeders can prevent the scaling from becoming a problem.

A similar scaling problem can result when lime is added in conjunction with alum. This is done to make sure the pH is in the proper range (6.0–7.8) for effective alum coagulation. However, if too much lime is added and the pH goes above 7.8, the alum will remain in solution and pass into the distribution system, where it will precipitate and form a scale. This problem can be prevented by monitoring the pH and closely controlling the amount of lime fed.

Persistence of Red-Water Problems

A frustrating problem is the persistence of red water and other corrosion-related problems even when stabilization is being practiced. This can occur for a number of reasons, but in most cases the problems relate to poor flow velocity, tuberculation on the pipe surface, and the presence of iron bacteria.

Flow velocity. Flow velocity in the distribution system plays an important role in stabilization. There must be enough velocity to carry the chemicals throughout the system and bring them into contact with the pipe surfaces. Where the velocity is low, such as in dead ends and in areas where water use is low, the stabilization chemicals will not form a protective coating or react with the corrosion by-products. A regular main-flushing program should concentrate on such problem areas to prevent the buildup of corrosion by-products until the dead ends and other problem areas can be eliminated.

Tuberculation. Problems can occur if the mains have considerable tuberculation. The rough, uneven surface will prevent a uniform protective coating from forming. As velocity and pressure changes occur, the tubercules can break away, which will also destroy the protective layer. The addition of polyphosphates can actually increase the problems since they can loosen the deposits and cause them to break away from the pipe. A better approach would be to clean the mains and then start the stabilization process.

Iron bacteria. Iron bacteria can interfere with stabilization. The slimy covering that these bacteria produce can prevent the stabilization chemicals from reaching the pipe surface. Any protective layer becomes stuck to the slime and is destroyed as the slime sloughs off during pressure and velocity changes. The pipes must be cleaned to rid them of the slime deposits. In addition, proper disinfection to prevent the bacteria from thriving in the pipe must be performed. Once these measures are accomplished, stabilization can be effective.

9-5. Operational Control Tests and Record Keeping

Once a stabilization program is started, monitoring and record keeping are essential to control the process, to determine if it is effective, and to ensure that public health is being protected from high levels of corrosion products such as lead. The monitoring program should consist of water quality analyses along with pipe and coupon testing to determine the effectiveness of the stabilization process.

Water Quality Analyses

The water quality analyses must include those parameters needed to control the stabilization process being used. In most cases this will include pH and total alkalinity as a minimum. However, the data needed to calculate the Langelier Index (or one of the other indices) should also be collected. This data includes the concentration of the calcium ion and total dissolved solids (total filterable residue) and the water temperature.

In-plant monitoring. The pH of the treated water entering the distribution system should be monitored with a suitable continuously recording pH meter. The pH must be kept within the proper range at all times to prevent excessive corrosion or scaling. The total alkalinity concentration should be determined at least every 8 hours, more frequently if the quality of the source water is changing quickly. The LI (or other index) of the raw and treated water should be calculated daily.

To calculate the chemical dose necessary for proper pH and alkalinity adjustment, the pH and total alkalinity of the water entering the stabilization process must also be monitored. Samples should be tested at least every 8 hours. It is important that the samples be taken just before the application point of the stabilization chemicals. The raw water entering the plant cannot provide meaningful data, since other water treatment chemicals such as alum, chlorine, and hydrofluosilicic acid lower the alkalinity and pH.

Distribution system monitoring. Water quality analyses should be conducted on samples from throughout the distribution system to help determine if the stabilization is effective. These analyses should take into consideration the materials in the distribution and plumbing systems, particularly if corrosion is a problem. Table 9-7 lists some of the parameters that should be included for various types of pipe material.

The pH and alkalinity are important characteristics to measure if cement-lined pipe or asbestos-cement pipe is used. If leaching of the cement is occurring due to aggressive water, the pH and alkalinity will increase through the distribution system, since cement contains lime. Lead and cadmium levels should be tested since these metals are toxic to humans. Samples from household taps should be routinely analyzed to determine if lead or cadmium are below the established maximum contaminant levels, particularly if lead or galvanized steel service lines or lead solder is part of the plumbing.

Pipe and Coupon Testing

A useful method for determining what is happening in the distribution system is the inspection of pipe specimens taken from the systems. This inspection allows the operator to determine the extent of corrosion or scaling taking place. A number of procedures can be used to determine the corrosion/scaling rate from pipe specimens, including calculation of the depth of pits in the pipe or the loss in weight of the pipe section. The scale on the pipe can also be analyzed to help determine why a pipe is being protected or corroded or why excessive scale is being deposited.

Table 9-7. Water Characteristics of Importance to Water-Main Materials

Water-Main Material	*Water Characteristic**
Ductile and cast iron	Color, conductivity, dissolved oxygen, iron, manganese, pH (alkalinity and calcium if main is cement-lined)
Steel	Color, conductivity, dissolved oxygen, iron, manganese, pH
Concrete cylinder	Alkalinity, calcium, conductivity, pH
Asbestos-cement	Alkalinity, asbestos fibers, calcium, conductivity, pH
Galvanized steel	Alkalinity, cadmium, color, conductivity, dissolved oxygen, iron, lead, pH, zinc

*See Volume 4, *Water Quality Analyses* for further discussion of many of these characteristics.

Courtesy of Virginia Chemicals Inc.

Figure 9-8. Coupons After They Have Been Cleaned

Another valuable technique is the COUPON TEST. This test measures the effects of the water on a small section of metal (coupon) inserted in a water line. After a minimum of 120 days, the inserts are removed, cleaned, weighed, and examined. The weight loss or gain of the coupon can provide an indication of the corrosion or scaling rate. Figure 9-8 shows some coupons after they have been cleaned. Coupons cannot provide the day-to-day information needed to adjust the chemical feed rates. However, they can be used with the water quality data to provide valuable information for long-term stability control.

Record Keeping

Record keeping is especially important for stabilization since a variety of information from different sources should be used to monitor the process. For example, records at the treatment plant should be used in conjunction with records from the distribution system to provide an understanding of how the stabilization program is working.

At the treatment plant, records should be kept of the following:

- Current inventory of stabilization chemicals
- Amount of chemical being fed (mg/L and lb/day)
- Quantity of water being treated
- pH, alkalinity, other control tests, and the LI (or other index)
- Maintenance on feeders or solution lines and diffusers.

The distribution system records should include:

- Water quality analyses
- Results of coupon tests

- Results of examination of pipe specimens or valves removed from the system
- Customer complaints, particularly as they relate to corrosion or scaling
- Results of examination of pipe specimens taken from household plumbing.

None of this information taken alone will be useful. However, good records of all the information will provide excellent indications of the condition of the distribution system and what changes might be necessary in the stabilization process to improve control of corrosion or scaling.

9-6. Safety and Stabilization

In operating the stabilization process, the operator will be exposed to caustic or acidic chemicals and solutions, depending on the process used.

When handling dry chemicals such as lime, soda ash, and sodium bicarbonate, care should be taken to minimize dust, since chemical dusts are irritating to the respiratory system, skin, and eyes. Dust collectors should be installed on storage hoppers and cleaned regularly. Dust around the feeders and hoppers should be cleaned routinely using a vacuum cleaner. When handling dry chemicals, operators should wear proper protective clothing, including a close-fitting respirator and tight-fitting safety glasses with side shields. In hot weather, when workers are perspiring, chemical burns become more of a problem, especially where quicklime dust is present. A long-sleeve shirt with collar buttoned and pants with the legs over the top of the shoes or boots are recommended. Gloves and headgear should also be worn. A protective cream should be placed on exposed skin near the face, neck, and wrists. After handling dusty, dry chemicals, the operator should shower immediately. If clothing is dusty, it should be laundered before wearing again.

Proper storage of chemicals can prevent many safety problems by minimizing dust and reducing opportunities for dangerous chemical reactions. The chemicals should be stored in dry areas with adequate ventilation. Each chemical should be stored in its own area, as shown in Figure 9-9, not together with other chemicals and equipment. If bags are used they should be stored on pallets and as close to the feeder hopper as possible.

The chemical solutions made from the dry chemicals (such as lime slurry) or liquid chemicals (acid or caustic soda) must also be handled with care. Most of these solutions can cause burns and serious damage to the eyes; therefore, face shields, rubber aprons, rubber boots, and rubber gauntlets should always be worn when working with the solutions. Pump heads should have splatter-proof shields installed to prevent problems in case of leaks. Spills should be cleaned up immediately to prevent accidental contact with other chemicals and damage to equipment. Storage tanks, dilution tanks, pumps, and piping should be inspected routinely to check for leaks.

Figure 9-9. Storage of Stabilization Chemicals

Operators should be familiar with the proper first-aid procedures to follow for chemical burns from dust or solutions. These procedures should be posted and first-aid training conducted using the procedures. The plant should have safety showers and eye-washing facilities available for emergencies.

Selected Supplementary Readings

Controlling Corrosion Within Water Systems. AWWA Seminar Proc. AWWA, Denver, Colo. (June 1978).

Curry, M. D. Is Your Water Stable and What Difference Does it Make? *Jour. AWWA*, 70:9:506 (Sept. 1978).

Kirmeyer, G. J. & Logsdon, G. S., Principles of Internal Corrosion and Corrosion Monitoring. *Jour. AWWA*, 75:2:78 (Feb. 1983).

Lime Handling, Application and Storage. National Lime Institute, Arlington, Va. (1982).

Mullon, E. D. & Ritter, J. A. Monitoring and Controlling Corrosion by Potable Water. *Jour. AWWA*, 72:5:286 (May 1980).

Ryder, R. A. The Costs of Internal Corrosion in Water Systems. *Jour. AWWA*, 72:5:267 (May 1980).

Safe Handling of Chemicals: Lime & Soda Ash. *OpFlow*, 6:2:3 (Feb. 1980).

Safe Storage and Handling of Lime. *OpFlow*, 2:4:6 (Apr. 1976).

Safety, Care Essential in Lime Handling. *OpFlow*, 2:1:1 (Jan. 1976).

Sawyer, Clair & McCarty, P.L. Chemistry for Environmental Engineering McGraw-Hill Book Company, New York. (3rd ed., 1978).

Swayze, James. Corrosion Study at Carbondale, Illinois. *Jour. AWWA*, 75:2:101 (Feb. 1983).

Water Quality and Treatment. AWWA Handbook. McGraw-Hill Book Company, New York. Chap. 8. (3rd ed., 1971).

Glossary Terms Introduced in Module 9

(Terms are defined in the Glossary at the back of the book.)

Aggressive
Anode
Cathode
Concentration cell corrosion
Corrosion
Corrosive
Coupon test
Erosion
Galvanic corrosion
Galvanic series
Iron bacteria
Langelier index (LI)
Localized corrosion
Milk of lime
Red water
Saturation point
Sequestering agent
Slaker
Stabilization
Tubercules
Uniform corrosion
Unstable

Review Questions

(Answers to the Review Questions are given at the back of the book.)

1. What is the purpose of stabilization?

2. In what three areas does unstable water cause major problems?

3. What is corrosion?

4. List at least five factors that affect water stability.

5. Explain how TDS affects stability.

6. What is galvanic corrosion?

7. How do bacteria affect corrosion?

8. Compare the two major types of corrosion.

9. How is scale formed?

10. What are some common forms of scale?

11. How does temperature affect scale formation?

12. What are the basic stabilization techniques?

13. What is the chemical most commonly used for pH adjustment?

14. Why is carbon dioxide added to lime-softened water?

15. How do sequestering agents prevent scale formation?

16. What type of feed equipment is necessary if quicklime is used?

17. What common operating problem is associated with lime feeding?

18. What does a negative Langelier Index indicate?

19. What types of information are needed to select a stabilization process?

20. What pH range should be maintained for a good $CaCO_3$ coating?

21. What is the chief cause of excessive scaling?

22. What are the two most important parameters to monitor for control of stabilization?

23. In addition to water quality, what other methods can be used to monitor the stabilization proccess?

24. What is a good way to minimize dust?

Study Problems and Exercises

1. The town council has asked you to investigate the rash of red-water complaints that have occurred since the town started using a new surface source. Based on the following water quality data, determine what the problem is and discuss how you would solve it.

TDS, mg/L	150
Total Hardness, mg/L*	76
Calcium, mg/L*	20
Magnesium, mg/L*	8
Alkalinity, mg/L*	30
pH	7.0
Temperature	39° F (4° C)

*As $CaCO_3$

2. Using water quality data from a local water supply, determine if it has corrosive potential based on the Langelier, Ryznar, and Aggressive Indices. Discuss what other data are needed to determine water stability.

3. You are to make recommendations as to which chemicals your water system should use to raise the pH and alkalinity. Discuss what factors should be considered in making the selection.

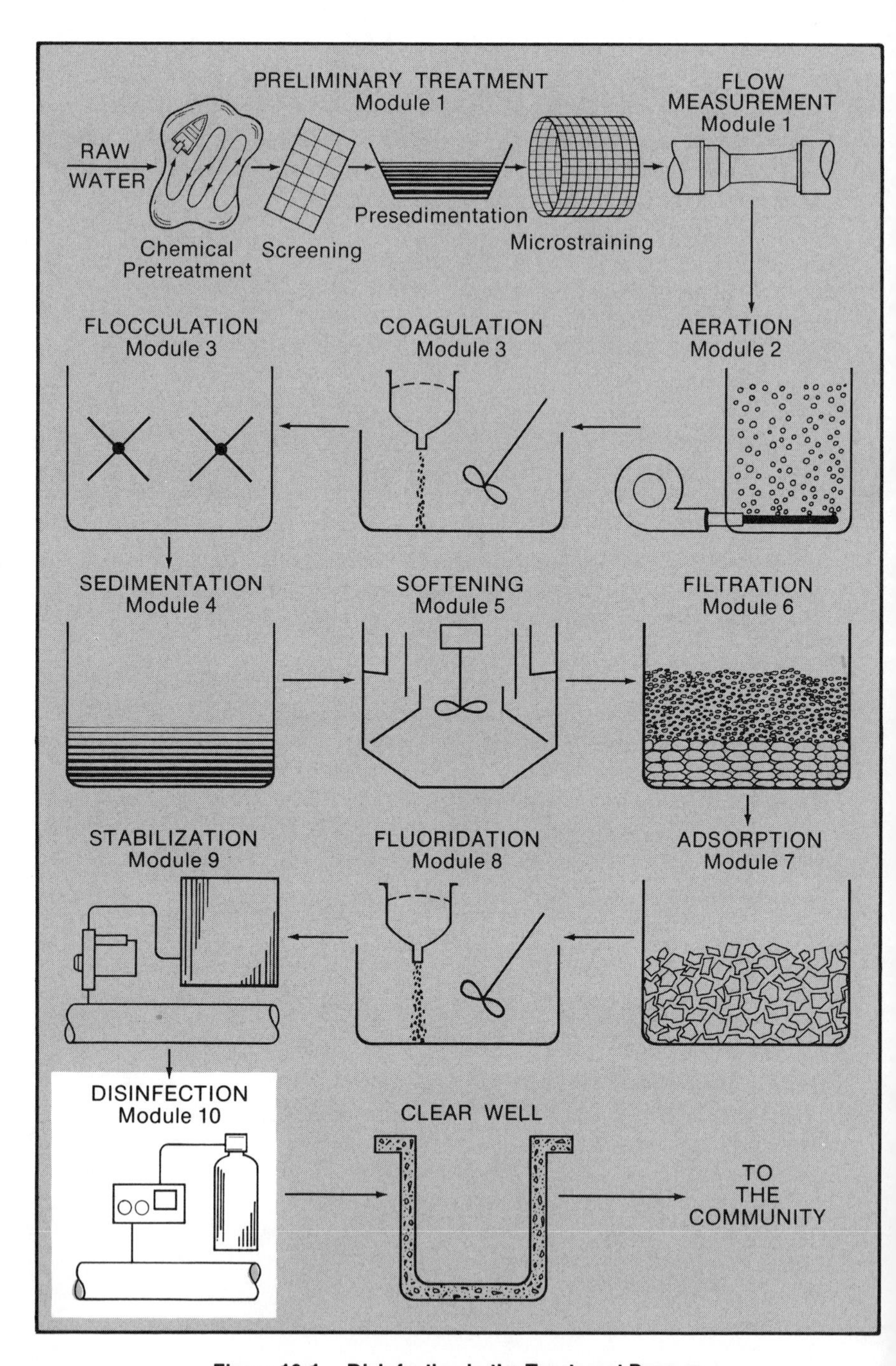

Figure 10-1. Disinfection in the Treatment Process

Module 10

Disinfection

One of the operator's most important roles in water treatment is to deliver water that is free from disease-causing (PATHOGENIC) organisms. DISINFECTION is the treatment process used to destroy these organisms, usually through the addition of chlorine to the water. Diseases caused by pathogenic organisms in water are called WATERBORNE DISEASES; the more common ones are summarized in Table 10-1. As the table indicates, the consequences of waterborne disease can range from mild illness to death. Clearly, the importance of disinfection cannot be overemphasized. (Disinfection—the destruction of disease-causing microorganisms—should not be confused with STERILIZATION, which is the destruction of all living microorganisms.) The location of the disinfection process in a typical treatment plant is shown in Figure 10-1.

After completing this module you should be able to

- Define the three general types of disinfection.
- Discuss the different chemicals used for disinfection along with their advantages and disadvantages.
- Understand the general chlorination chemical reactions.
- Describe and plot a typical chlorination "breakpoint" curve.
- Identify and discuss the five factors important to successful chlorination.
- Describe the three chemicals used for chlorination.
- Identify the equipment needed with each of the three chlorine chemicals.
- Describe general procedures for using chlorine cylinders and ton containers.
- Discuss typical operating problems with chlorination and possible solutions.

Table 10-1. Waterborne Diseases

Waterborne Disease	*Causative Organism*	*Source of Organism in Water*	*Symptom*
Gastroenteritis	*Salmonella* (bacteria)	Animal or human feces	Acute diarrhea and vomiting
Typhoid	*Salmonella typhosa* (bacteria)	Human feces	Inflamed intestine, enlarged spleen, high temperature—fatal
Dysentery	*Shigella* (bacteria)	Human feces	Diarrhea—rarely fatal
Cholera	*Vibrio comma* (bacteria)	Human feces	Vomiting, severe diarrhea, rapid dehydration, mineral loss—high mortality
Infectious hepatitis	Virus	Human feces, shellfish grown in polluted waters	Yellowed skin, enlarged liver, abdominal pain—low mortality, lasts up to 4 months
Amebic dysentery	*Entamoeba histolytica* (protozoa)	Human feces	Mild diarrhea, chronic dysentery
Giardiasis	*Giardia lamblia* (protozoa)	Animal or human feces	Diarrhea, cramps, nausea and general weakness—not fatal, lasts 1 week to 30 weeks

- Describe the operational control tests and record keeping suggested for effective chlorination.
- Describe the safety practices that must be followed when using each chlorine chemical.

10-1. Destroying Pathogens in Water

CHLORINATION, usually performed as the final treatment process (Figure 10-1), is the most common means of disinfecting drinking water. Other processes, natural and artificial, aid in the destruction or removal of pathogens (Table 10-2).

Most pathogens are accustomed to living in the temperatures and conditions found in the bodies of humans and animals; they do not survive well outside the body. Nonetheless, significant numbers can survive in potable water. Some pathogens, particularly certain viruses and those organisms that form cysts, can survive for surprisingly long periods even under the most adverse conditions. Because such organisms also tend to be resistant to chlorine doses normally used in water treatment, chlorination alone cannot always ensure safe drinking water.

Table 10-2. Pathogen Reduction from Various Treatment Processes

Unit Process	*Percent Reduction*
Storage*	Significant amounts
Sedimentation*	0–99
Coagulation*	Significant amounts
Filtration*	0–99
Chlorination	99

*These methods do not, in themselves, provide adequate pathogen reduction. However, their use prior to disinfection may significantly lower the costs associated with disinfection.

Storing water for extended periods in open tanks or reservoirs prior to treatment can accomplish some destruction of pathogens through sedimentation and natural die-off of the organisms. Significant pathogen removal also occurs during the conventional treatment processes of coagulation, flocculation, sedimentation, and filtration.

All drinking water should receive some type of disinfection to help ensure that pathogens are not present. Because surface waters are more susceptible to contamination than ground waters, they should always receive treatment (such as coagulation, flocculation, sedimentation, and filtration) before disinfection and delivery to consumers.

Methods of Disinfection

Although chlorination is the most common technique for disinfection, other methods are available and may be useful in some situations. This section briefly summarizes the methods available.

Three general types of disinfection are:

- Heat treatment
- Radiation treatment
- Chemical treatment.

Heat treatment. Heat was probably the first method used for disinfection; it is still a good emergency procedure for small quantities of water. To disinfect with heat, bring water to a boil and continue to boil it for 5 to 20 min. The boiling time varies with altitudes—at higher altitudes more time will be needed because boiling temperature is lower. Boiling can be used to disinfect drinking water at a campsite or to disinfect household drinking water supplies when a local health agency has issued a "boil order." Such an order could be issued whenever the bacteriological quality of the water does not meet drinking water standards. Boiling is not suited for large-scale use, primarily because of the high cost of energy required.

Radiation treatment. Sunlight is a natural means of disinfection—the ultraviolet portion of the sun's light destroys microorganisms. Ultraviolet radiation can also be produced from specially designed lamps; these have had limited use in water disinfection. The use of ultraviolet radiation to destroy pathogens, called UV DISINFECTION, is costly. Since UV light is readily absorbed and scattered by the turbidity-causing impurities in water, the light's killing power is greatly diminished in water that does not pass close to the lamp. These disadvantages and the lack of measurable residual have discouraged widespread use of UV disinfection.

Chemical treatment. The chemicals used as disinfectants are:

- Bromine
- Iodine
- Ozone
- Chlorine and chlorine compounds.

Bromine. Bromine is a dark reddish-brown liquid. It vaporizes at room temperature and has a penetrating, suffocating odor. The vapor is extremely irritating to the eyes, nose, and throat, and it is very corrosive to most metals. When spilled or splashed onto the skin, bromine causes painful burns that are slow to heal.

The residual formed when bromine is added to water is as effective a disinfectant as chlorine, but not as stable. Consequently, depending on the constituents in the water being treated, it may be necessary to add bromine at two or three times the concentration required for chlorine. Because of the added cost factor and the handling hazards, liquid bromine is not used to disinfect municipal water supplies; it is used in the safer, but more costly, solid "stick form" (ORGANOBROMINE COMPOUND) to disinfect swimming pools.

Iodine. Iodine is a lustrous, blue-black solid, about five times the density of water, with a peculiar chlorine-like odor. The solid can quickly change to a gas, releasing a characteristic violet vapor. Iodine has been used extensively for medicinal purposes and as a disinfectant for small drinking-water supplies. Iodine, like bromine, is more costly than chlorine; although it is an effective disinfectant, it is usually too costly to use on a municipal scale. Due to possible health effects associated with long-term consumption, iodine is not recommended as a disinfectant for water supplies serving permanent populations. However, iodine can be used economically to disinfect water serving non-permanent populations, as in campgrounds; and it can be used for emergency water disinfection. Iodine is available in crystals for use in saturator-type feeders.

Ozone. Ozone (O_3) is formed when a high-voltage arc passes through the air between two electrodes. It is also formed PHOTOCHEMICALLY in the atmosphere and is one of the constituents in smog. Ozone is a bluish, toxic gas with a pungent odor. It is considered hazardous to health at a concentration in air of 0.25 mg/L by volume and extremely hazardous at levels of 1.0 mg/L and greater.

Ozone is a powerful oxidizing agent used in water treatment for disinfection and for control of color, taste, and odor. The gas is chemically unstable and disappears in minutes, leaving no residual disinfectant to continue disinfection within the water system. Furthermore, because of this lack of residual, the operator has no simple way to determine whether enough ozone has been added to destroy all pathogens. Bacteriological analysis of samples collected before and after the point of application is the only method available for measuring the effectiveness of ozone.

Ozone has been used in a few US water plants for disinfection and taste-and-odor removal, but it has found use as a disinfectant primarily in Europe, mostly in France. Manufacturing ozone requires a great deal of energy, and rising energy costs are making ozonation more expensive to use. Chlorination, with its lower cost, greater flexibility, and greater familiarity, continues to be the preferred alternative in the United States.

Chlorine and chlorine compounds. Chlorine and chlorine compounds (such as calcium hypochlorite and sodium hypochlorite) are the disinfectants most widely used for water treatment in the United States. The remainder of this module describes chlorine and the chlorination of drinking water.

10-2. Description of Chlorination

Chlorination, the addition of chlorine to water, is the most common form of disinfection practiced in the United States today. When properly understood and correctly operated, the chlorination process is a safe, practical, and effective way to destroy disease-causing organisms.

There are several secondary benefits gained from using chlorine as the disinfectant for treated water, and chlorine may also be used as part of other treatment processes. Chlorine is useful for disinfecting storage tanks and pipelines; for oxidizing iron, manganese, and hydrogen sulfide; and for controlling tastes, odors, algae, and slime. These uses are discussed in greater detail in other modules.

Chemistry of Chlorination

To understand reactions of chlorine in natural water, consider the reaction of chlorine in distilled water. As shown in Figure 10-2, the amount of FREE CHLORINE RESIDUAL is directly related to the dose or amount of chlorine added.

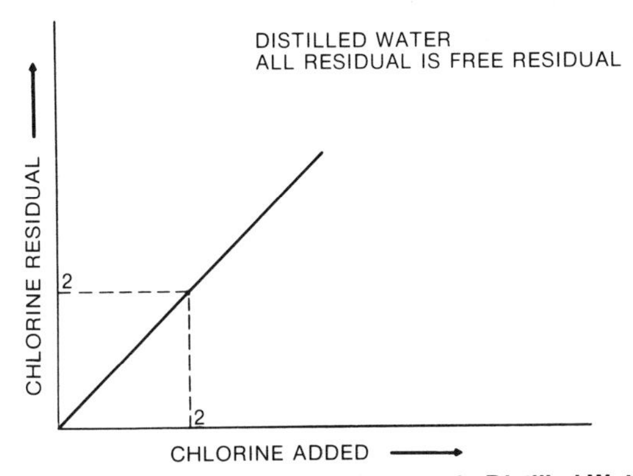

Figure 10-2. The Reaction of Chlorine in Distilled Water

For example, if 2 mg/L of chlorine is added, 2 mg/L of free residual is produced. The reactions occurring are as follows (Equation numbers are the same as those used in *Basic Science Concepts and Applications*.):[1]

Cl_2	+	H_2O	→	HOCl	+	HCl	*Eq 52*
Chlorine		Water		Hypochlorous acid		Hydrochloric acid	

The products are weak compounds that dissociate as follows:

HOCl	→	H^+	+	OCl^-	*Eq 53*
Hypochlorous acid		Hydrogen ion		Hypochlorite ion	
HCl	→	H^+	+	Cl^-	*Eq 54*
Hydrochloric acid		Hydrogen ion		Chlorine ion	

Hypochlorous acid, one of two forms of free chlorine residual, is the most effective disinfectant available. When it dissociates as in Equation 53, the hypochlorite ion (the second form of free chlorine residual) is formed. The hypochlorite ion as a disinfectant is only 1/100 as effective as hypochlorous acid (as indicated in Table 10-3).

Table 10-3. Estimated Effectiveness of Residual Types

Type	*Chemical Abbreviation*	*Estimated Effectiveness Compared to HOCl*
Hypochlorous acid	HOCl	1
Hypochlorite ion	OCl^-	1/100
Trichloramine*	NCl_3	†
Dichloramine	$NHCl_2$	1/80
Monochloramine	NH_2Cl	1/150

*Commonly called nitrogen trichloride.
†No estimate; possibly more effective than dichloramine.

Natural water is not pure, and the reaction of chlorine with the impurities it contains interferes with the formation of a free chlorine residual. For example, if the water contains organic matter, nitrites, iron, manganese, and ammonia, then the chlorine added will react as shown in Figure 10-3. Between points 1 and 2, the chlorine added combines immediately with iron, manganese, and nitrites. Iron, manganese, and nitrites are REDUCING AGENTS, and no residual can be formed until all reducing agents are completely destroyed by the chlorine.

As more chlorine is added between points 2 and 3, the chlorine begins to react with ammonia and organic matter to form chloramines and chloroorganic compounds. These are COMBINED CHLORINE RESIDUALS. Since the chlorine is combined with other compounds, this residual is not as effective as a free chlorine residual. (The effectiveness of residuals is compared in Table 10-3.)

[1] *Basic Science Concepts and Applications*, Chemistry Section, Chemistry of Treatment Processes (Chemistry of Chlorination).

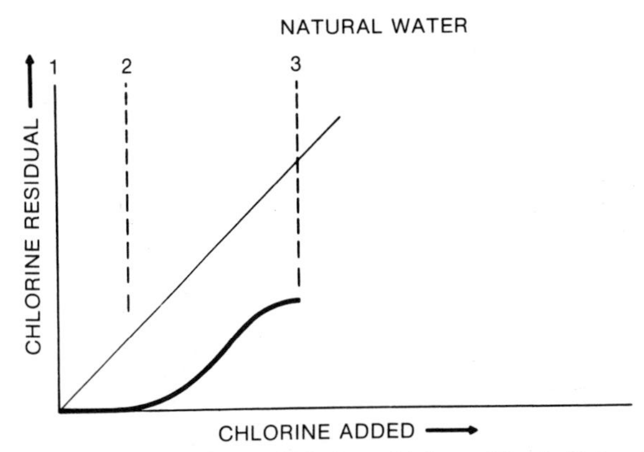

Figure 10-3. The Reaction of Chlorine with Impurities in Natural Water

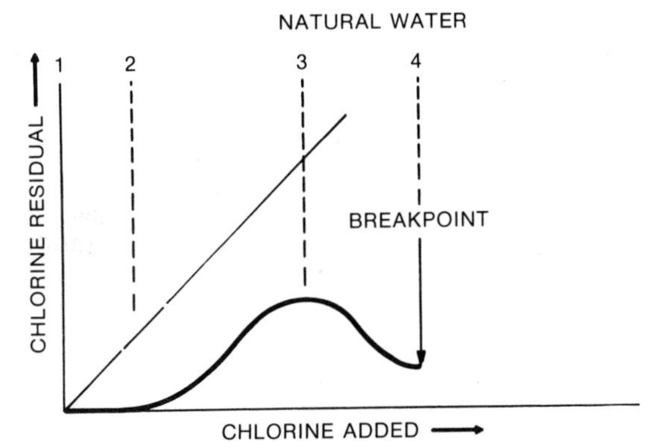

Figure 10-4. Decrease of Chlorine Residual

Between points 2 and 3, the combined residual is primarily monochloramine:

NH_3	+	HOCl	→	NH_2Cl	+	H_2O	*Eq 55*
Ammonia		Hypochlorous acid		Monochloramine		Water	

Adding more chlorine to the water (Figure 10-4) actually decreases the residual. The decrease (shown from point 3 to point 4) results because the additional chlorine oxidizes some of the chloroorganic compounds and ammonia. The additional chlorine also changes some of the monochloramine to dichloramine and trichloramine.

NH_2Cl	+	HOCl	→	$NHCl_2$	+	H_2O	*Eq 56*
Monochloramine		Hypochlorous acid		Dichloramine		Water	
$NHCl_2$	+	HOCl	→	NCl_3	+	H_2O	*Eq 57*
Dichloramine		Hypochlorous acid		Trichloramine		Water	

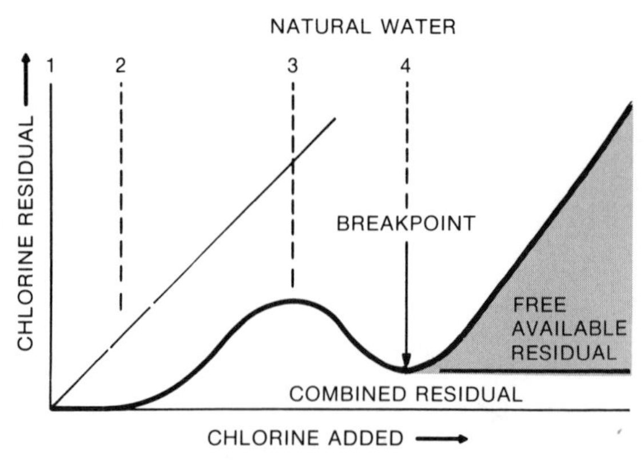

Figure 10-5. The Chlorine Breakpoint

As additional chlorine is added, the amount of chloramine reaches a minimum value. Beyond this point, a point is reached where further addition of chlorine produces free residual chlorine. The point at which this occurs (point 4, Figure 10-5) is known as the BREAKPOINT. To the right of the breakpoint (Figure 10-5) an increase in the chlorine dose will usually produce a proportionate increase in the free chlorine residual. Beyond the breakpoint, the free chlorine residual should be 85 to 90 percent of the total chlorine residual. The remaining 10 to 15 percent is combined residual consisting of dichloramines, trichloramines, and chloroorganic compounds. A group of the chloroorganic compounds, TRIHALOMETHANES, are discussed later in this module.

Principle of Disinfection by Chlorination

There are five factors important to the success of chlorination:

- Concentration
- Contact time
- Temperature
- pH
- Substances in the water.

Concentration and contact time. The effectiveness of chlorination depends primarily on two factors:

- Concentration (C)
- Contact time (t).

The destruction of organisms, often referred to as the "kill," is directly related to these two factors as follows:

Kill is proportional to $C \times t$
or
$$\text{Kill} \propto C \times t$$

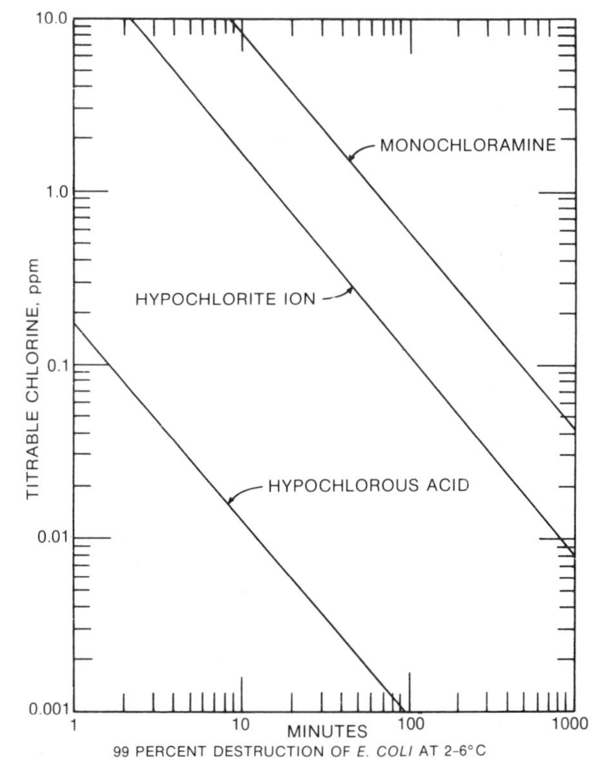

From Handbook of Chlorination *by Clifford White, copyright © 1972 by Van Nostrand Company. Reprinted by permission of the publisher.*

Figure 10-6. Efficiency of Hypochlorous Acid, Hypochlorite Ion, and Monochloramine as Disinfectants

This means that if the chlorine concentration is decreased, then the contact time—the length of time the chlorine and the organisms are in physical contact—must be increased to ensure that the kill remains the same. Similarly, as the chlorine concentration increases, the contact time needed for a given kill decreases.

A combined chlorine residual, which is a weak disinfectant, requires a greater concentration acting over a longer period of time than is required for a free chlorine residual. Therefore, when the contact time between the point of application of chlorine and the consumption of water is short (for example, 10 min), only a free residual will provide effective disinfection. It is important to know the contact time and type of residual chlorine available so the proper concentration can be provided. Figure 10-6 illustrates this point. Usually, a minimum free chlorine residual of 0.3 mg/L should be maintained at the extremities of the distribution system.

Temperature. The effectiveness of chlorination is also related to the temperature of the water—at lower temperatures, bacterial kill tends to be slower. However, chlorine is more stable in cold water, and the residual will remain for a longer period of time, compensating to some extent for the lower rate of disinfection. Other factors being equal, chlorination is more effective with higher water temperatures. It is important for the operator to maintain a record of water temperatures. As temperatures change seasonally, the chlorine dosage will also need to be changed. The effectiveness of combined chlorine residuals is affected more by low temperatures than that of free chlorine residuals.

pH. The pH of the water affects the disinfecting action of chlorine since it determines the ratio of HOCl to OCl^-—depending on the pH, either more hypochlorite ion or more hypochlorous acid could be present. As shown in Figure 10-7, the dissociation can go either direction, and the ratio of the ions will shift as the pH changes.

Hypochlorous acid dissociates poorly at low pH levels; the dominant residual is HOCl. On the other hand, HOCl will dissociate almost completely at high pH levels, leaving OCl^- as the dominant residual. The most dramatic dissociation occurs between pH 6.0 and pH 8.5, as shown in Figure 10-7. Note the slight effect of temperature on the distribution of HOCl and OCl^- at the various pH levels. Figure 10-8 summarizes the effect of pH on free and combined residuals. It is essential that the operator understand and use these relationships in order to obtain the most effective disinfectant. The operator should routinely check the pH. This is most important if the pH of the water is being raised to control corrosion, since the chlorine dosage will also have to be raised to maintain an effective level. Addition of chlorine gas lowers the pH of water; the use of hypochlorites raises the pH slightly.

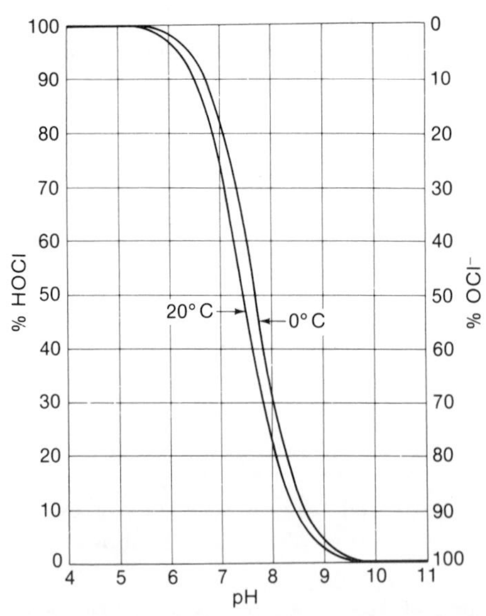

Figure 10-7. Distribution of HOCl and OCl^- in Water at Indicated pH Levels

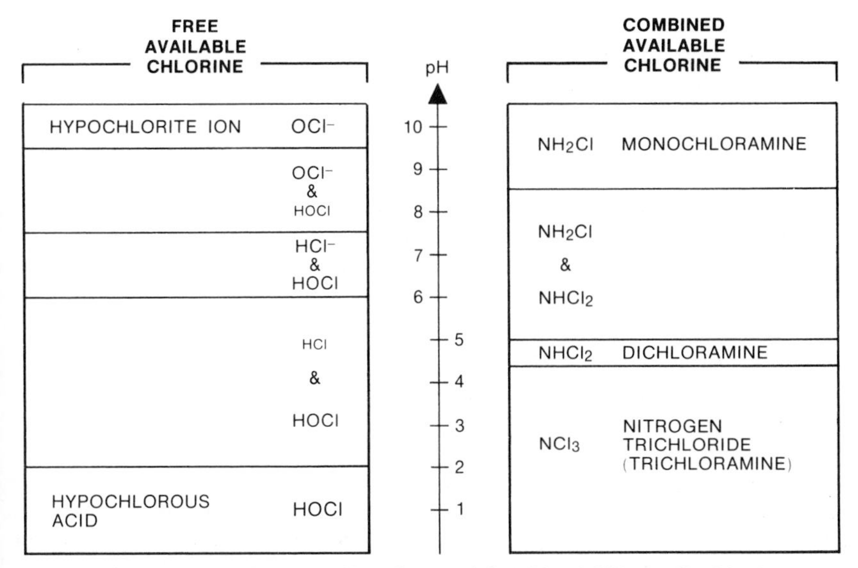

Figure 10-8. Effect of pH on Free and Combined Chlorine Residual

Substances in the water. Chlorine acts as an effective disinfectant only if it comes in contact with the organisms to be killed. TURBIDITY, caused by tiny particles of dirt and other impurities suspended in water, can prevent good contact and protect the pathogens. Therefore, for chlorination to be effective, turbidity must be reduced to the greatest extent possible by treatment methods such as coagulation, flocculation, and filtration.

As discussed earlier, chlorine reacts with other substances in water, such as organic matter and ammonia. Since these compounds result in the formation of the less-effective combined residuals, their concentrations are an important factor in determining chlorine dosages. Breakpoint chlorination should be practiced to ensure the formation of free chlorine residual. Failing to practice breakpoint chlorination can allow the survival of pathogens and can cause serious taste-and-odor problems.

10-3. Chlorination Facilities

Chlorine is available in gaseous, liquid, and solid form. The chemicals and equipment used for chlorination depend primarily on the type of chlorine used.

Chlorine Chemicals

Hypochlorous acid (HOCl) and hypochlorite ion (OCl^-) are the most effective residuals. They can be derived from three chemicals:

- Chlorine, Cl_2
- Calcium hypochlorite, $Ca(OCl)_2$
- Sodium hypochlorite, NaOCl.

Chlorine. Chlorine gas, Cl_2, is about 2.5 times as dense as air. It has a pungent, noxious odor and greenish-yellow color, although it is visible only at high concentrations (above 100 ppm by volume). The gas is highly irritating to the eyes, nasal passages, and the respiratory tract, and can kill in a few breaths at concentrations as low as 0.1 percent (1000 ppm) by volume. Its odor can be detected at concentrations above 0.3 ppm (by volume).

Chlorine liquid, Cl_2, is created by compressing chlorine gas. The liquid, which is about 99.5-percent pure chlorine, is amber in color and is about 1.5 times as dense as water. It can be purchased in a wide range of pressure containers, tank trucks, and railroad tank cars (Figures 10-9 through 10-12).

Liquid chlorine changes easily to a gas at room temperatures and pressures. One volume of liquid chlorine will expand to about 460 volumes of gas. Dry chlorine gas will not corrode steel or other metals, but it is extremely corrosive to most metals in the presence of moisture.

Chlorine will not burn; but, like oxygen, it will *support combustion*—that is, it takes the place of oxygen in the "burning" of combustible materials. Chlorine is not explosive, but it will react violently with greases, turpentine, ammonia, hydrocarbons, metal filings, and other flammable materials. Chlorine will not conduct electricity. Because of the inherent hazards involved, chlorine requires special care in storage and handling, as will be described throughout this module.

Calcium hypochlorite. Calcium hypochlorite, $Ca(OCl)_2$, is a dry, white or yellow-white, granular material; it is also available in compressed tablets. It contains 65-percent available chlorine by weight. This means that when 1 lb of 65-percent calcium hypochlorite powder is added to water, only 0.65 lb of pure chlorine is added. To add 1 lb of chlorine, 1.54 lb of calcium hypochlorite must be added. (See Table 10-4.)

Calcium hypochlorite requires special storage to avoid contact with organic material—the reactions possible between calcium hypochlorite and organics can generate enough heat and oxygen to start and support a fire. When calcium hypochlorite is mixed with water, heat is given off. To provide adequate dissipation of the heat generated, the dry chemical should be added to the correct volume of water, rather than adding water to the chemical.

Table 10-4. Chlorine Content of Common Disinfectants

Compound	*Percentage Cl*	*Amount of Compound Needed to Yield 1 lb of Pure Cl*
Chlorine gas or liquid (Cl_2)	100	1 lb (0.454 kg)
Sodium hypochlorite (NaOCl)	15	0.8 gal (3 L)
	12.5	1.0 gal (3.8 L)
	5	2.4 gal (9.1 L)
	1	12.0 gal (45.4 L)
Calcium hypochlorite [$Ca(OCl)_2$]	65	1.54 lb (0.7 kg)

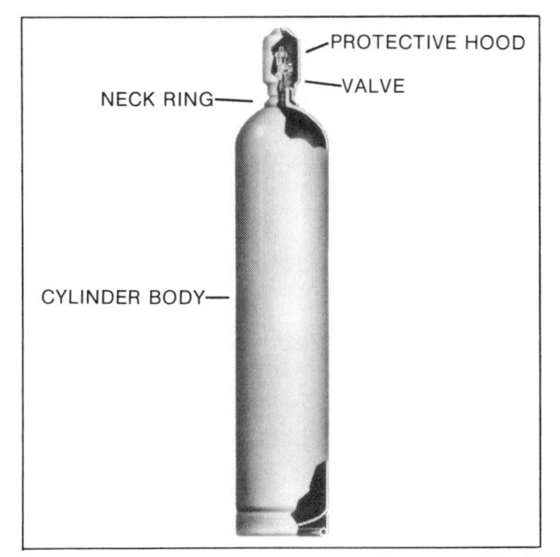

Courtesy of the Chlorine Institute, Inc.

Figure 10-9. Chlorine Cylinder

Courtesy of the Chlorine Institute, Inc.

Figure 10-10. Chlorine Ton Container

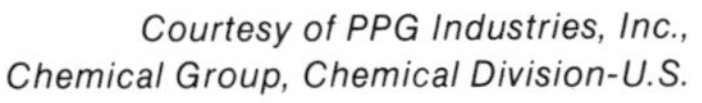
Courtesy of PPG Industries, Inc.,
Chemical Group, Chemical Division-U.S.

Figure 10-11. Chlorine Ton Container Truck

Courtesy of the Chlorine Institute, Inc.

Figure 10-12. Chlorine Tank Car

Sodium hypochlorite. Sodium hypochlorite, NaOCl, is a clear, light-yellow liquid commonly used for bleach. Ordinary household bleach is an example of sodium hypochlorite. Household bleach contains 5 percent sodium hypochlorite; industrial bleaches are stronger, containing from 9 to 15 percent. The sodium hypochlorite solution is alkaline, with a pH of 9 to 11, depending on the available chlorine content. For common strengths of sodium hypochlorite, Table 10-4 shows the amount of solution needed to supply 1 lb of pure chlorine. The solution can be purchased in 5-gal carboys, 55-gal rubber-lined steel drums, and in railroad tank cars. However, for use in small water systems, it is often purchased in 1-gal opaque plastic jugs. There is no fire hazard when storing sodium hypochlorite. The chemical is quite corrosive, however, and should be kept away from equipment susceptible to corrosion damage. Sodium hypochlorite solution can lose 2 to 4 percent of its available chlorine content per month at room temperature. Therefore, manufacturers recommend a maximum shelf life of 60 to 90 days.

Of the three forms of chlorine, chlorine gas is the most common, used in 90 to 95 percent of all disinfection applications. The remainder of this section focuses on facilities required when using chlorine gas, with a brief description of hypochlorination facilities.

Handling and Storing Chlorine Gas

Safe handling and storage of chlorine is vitally important to the operator and to the communities immediately surrounding and downwind of the plant. An error or accident in handling or storage of chlorine pressure vessels can cause serious injuries, even fatalities.

The containers commonly used to supply chlorine in small-to-medium-sized water treatment plants are 150-lb cylinders and ton containers. Handling and storage requirements differ depending on the container size used.

Cylinders. Dimensions and weights of the various cylinders differ depending on whether 100-, 105-, or 150-lb cylinders are used. The most commonly used cylinder, having a total filled weight of 250–285 lb (110–130 kg), is the 150-lb unit, which holds 150 lb (68 kg) of chlorine. The 150-lb cylinders are about 10½ in. (0.27 m) in diameter and 56 in. (1.42 m) high. As illustrated in Figure 10-9, each cylinder is equipped with a protective hood, protecting the cylinder valve against shipping damage that could lead to a serious chlorine leak. The hood should be properly screwed into place whenever a cylinder is handled.

Cylinders are usually delivered by truck. Each cylinder should be unloaded to a dock at truck-bed height. If a hydraulic tailgate is used, the cylinders should be secured to keep them from falling. The cylinders should never be dropped (full or empty) and must be protected from any forceful impact. The easiest and safest way to move cylinders in the plant is with a hand truck. As shown in Figure 10-13, the hand truck should be equipped with a restraining chain that snugly fastens around the cylinder about two-thirds of the way up. Slings should never be used to lift cylinders, and a cylinder should never be lifted by the protective hood—the neck ring that holds the hood to the cylinder cannot support the weight of the cylinder.

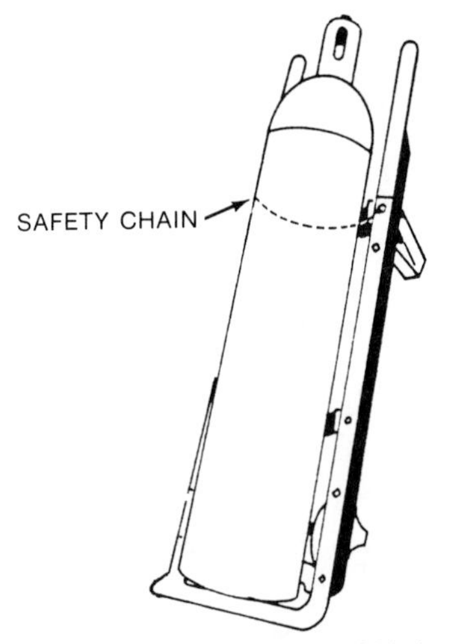

Courtesy of the Chlorine Institute, Inc.

Figure 10-13. Hand Truck for Moving Chlorine Cylinders

Cylinders may be stored indoors or outdoors. If indoors, the building should be fire resistant, have multiple exits with outward opening doors, and be adequately ventilated. Outdoor storage areas must be fenced and protected from direct sunlight, and they should be protected from vehicles or falling objects that might strike the cylinders. If standing water accumulates in the outdoor storage area, the cylinders should be stored on elevated racks. Avoiding contact with water will help to minimize cylinder corrosion.

Some operators find it convenient to hang "full" or "empty" ID tags on cylinders in storage, so that the status of the chlorine inventory can be determined at a glance. Full and empty cylinders should receive the same care while in storage. Protective hoods should be replaced on empty cylinders. Even though "empty," small amounts of chlorine gas remain and could escape if the valve were damaged. Full or empty, cylinders should always be stored upright and secured.

Ton containers. The ton container is a reusable, welded tank. It holds 2000 lb (910 kg) of chlorine and weighs about 3700 lb (1700 kg) when full. Containers are generally 30 in. (0.76 m) in diameter and 80 in. (2.03 m) long. As shown in Figure 10-10, the ends are concave and the container is crimped around the perimeter of the ends, forming good gripping edges for the hoists used in lifting and moving the containers. The ton container is designed to rest horizontally both in shipping and in use. It is equipped with two valves and two EDUCTORS, so the operator has the option of withdrawing liquid or gaseous chlorine.

Handling these heavy containers is, by necessity, far more mechanized than handling cylinders. Containers are loaded or unloaded using a lifting beam in combination with a manual or motor-operated hoist mounted on a monorail (Figure 10-14) with at least a 2-ton (1815-kg) capacity. The containers are always stored horizontally. To prevent accidental rolling, they are stored on TRUNNIONS

Figure 10-14. Lifting Beam for Ton Containers with Motorized Hoist

Figure 10-15. Ton Containers Stored on Trunnions

(Figure 10-15). The trunnions allow the container to be rotated so it can be correctly positioned for connection into the chlorine supply line.

Ton containers may be stored indoors or outdoors and require the same precautions as 150-lb cylinders. Full and empty ton containers should receive the same care. Often, "empty" containers still contain enough chlorine to pose a danger if a leak were to occur. The bowl-shaped hood that covers the two valve assemblies when the tank is delivered should be replaced each time the container is handled, even when it is "empty" and being moved to the storage area.

The storage area serves as a stockpile area, providing space for a 30- to 60-day supply of chlorine. Chlorine may also be fed directly from this area. When ton

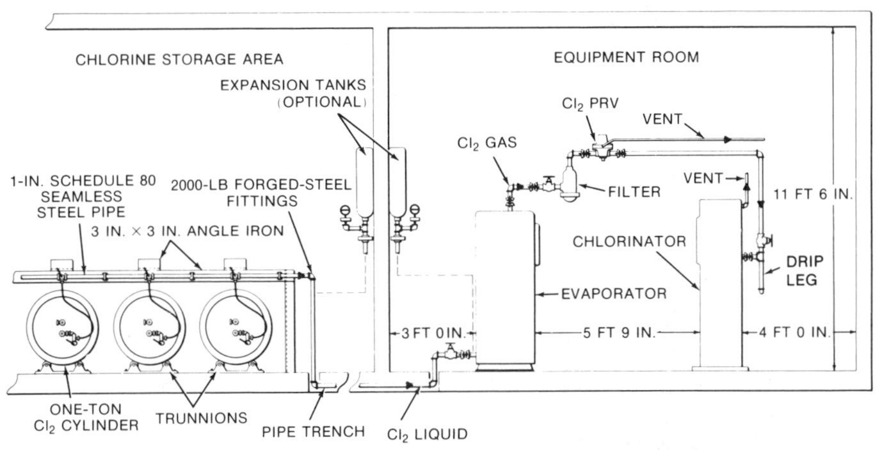

From Handbook of Chlorination *by Clifford White, copyright © 1972 by Van Nostrand Company. Reprinted by permission of the publisher.*

Figure 10-16. Chlorination Feed Equipment Located in a Separate Room

containers are used, the chlorination feed equipment is usually in a separate room (Figure 10-16).

The decision of whether to use 150-lb cylinders or ton containers should be based on cost and capacity. The cost per pound of chlorine in 150-lb cylinders is usually more than that of chlorine in ton containers. If a plant's needs for chlorine are less than 50 lb/day (23 kg/d), 150-lb cylinders usually should be selected. For systems using more than 50 lb/day (23 kg/d), ton containers are the most economical size.

Feeding Chlorine Gas

Chlorine feeding begins where the cylinder or ton container connects to the chlorinator (or to the chlorine-supply manifold if more than one container is connected). The feed system ends at the point where the chlorine solution mixes into the water being disinfected. The main components of the system are:

- Weighing scale
- Valves and piping
- Chlorinator
- Diffuser or injector.

Weighing scales. It is important that an accurate record be kept of the amount of chlorine used and the amount of chlorine remaining in a cylinder or container. A simple way to do this is to place the cylinders or ton containers on weigh scales. The weigh scales can be calibrated to read the amount used or the amount remaining. By recording weight readings at regular intervals, the

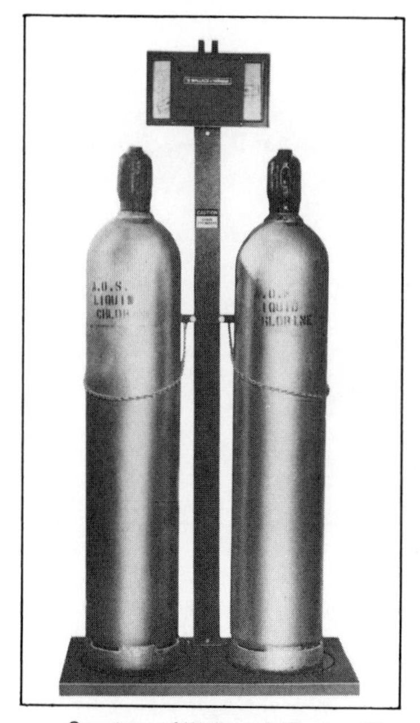

Courtesy of Wallace & Tiernan Div., Pennwalt Corp.

Figure 10-17. Two-Cylinder Scale

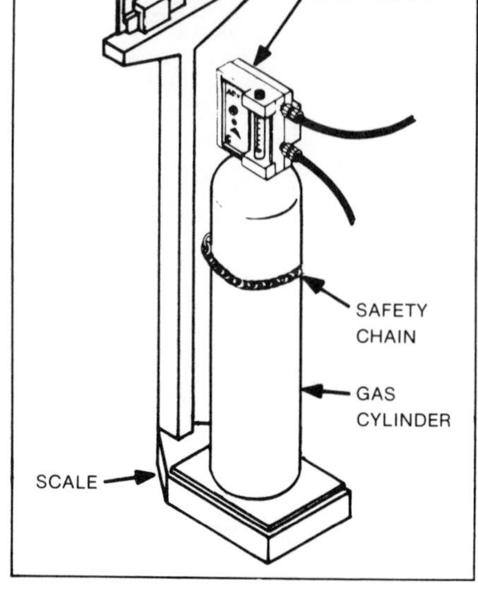

Courtesy of Capital Controls Company

Figure 10-18. Portable Beam Scale

operator can develop a record of chlorine-use rates. Figure 10-17 shows a common type of two-cylinder scale. Figure 10-18 shows a portable beam scale. Figure 10-19 shows a combination trunnion-and-scale for use with ton containers. The scale operates hydraulically and has a dial readout.

Valves and piping. Chlorine cylinders and ton containers are equipped with valves. These CYLINDER VALVES or CONTAINER VALVES, shown in Figures 10-20 and 10-21, comply with the standards set by the Chlorine Institute. It is standard practice for the operator to connect an AUXILIARY TANK VALVE directly to the container valve, as shown in Figure 10-22. The connection should be made using either a union-type or yoke-type connector. The auxiliary valve can be used to close off all downstream piping, minimizing gas leakage during container changes. The auxiliary tank valve will also serve as an emergency shut-off if the container valve fails. If a direct-mounted chlorinator is used (Figures 10-23 and 10-24), an auxiliary tank valve is not required.

A typical valve assembly is pictured in Figure 10-22. The figure shows that the assembly is connected to the chlorine-supply piping by flexible tubing, commonly used for this purpose. The tubing is usually 3/8-in. (10-mm) copper, rated at 500 psig (3500 kPa).

When more than one container is connected, a manifold such as the one shown in Figure 10-22 must be used. The manifold channels the flow of chlorine from two or more containers into the chlorine-supply piping. The manifold and

Courtesy of Force Flow Equipment

Figure 10-19. Combination Trunnion-and-Scale

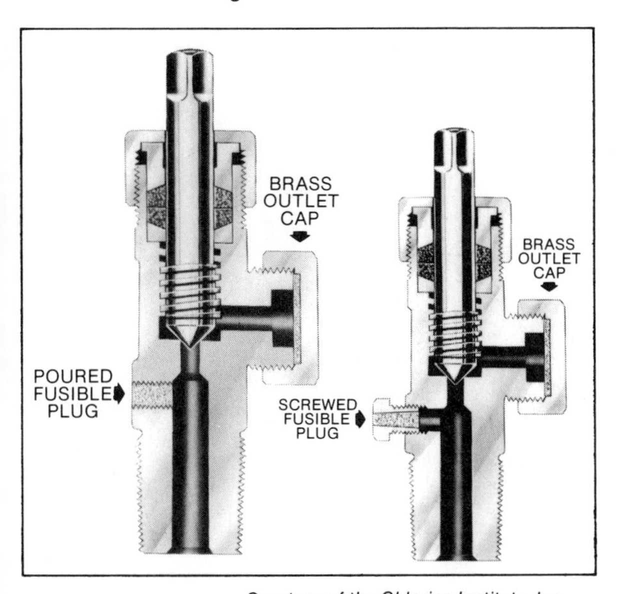

Courtesy of the Chlorine Institute, Inc.

Figure 10-20. Standard Cylinder Valve: Poured-Type Fusible Plug (left) and Screwed-Type Fusible Plug (right)

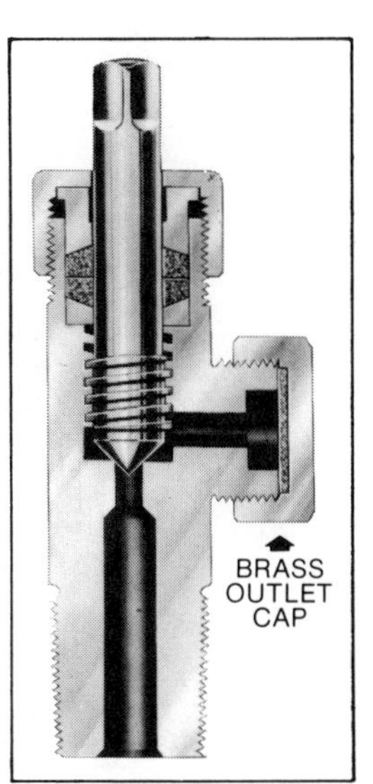

Courtesy of the Chlorine Institute, Inc.

Figure 10-21. Standard Ton Container Valve

supply piping must meet the specifications of the Chlorine Institute. Manifolds may have from 2 to 10 connecting points. Each connecting point is a union nut suitable for receiving flexible connections. Notice in Figure 10-22 that a header valve is connected at the manifold discharge end, providing another shut-off point. Additional valves are used along the chlorine supply line for shutoff and isolation.

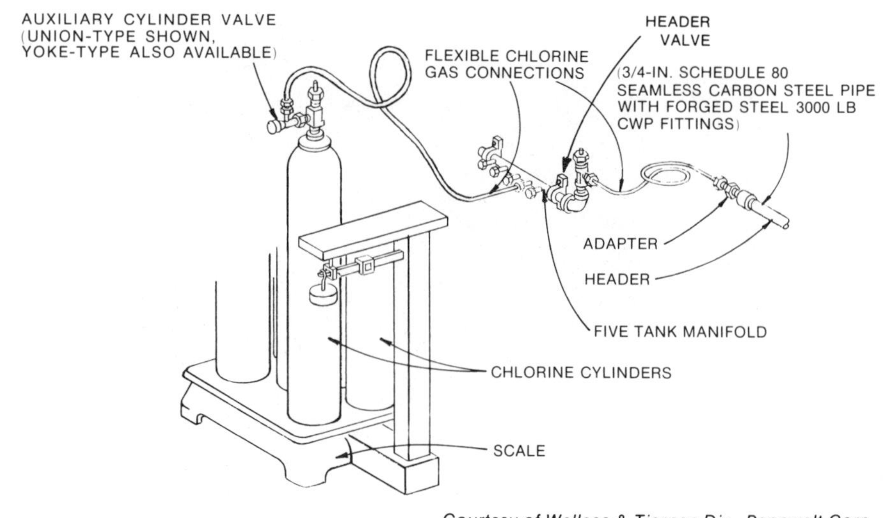

Courtesy of Wallace & Tiernan Div., Pennwalt Corp.

Figure 10-22. Auxiliary Tank Valve Connected Directly to Container Valve

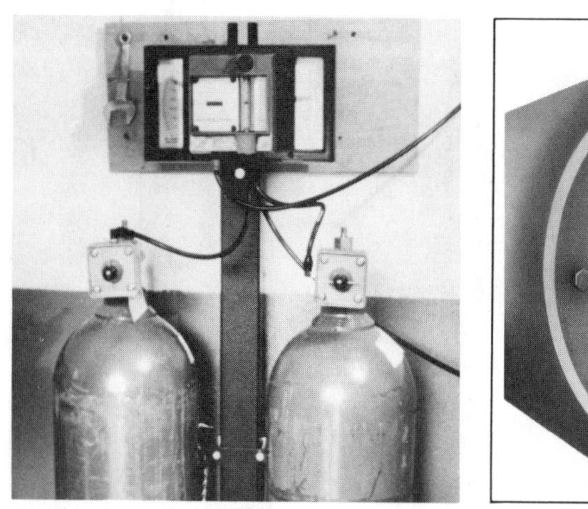

Figure 10-23. Direct-Mounted Chlorinator

Courtesy of Fischer & Porter Co., Warminster, PA 18974.

Figure 10-24. Direct-Mounted Chlorinator

Chlorinators. The CHLORINATOR can be a simple direct-mount unit on a cylinder or ton container (Figures 10-23 and 10-24) feeding chlorine gas directly to the water being treated; or it can be a free-standing cabinet (Figure 10-25). Figures 10-26 and 10-27 are schematic diagrams of two typical chlorinators. The purpose of the chlorinator is to safely and accurately meter chlorine gas from the cylinder or container, and accurately deliver the set dosage. To do this, a chlorinator is equipped with pressure and vacuum regulators, actuated by

Courtesy of Wallace & Tiernan Div., Pennwalt Corp.

Figure 10-25. Free-Standing Chlorinator Cabinet

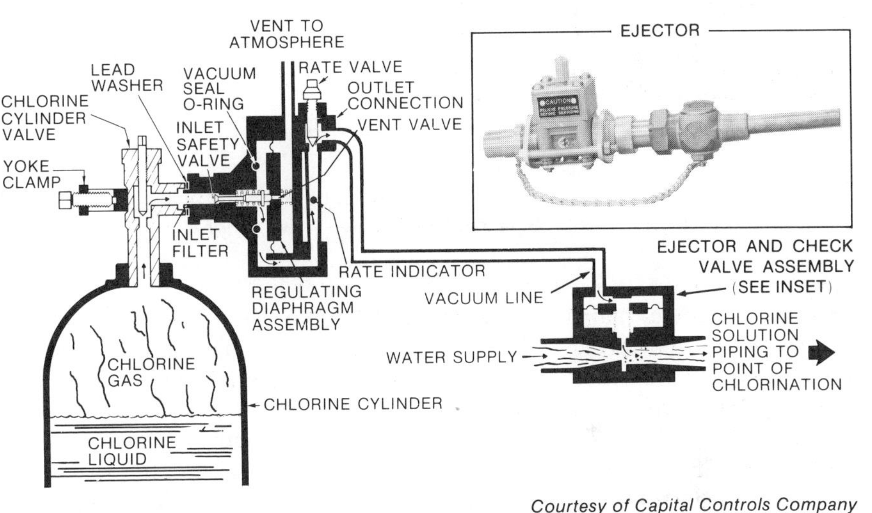

Courtesy of Capital Controls Company

Figure 10-26. Schematic of Direct-Mounted Gas Chlorinator

diaphragms and orifices that reduce the chlorine gas pressure. The reduced pressure allows a uniform gas flow, accurately metered by the ROTAMETER (feed rate indicator). In addition, a vacuum is maintained in the line to the injector for safety purposes. If a leak develops in the vacuum line, air will enter the atmospheric vent, causing the vacuum relief valve to close and stopping the flow of chlorine gas. To vary the chlorine dosage, the operator manually adjusts the setting of the rotameter.

Injector. An INJECTOR, or EJECTOR (Figure 10-26) located within or downstream of the chlorinator, is a venturi device that pulls chlorine gas into a passing stream of dilution water, forming a strong solution of chlorine and water. The injector also creates the vacuum needed to operate the chlorinator. The highly

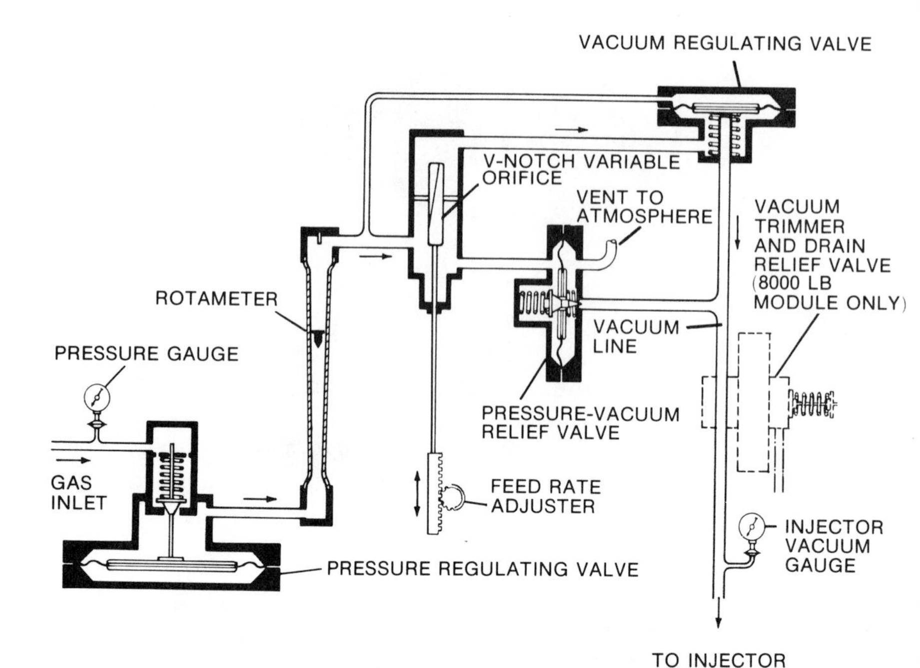

Courtesy of Wallace & Tiernan Div., Pennwalt Corp.

Figure 10-27. Schematic of Cabinet-Style Chlorinator

corrosive chlorine solution (pH of about 2 to 4) is carried to the point of application with the main flow of water in PVC or fiberglass pipe or in steel pipe lined with PVC or rubber. A strainer should be installed on the water line before the injector. This prevents any grit, rust, or other material from entering and blocking the injector or causing wear of the injector throat.

Diffusers. A DIFFUSER is one or more short lengths of pipe, usually perforated, that quickly and uniformly disperses the chlorine solution into the main flow of water. There are two types of diffusers: those used in pipelines and those used in facilities such as open channels and tanks. A properly designed and operated diffuser is necessary for the complete mixing needed for effective disinfection.

The diffuser used in pipelines 3 ft (0.9 m) or less in diameter is simply a pipe protruding into the center of the pipeline. Figure 10-28 shows a diffuser made from Schedule 80 PVC, and Figure 10-29 shows how the turbulence of the flowing water completely mixes the chlorine solution throughout the water. Complete mixing should occur within 10 pipe diameters.

Figure 10-30 shows a perforated diffuser for use in pipelines larger than 3 ft (0.9 m) in diameter. A similar design is used to introduce chlorine solution into a tank or open channels, as shown in Figure 10-31. In the figure, the water level has been dropped to show the chlorine solution passing out each perforation. Normally, the perforations in the diffuser would be completely submerged.

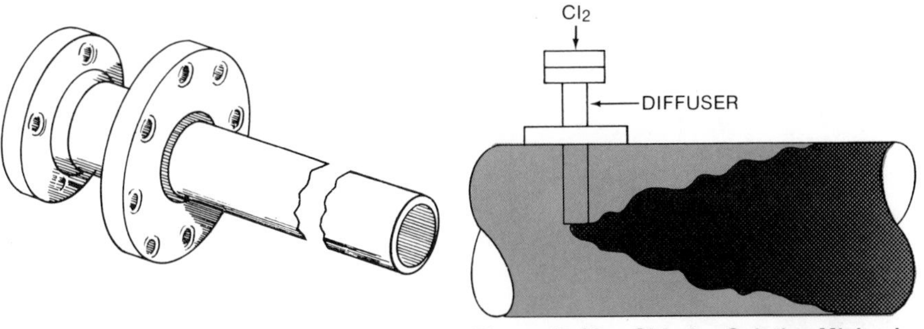

Figure 10-28. Diffuser Made From Schedule 80 PVC

Figure 10-29. Chlorine Solution Mixing in a Pipeline 2 ft or Less in Diameter

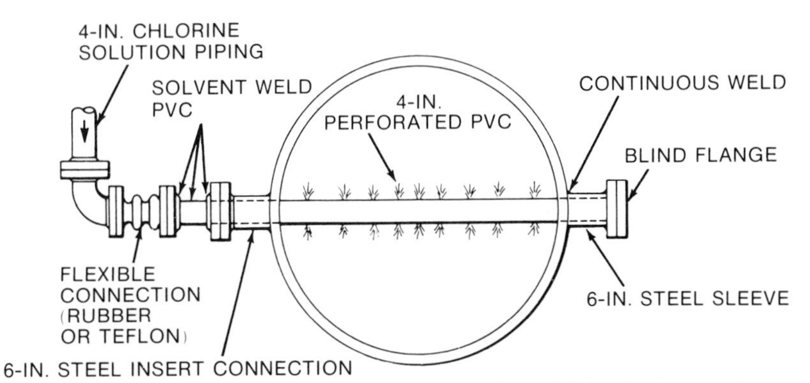

From Handbook of Chlorination *by Clifford White, copyright © 1972 by Van Nostrand Company. Reprinted by permission of the publisher.*

Figure 10-30. Perforated Diffuser for Pipelines Larger Than 3 ft

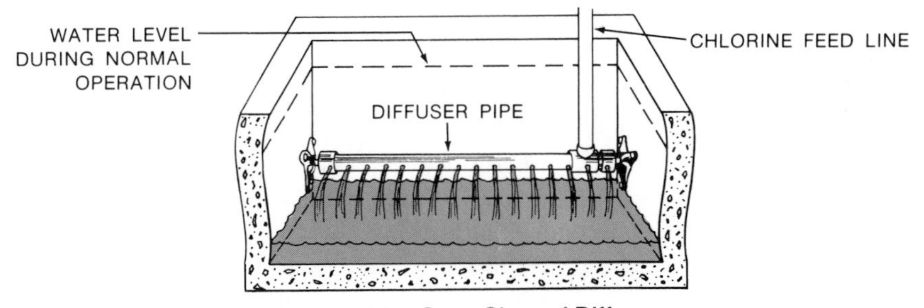

Figure 10-31. Open Channel Diffuser

Auxiliary Equipment

A variety of auxiliary equipment may be used in chlorination. The following discussion describes the functions of the more commonly used items.

Booster pumps. Many times a booster pump (Figure 10-32) is needed to provide the proper flow rate and velocity of water to ensure that the injector

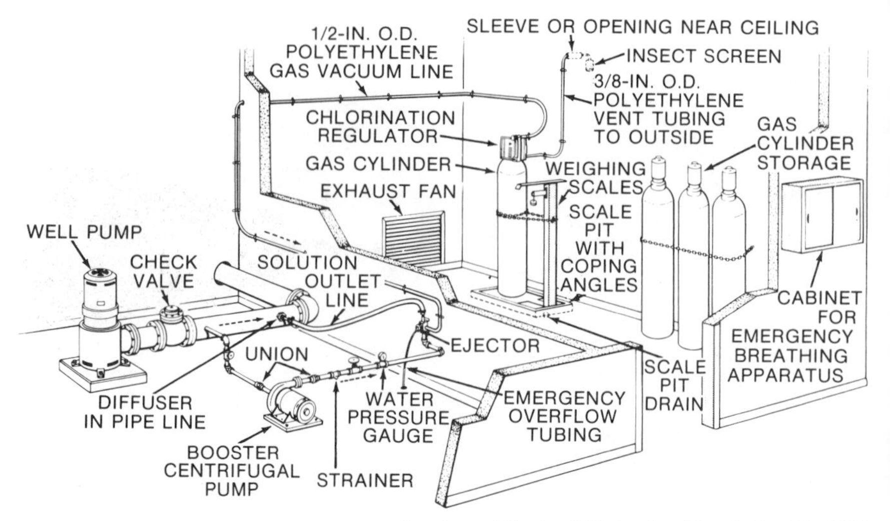

Courtesy of Fischer & Porter Co., Warminster, PA 18974.

Figure 10-32. Typical Chlorinator Deep Well Installation Showing Booster Pump

operates properly. The booster pump is usually a low-head, high-capacity centrifugal type. It must be sized to overcome the pressure in the line carrying the main flow of water being treated.

Automatic controls. The equipment described above is sufficient for a manual chlorination system. In such a system, to vary the amount of chlorine added, the rotameter is adjusted manually. Adjustments must be made each time the flow rate changes. For constant or near-constant flow rate situations, a manual system is suitable; however, when flow rates vary continuously, the system would require the operator to continuously change the rotameter settings. In such a situation, automatic controls are valuable. Although many automatic control arrangements are possible, there are two common types:

- Flow proportional control
- Residual flow control (also called compound loop control).

Flow proportional control. FLOW PROPORTIONAL CONTROL automatically increases or decreases the chlorine feed rate as the flow rate of the water increases or decreases. Equipment needed includes a flow-meter primary element for the treated water (discussed in Module 1, Preliminary Treatment and Flow Measurement); a transmitter to sense the flow rate and send a signal to the chlorinator; and a receiver at the chlorinator, which responds to the transmitted signal by opening or closing the valve controlling the chlorine flow rate.

As long as every gallon of water being treated requires the same dosage of chlorine, flow proportional control works well. However, if water quality changes, either gradually or suddenly, the required chlorine dosage also changes. Flow proportional control cannot sense or correct for this change in chlorine demand.

Residual flow control. RESIDUAL FLOW CONTROL automatically maintains a constant chlorine residual, no matter how chlorine demand (based on water quality) or water flow rate changes. The system uses an automatic chlorine-residual analyzer (Figure 10-33) in addition to a primary flow-rate element. The analyzer uses an electrode to determine the chlorine residual in the treated water. Signals from the residual analyzer and flow element are sent to a receiver in the chlorinator. That receiver combines the two signals to control a motor-operated valve, which opens or closes to adjust chlorine feed rate.

Evaporator. An EVAPORATOR is a heating device used to convert liquid chlorine to chlorine gas. As noted above, ton containers are equipped with both gas and liquid eductor pipes. At 70° F (21° C), the maximum gas withdrawal rate from ton containers is 400 lb/day (180 kg/d). If higher withdrawal rates are needed, the liquid eductor pipe is connected to an evaporator. The evaporator speeds the change of liquid chlorine to gas, so withdrawal rates up to 9600 lb/day (4400 kg/d) can be obtained.

An evaporator (Figure 10-34) is a water bath heated by electric immersion heaters to a temperature of 170 to 180° F (77 to 82° C). The pipes carrying the liquid chlorine pass through the water bath, and liquid chlorine is converted to chlorine gas by the heat.

Automatic switchover system. In many small water systems it is either impossible or uneconomical to have an operator available to monitor operation of the chlorination system 24 hours a day. An automatic switchover system makes 24-hour operation of the chlorinator possible without full-time staff attention. The system provides automatic switchover to a new chlorine supply when the on-line supply runs out. The switchover from one tank to another is either pressure or vacuum activated. The vacuum type of installation is shown in Figure 10-35. The automatic changeover mechanism has two inlets and one

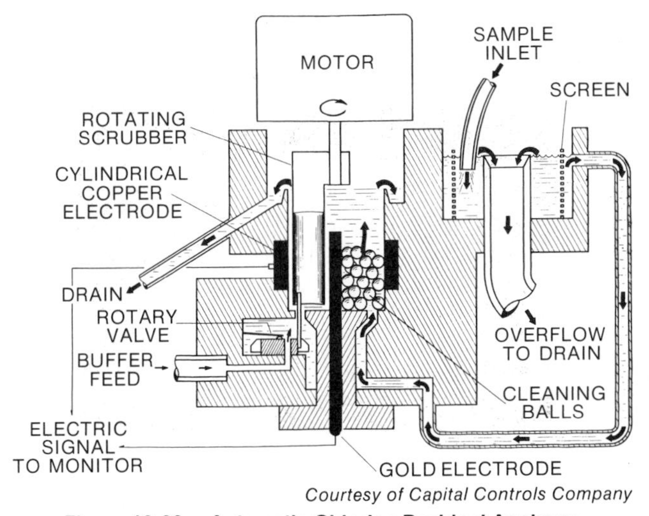

Courtesy of Capital Controls Company

Figure 10-33. Automatic Chlorine Residual Analyzer

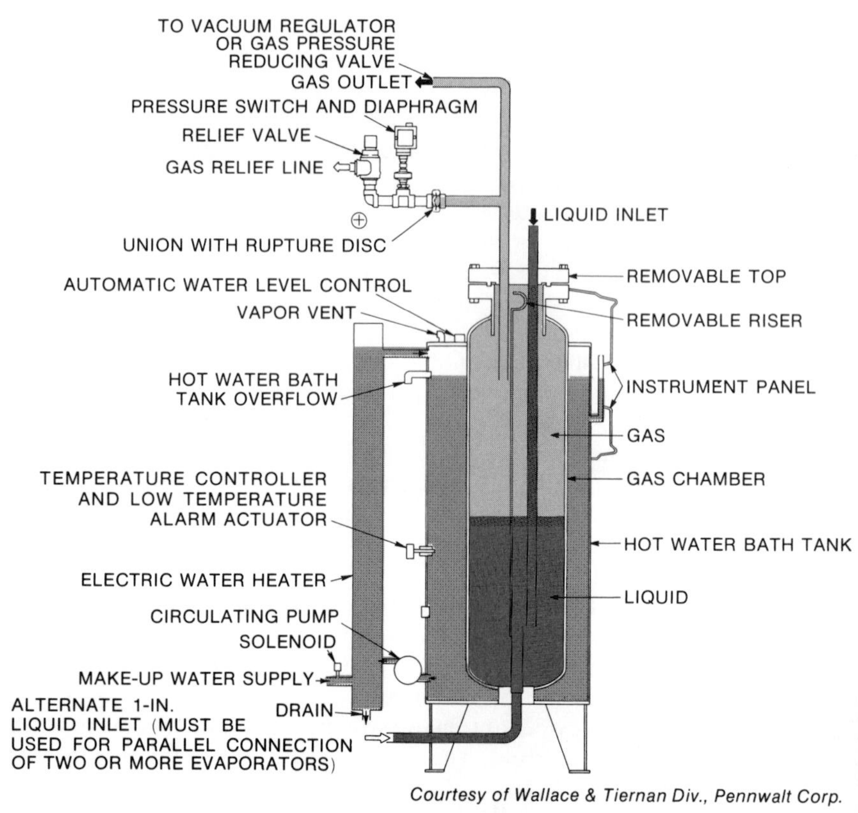

Courtesy of Wallace & Tiernan Div., Pennwalt Corp.

Figure 10-34. Chlorine Evaporator

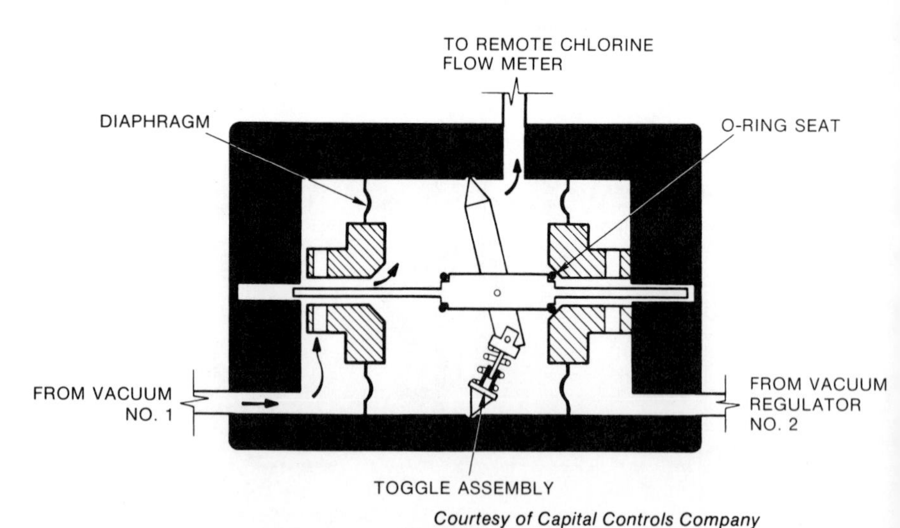

Courtesy of Capital Controls Company

Figure 10-35. Automatic Switchover Unit

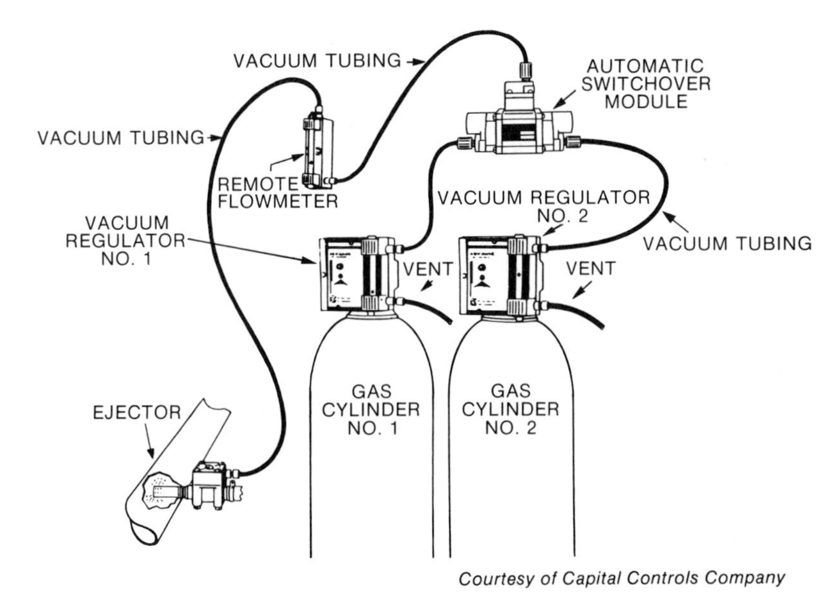

Figure 10-36. Typical Installation of Switchover System

outlet. As the on-line supply is exhausted, the vacuum increases, causing the changeover mechanism to close on the exhausted supply and open the new chlorine supply. The unit can send a signal to notify operating personnel that the on-line source is empty and should be replaced when convenient. Figure 10-36 shows a typical installation. This system is ideal for remote locations to ensure uninterrupted chlorine feeding.

Chlorinator alarm. Chlorinators are often equipped with a vacuum switch that triggers an alarm when it senses abnormally low or high vacuum. A low-vacuum condition can mean an injector failure, vacuum-line break, or booster-pump failure. A high-vacuum condition can be caused by a plugged chlorine-supply line or by empty chlorine tanks.

Safety equipment. Safety in and around the chlorination process is important in preventing serious accidents and equipment damage. Safety equipment is discussed in detail in Section 10-7. There are certain items of equipment essential for safe operation of a chlorination facility, including:

- Chlorine detectors
- Self-contained breathing apparatus
- Emergency repair kits.

Hypochlorination Facilities

HYPOCHLORINATION is a method for chlorination that is well suited to smaller water-supply facilities. As a rule, plants using less than 3 lb/day (1.4 kg/d) of available chlorine will use hypochlorinators. However, hypochlorinators may be used with any size system.

Table 10-5. Properties of Hypochlorites

Property	*Sodium Hypochlorite*	*Calcium Hypochlorite*
Symbol	NaOCl	$Ca(OCl)_2$
Form	Liquid	Dry granules, powder, or tablets
Strength	Up to 15% available chlorine	65–70% available chlorine, depending on form
pH	9–11, depending on percentage of available chlorine	———

Hypochlorite compounds. The two most commonly used compounds are calcium hypochlorite and sodium hypochlorite. Table 10-5 lists the properties of both compounds.

Common equipment. Plants using calcium hypochlorite are usually equipped with:

- A cool, dry storage area to stockpile the compound in their shipping containers.
- A variable-speed chemical feed pump (hypochlorinator)—usually a positive-displacement diaphragm type.
- A mix tank and day tank (Figure 10-37). After mixing calcium hypochlorite with water, impurities and undissolved chemicals settle to the bottom of the mix tank. The clear solution is then transferred to the day tank. This prevents the undissolved material from reaching and plugging the hypochlorinator or rupturing the diaphragm.

Since sodium hypochlorite is a liquid, it is simpler to use than calcium hypochlorite. Sodium hypochlorite may be fed full strength or it may be diluted with water, usually to a 1-percent solution. In either case, only one feed tank is required.

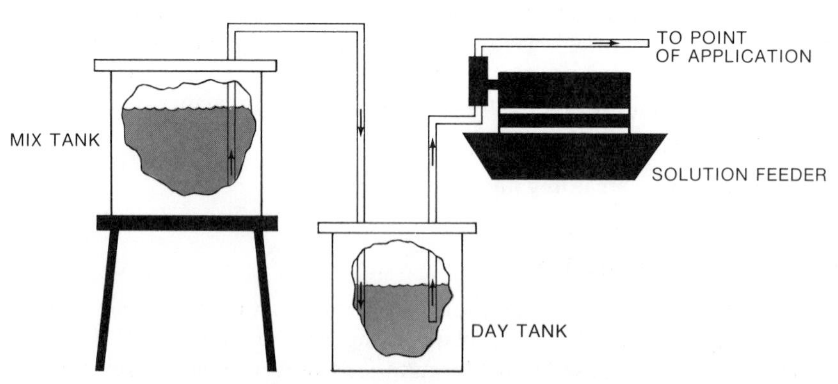

Figure 10-37. Mix Tank and Day Tank

10-4. Operation of the Chlorination Process

Successful operation of the chlorination process requires an understanding of how each of the system components operates. In addition, the operator must be aware of the safety procedures that must be followed when changing cylinders and when dealing with leaks or breakdowns.

Using Cylinders and Ton Containers

Cylinders. The 100-lb or 150-lb chlorine cylinders should always be stored and used in an upright position. In this position, the cylinders will deliver chlorine gas continuously (without frost formation) at a maximum rate of about 42 lb/day (19 kg/d) against a water pressure of 35 psig (249 kPa) at 70° F (21° C).

If chlorine gas is withdrawn at a rate greater than 42 lb/day (1.75 lb/hour, or 19 kg/d), the pressure in the cylinder will drop very quickly, causing a sudden decrease in cylinder temperature. If the high withdrawal rate continues, frost will form on the outside of the cylinder. This reduces the withdrawal rate because the cooler temperature retards the vaporization of the liquid chlorine.

To eliminate frosting, air circulation around the cylinder should be improved by placing a fan in the chlorinator room. If the problem continues, the withdrawal rate must be reduced. Heat should never be applied directly to the container, since the fusible plug may melt or pressure may increase to the point where the valve fails, causing a serious chlorine leak.

Ton containers. Ton containers are transported, hoisted, stored, and used in the horizontal position. As shown in Figure 10-10, the containers must be positioned so the two valves are oriented one above the other. In this position, the top valve delivers chlorine gas, and the bottom delivers liquid chlorine.

Ton containers can deliver chlorine gas at rates up to 400 lb/day (180 kg/d) against a back pressure of 35 psig (240 kPa) at 70° F (21° C) without frost appearing on the container wall. Liquid chlorine can be delivered at rates up to 9600 lb/day (4400 kg/d) if an evaporator is being used. Gas withdrawal can be used in any situation if enough containers are available at the same time. For example, a plant's requirement of 1400 lb/day (640 kg/d) could be supplied by using four ton containers (4 × 400 lb/day = 1600 lb/day) connected to a common supply line or manifold.

The exact maximum withdrawal rate of gas for vacuum systems can be determined with the following formula:

Chlorine-Room Temp °F		Threshold Temp °F		Withdrawal Factor		Maximum Withdrawal Rate lb/day
	−		×		=	

The withdrawal factor depends on the size and shape of the cylinder or container. For a 150-lb cylinder, the factor is 1.0; for a ton container, the factor is 8.0.

The threshold temperature is the temperature at which the minimum gas pressure required to operate a gas chlorinator at the point of withdrawal (line pressure at point of application) is reached. Threshold temperature is determined

Table 10-6. Values from the Vapor Pressure Curve for Liquid Chlorine

Minimum Pressure psig (kPa)	*Threshold Temperature* °F (°C)	*Minimum Pressure* psig (kPa)	*Threshold Temperature* °F (°C)
9 (62)	−10 (−23)	47 (324)	40 (4)
14 (97)	0 (−18)	59 (407)	50 (10)
21 (145)	10 (−12)	71 (490)	60 (16)
28 (193)	20 (−7)	86 (593)	70 (21)
37 (255)	30 (−1)	102 (703)	80 (27)

from the vapor pressure curve of liquid chlorine. Table 10-6 is a chart of values from that curve. For example, given a 60° F (15° C) room temperature, a 150-lb cylinder having a withdrawal factor of 1, and a minimum gas pressure of 14 psig (97 kPa), the threshold temperature is 0° F (−17° C).

The maximum withdrawal rate under these conditions is:

$$(60 - 0)\ 1 = 60 \text{ lb/day maximum withdrawal rate.}$$

Weighing Procedures

The only reliable method of determining the amount of chlorine remaining in a cylinder or container is to weigh the unit and check its weight against the TARE WEIGHT (empty weight) stamped on the shoulder of the cylinder or container. This information can be used to determine the feed rate and to decide when to change cylinders or containers. Since the pressure in a cylinder depends on the liquid chlorine temperature, not upon the amount of chlorine in the container, the pressure cannot be used to determine when a cylinder or container needs changing.

Weighing procedures depend on the scale in use. Simple scales show the combined weight of container and chlorine. On other scales, the operator sets the tare weight into the scale and the scale displays only the weight of chlorine remaining in the cylinder or container.

Connecting Cylinders and Ton Containers

When connecting a cylinder to the chlorine supply line or manifold (using yoke and adapter), the following procedure should be observed.

1. Always wear personal respiratory protection when changing cylinders or containers.
2. Never lift cylinder by the protective hood. The protective hood on the 150-lb cylinder is screwed onto a threaded neck ring. Despite its appearance, the neck ring is not part of the cylinder, and is often not securely attached to the cylinder.
3. Secure the cylinder with a safety chain or steel strap in a solid, upright position.
4. Remove the protective hood. If the cylinder has been exposed to the weather for a long time, the threads at the base of the hood may have been corroded, in which case a few sound raps with a wooden or rubber mallet on opposite sides of the hood will loosen it so it can be unscrewed easily.

5. Remove the brass outlet cap (Figures 10-20 and 10-21) and any foreign matter that may be in the valve outlet recess. Use a wire brush to clear out any pieces of the old washer, being careful not to scratch the threads or gasket-bearing surface.

6. Place a new lead washer (Figure 10-26) in the outlet recess. Do not reuse washers.

7. Place the yoke over the valve. Insert the adapter in the outlet recess and then, fitting the adapter in the yoke slot, tighten the yoke screw. Make sure the end of the adapter seats firmly against the washer. Use only the cylinder-valve wrench provided by the chlorine supplier for all chlorine cylinder or container valve connections.

8. Install the flexible connector, sloping it back toward the chlorine cylinder so that any liquid chlorine droplets will flow back to the cylinder and not to the chlorinator unit (Figure 10-38).

The procedure for connecting ton containers to the piping or the manifold is basically the same as described for the 150-lb cylinders. Note that when the cylinder or the ton container is being connected, both the container valve and auxiliary valve should be closed. Ton containers should have a DRIP LEG (also called a liquid chlorine trap) with a heater installed before the chlorinator to vaporize any liquid chlorine initially coming from the eductor at the start of gas operation and any liquid droplets that may flow out of the container or evaporator during normal operation (Figure 10-16).

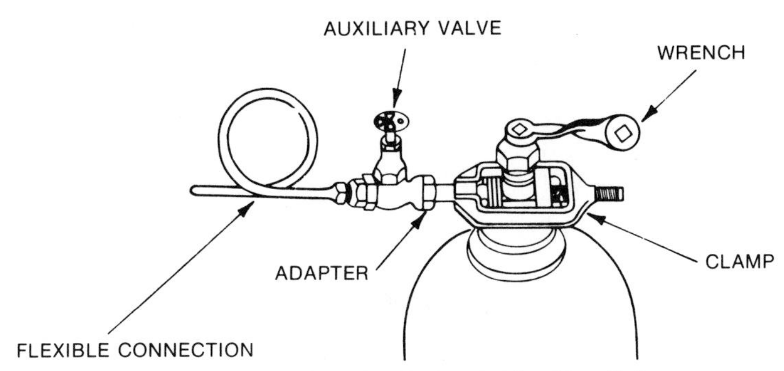

Figure 10-38. An Installed Yoke and Auxiliary Valve

Opening the Valves

Once the cylinder or container has been installed and the flexible connector attached, the valves should be opened and the lines checked for leakage according to the following procedure:

1. Place the valve wrench provided by the chlorine supplier on the cylinder or container valve stem. Stand behind the valve outlet. Grasp the valve firmly with one hand and hit the wrench a sharp blow in a counterclockwise direction with the palm of the other hand. Do not pull or tug at the wrench as this may bend the stem, causing it to stick or to fail to close properly.

To open a stubborn valve, follow the normal opening procedure, but use a small block of wood held in the palm of the hand when striking the wrench. If the valve continues to resist opening, return the cylinder to the supplier. Do not under any circumstances use a pipe wrench or an ill-fitting wrench, since these wrenches will round the corners of the squared-end valve stem. Avoid the use of wrenches longer than 6 in. (150 mm) for opening stubborn valves because of the danger of bending or breaking the valve stem.

2. Open the valve and close it immediately.

3. The line is now pressurized and all new joints and connections can be checked for leaks using an ammonia solution. Use only commercial 26° BAUME' ammonia, which can be obtained from a supplier of chemicals or chlorine. Common household ammonia is not strong enough. Hold an opened plastic squeeze bottle of the ammonia beneath the valve, joint, or any possible leak, and allow the ammonia fumes to rise up around the suspected area. Ammonia fumes react with chlorine gas to form a white cloud of ammonium chloride, making small leaks easy to locate. Do not spray, swab, or otherwise bring ammonia liquid into contact with chlorination equipment—the chlorine gas will combine with the liquid ammonia and may start corrosion of the leaky connection.

4. If no leaks are found, open the cylinder valve. One complete turn will permit the maximum withdrawal rate.

5. Leave the wrench on the valve to allow for easy and rapid shutoff in an emergency. The wrench also indicates to other operators which cylinder is being used. A sign or disk should be installed on the stem of the cylinder or container valve (Figure 10-39) to show the direction of closure in case an emergency should arise.

6. If the injector and chlorinator are already operating, open the auxiliary valve and the newly connected cylinder or container will start feeding chlorine to the system. Do not open the auxiliary valve until the injector is operating since the necessary vacuum will not be developed and the regulating valves will not function.

Courtesy of Capitol Controls Company

Figure 10-39. Sign to Show Direction of Valve Closure in Case of an Emergency

Closing the Valves

When all the chlorine has been released from the cylinder or container (as indicated by the weigh-scale reading), the container should be disconnected and replaced according to the following procedures:

1. Close the cylinder or container valve. After about 2 min, close the auxiliary valve. The 2-min delay allows the remaining chlorine gas in the line to be drawn into the injector. To close the valve, use the wrench provided, grasping the valve in one hand and tapping the wrench in a clockwise direction with the palm of the other. If the valve does not close tightly on the first try, open and close it lightly several times until the proper seating is obtained. Never use a hammer or any other tool to close the cylinder valve tightly.
2. Disconnect the flexible tubing from the cylinder or container, and replace the brass outlet cap on the cylinder valve immediately so that the valve parts will be protected from moisture in the air.
3. Screw the protective hood in place.
4. The outlet cap of each valve is fitted with a gasket that is designed to fit against the valve outlet face. If a valve leaks slightly after closing, the leak can often be stopped by drawing up the valve cap tightly.
5. The flexible copper tubing from the manifold to the container should be supported while the empty container is being replaced. Support it on another container, a wall hook, or a block, in order to prevent any kinking or weak spots from developing in the pipe. If the flexible tubing is disconnected for any length of time, there is a danger of moisture forming in the line. Close the open end of the pipe with tape or plastic wrap and a rubber band.

Determining the Chlorine Dosage

The chlorine demand of the water being treated must be determined by performing a chlorine-demand test. Using the result from the demand test and knowing the residual needed, the operator can determine the chlorine dosage as follows:[2]

$$\text{Chlorine Dosage, mg/L} = \text{Chlorine Demand, mg/L} + \text{Chlorine Residual, mg/L}$$

Once calculated, the dosage should be converted into pounds per day and the chlorinator should be set to deliver that dosage. After setting the dosage rate, the water must be tested regularly to ensure the proper chlorine residual is maintained.

As noted previously, temperature, pH, and contact time are important variables affecting the success of chlorination. As these factors change, the amount of residual needed will also change. Table 10-7 shows recommended minimum residuals. Notice that recommendations are included for free residual

[2] *Basic Science Concepts and Applications*, Chemistry Section, Chemical Dosage Problems (Chlorine Dosage/Demand/Residual).

Table 10-7. Recommended Minimum Concentrations of Free Chlorine Residual Versus Combined Chlorine Residual to Ensure Effective Disinfection

pH Value	*Minimum Concentration of Free Chlorine Residual (disinfecting period is at least 10 min)* ppm	*Minimum Concentration of Combined Chlorine Residual (disinfecting period is at least 60 min)* ppm
6.0–7.0	0.2	1.0
7.0–8.0	0.2	1.5
8.0–9.0	0.4	1.8
9.0–10.0	0.8	Not recommended
10.0+	0.8 + (with longer contact)	Not recommended

as well as comparable values for combined residual. Free residual values are based on a contact time of at least 10 min; the combined residual concentrations require at least 60 min contact time.

10-5. Common Operating Problems

There are a variety of problems that can occur related to chlorination:

- Chlorine leaks
- Stiff container valves
- Hypochlorinator problems
- Tastes and odors
- Sudden change in residual
- Trihalomethane formation.

Other problems can occur (particularly when using chlorine gas) since more piping, valves, and equipment are involved. Proper maintenance can prevent many of these problems. Appendix B presents maintenance guidelines for gas chlorination equipment. If problems develop, the manufacturer's troubleshooting guides can be used to help locate and correct the problems.

Chlorine Leaks

A major concern in the operation and maintenance of the chlorination process is the prevention of chlorine leaks. The most common place for leaks to occur is the pressurized chlorine supply line between the containers and the chlorinator. Every joint, valve, fitting, and gauge in the line is a possible point of leakage.

Some chlorine leaks are obvious. Others are very slow, very small, partly hidden, or otherwise difficult to locate. The usual method of detection is to open a bottle of ammonia solution, place the bottle near a suspected leak, and allow the ammonia fumes to rise around the suspect area. If there is a sizeable leak, the chlorine will combine with the ammonia to form visible, white, ammonium-chloride vapor. Unfortunately, this method will not show the existence of a very small, slow leak (which also is usually too small to produce a noticeable odor).

Small leaks can go unnoticed for weeks unless the operators periodically look for two signs: (1) joint discoloration and (2) moisture.

Even the smallest leak will remove cadmium plating from chlorine tubing and fittings. The metal underneath (copper, brass, or bronze) will appear reddish, and a green copper-chloride scum may appear around the edges of the area affected.

Portions of the pressure piping system (the manifold, for example) are often painted. As a result, discoloration of the metal beneath the paint will not be apparent. To locate leaks in painted piping, look for small droplets of water that may form on the underside of joints. Small, almost-invisible leaks must be located early, or the corrosion they cause may, in a matter of months or even weeks, result in a sudden and massive chlorine leak.

The best and most reliable way to detect chlorine leaks is with a chlorine detector. The detector is sensitive to very small leaks—as small as 1 ppm chlorine in air. Such leaks are not normally detectable by the ammonia technique or by smell. If a major leak requires shutdown of the system for repair, then the tank valve should be closed, the yoke disconnected, and the injector left running with the auxiliary tank valve open until any remaining chlorine gas is purged from the line. Emergency kits for leak repair are discussed in Section 10-7. To prevent leaks, the operator should observe the following precautions:

- Replace the lead gasket every time a cylinder or container is changed.
- Each time a threaded fitting is opened, clean the threads with a wire brush and wrap with PTFE (sold under the trade name Teflon) tape or use one of the following pipe joint compounds: (1) linseed oil and graphite, (2) linseed oil and white lead, or (3) litharge and glycerine. If PTFE tape is used, remove any previous remnants of tape before remaking the joint.
- Replace all chlorine-supply-line valves annually. Refit and repack the old ones so they are ready for use the following year.

Stiff Container Valves

Container valves are carefully checked before leaving the manufacturer's plant, but occasionally a valve may be stiff to turn or difficult to shut off tightly. The problem is often caused by overly tight packing. Sometimes the valve can be freed by opening and shutting a few times. If the valve still does not operate, set the container aside and call the supplier.

Hypochlorinator Problems

Two problems commonly occur with hypochlorinators:

- Clogged equipment
- Broken diaphragms.

Clogged equipment. Clogging, due to scaling with calcium carbonate ($CaCO_3$), occurs primarily in two areas of a hypochlorinator: (1) the pump head and (2) suction and discharge hoses. Scale formation requires (1) the highly alkaline (about pH 10.6) hypochlorite solution, (2) calcium hardness, and (3) carbonate alkalinity in the make-up water. When all three factors are present, calcium carbonate forms and causes a deposition scale in the pump head, suction hose, and discharge hose; scale may also form in the solution injector or diffuser. This scale can be readily removed by pumping a dilute (5 percent) hydrochloric acid solution (also known as MURIATIC ACID) through the pump head, hoses, and diffuser. The hypochlorite solution should be completely flushed out of the system with water before the hydrochloric acid is used.

An associated problem affecting the pump head is the accumulation of dissolved calcium hypochlorite (lime sludge). When the solution tank level is low or the suction foot-valve is near the bottom of a one-tank installation, the suction foot-valve can draw the undissolved chemical up into the pump head and fill the area of the head. This can result in the pump not feeding hypochlorite solution and can cause diaphragm rupture.

The lime-sludge residue left after dissolving calcium hypochlorite can also be drawn into the pump head and cause the same results. To prevent these problems from occurring, a hypochlorite installation that uses calcium hypochlorite should use a two-tank setup. Clogging problems do not occur when sodium hypochlorite is used.

Broken diaphragms. The second most common problem with hypochlorinators is broken diaphragms. It is important for the operator to check the diaphragm regularly to ensure that the hypochlorinator is functioning properly. A visual inspection of the pump head may not reveal a broken diaphragm, but an outflow of solution from the discharge hose is a positive indication that the diaphragm is intact. Figures 10-40 and 10-41 indicate the points of diaphragm weakness in the two most common types of hypochlorinators.

Taste and Odor

There is a common misunderstanding about the cause of "swimming pool" tastes and odors in drinking water. It seems only natural to assume that water that tastes like chlorine must contain too much chlorine. Oddly enough, chlorine-like tastes and odors are usually caused by too little chlorine. In the

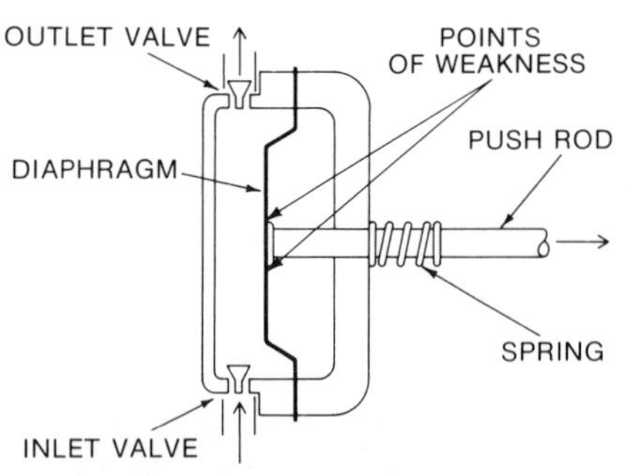

Figure 10-40. Point of Weakness in a Mechanically Actuated Diaphragm

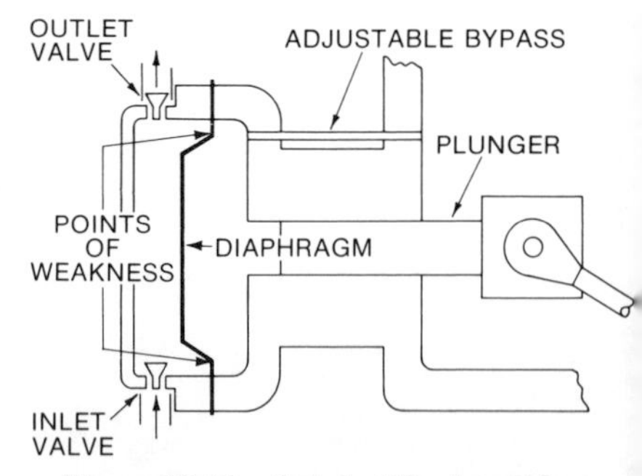

Figure 10-41. Point of Weakness in a Hydraulically Actuated Diaphragm

discussion of Figure 10-4, it was pointed out that the space between points 2 and 4 is predominantly combined chlorine residual. The combined residual available in this range is the weakest form of chlorine for disinfection, and the accompanying dichloramines, trichloramines and chloroorganic compounds cause taste and odor. By increasing the chlorine dosage beyond the breakpoint (beyond point 4 in Figure 10-4), the tastes and odors are reduced significantly or eliminated. Free available chlorine residual is free from taste and odor at the concentrations commonly used in disinfection.

Sudden Change in Residual

Chlorine offers a very valuable operating advantage over other disinfectants: measurable residual. Using a chlorine residual test procedure, the strength of disinfectant remaining in the water at any point in the distribution system can be determined. The test should take less than 5 min. Records of chlorine residual kept over several months or years can help predict the type and amount of residual that will be found at various locations in the distribution system. Any sudden drop in residual warns of a potentially dangerous situation (for example, a cross-connection that is allowing contaminated water into the drinking-water system).

When a sudden drop in residual occurs, chlorination levels should be increased immediately to bring the residual back to the desired level—this guards against the possibility of waterborne disease during the time that the problem is being analyzed and corrected. Samples should then be analyzed to identify the contaminants and the distribution system should be checked to locate the source of contamination or other cause of the drop in residual.

Formation of Trihalomethanes

All surface and ground waters contain natural organic compounds. These compounds come from decaying vegetation and are primarily humic and fulvic acids. They can cause color, taste, and odor problems in water. In addition, when organic compounds react with chlorine, complex chloroorganic compounds known as TRIHALOMETHANES (THMS) are formed. These compounds, the most common of which is chloroform, are considered potential cancer-causing substances. Because of the possible public health effects, the USEPA has set a maximum contaminant level of 0.1 mg/L (100 ppb) for trihalomethanes in drinking water. Further information on the regulation is included in Volume 4, *Introduction to Water Quality Analyses*.

Once trihalomethanes are formed, they are very difficult to remove. Therefore, the best way to keep THM levels low is to prevent them from forming. Most of the trihalomethanes are formed when untreated water is prechlorinated; the postchlorination disinfection process is generally not the major contributor.

One means of reducing THM levels is to discontinue prechlorination or move the prechlorination point to just ahead of the filters. This allows the coagulation/flocculation/sedimentation processes to remove a large part of the humic compounds before they have a chance to react with chlorine. The disinfection process itself should not be altered until comprehensive monitoring and engineering studies have been conducted.

10-6. Control Tests and Record Keeping

Control Tests

Two types of operational testing are used to monitor the disinfection process. First, the level of chlorine residual is checked regularly at points throughout the distribution system and in the treatment plant. Second, bacteriological tests (for total coliform bacteria) are performed on samples from selected points in the distribution system.

Chlorine residual test. The chlorine residual test is essential to successful and efficient operation of the chlorination process. The test results provide the operator with three important pieces of information:

- Whether or not a residual exists
- The type of residual (free or combined)
- The amount of residual (concentration).

The operator should also monitor temperature and pH. These two factors influence the amount and type of residual formed, which in turn controls the effectiveness of disinfection.

The chlorine residual test is one of the quickest and easiest to run of all water plant operational control tests. The recommended test, the DPD METHOD, can be performed using a field test kit, like any one of those shown in Figure 10-42. For laboratory use, the amperometric titrator is the most accurate method for measuring all forms of residual. The DPD method and the operational significance of chlorine residual are covered in detail in Volume 4, *Introduction to Water Quality Analyses*, Physical/Chemical Tests Module, Chlorine Demand and Residual.

As a general rule, it is best to maintain a free available chlorine residual, since it is a far more effective disinfectant than a combined residual. Usually, a free

Figure 10-42. Several Types of DPD Field Test Kits

residual of 0.3 mg/L is the minimum level to maintain at the extremities of the distribution system. The chlorinator dosage must be set to ensure that the desired level of free residual will exist at all times in all parts of the system.

Bacteriological test. All pathogenic organisms have a common source—the feces of humans and animals. One obvious way to determine the success of disinfection would be to test the treated water for the presence of pathogens. However, this is not practical since pathogens in water are few in number and quite difficult to measure, even with sophisticated laboratory equipment.

Fortunately, there is a group of bacteria, known as coliform bacteria, which are relatively easy to measure and whose presence indicates that pathogens might also be present. Since most pathogens are less resistant to chlorine than coliform bacteria, it is assumed that if no coliform bacteria are found in treated water, pathogens are not present. Therefore, routine samples are taken from the raw and treated water (including the distribution system) and tested for coliform bacteria (total coliform test). Discussion of the sampling and testing procedures for the total coliform test is included in Volume 4, *Introduction to Water Quality Analyses*, Microbiological Tests Module, Coliform Bacteria.

Record Keeping

Records for the chlorination process should show the type and amount of chlorine, and they should show bacteriological and other operational control test results. The following is a list of types of information that should be recorded as part of chlorination record keeping:

- Type of chlorine chemical in use
- Ordering information:
  —Manufacturers' name, address, and phone number
  —Shippers' name, address, and phone number
  —Type, size, and number of shipping containers
- Most recent costs
- Chemical Transportation Emergency Phone Number—available from CHEMTREC, a service of the manufacturing Chemists Association in Washington, D.C., for assistance with chlorine accidents
- Current dosage rate in milligrams per litre and pounds per day
- Bacteriological test results
- Chlorine-residual test results
- Water temperature
- pH
- Daily explanation of unusual conditions, mechanical problems, supply problems, emergencies, or unusual test results.

10-7. Safety and Chlorination

If used correctly by trained operators in adequately equipped plants, chlorine should never cause accident or injury. The key factors in safely operating a chlorination system are proper safety equipment and proper safety procedures.

Proper Safety Equipment

Without adequate equipment and the proper training in the use of such equipment, the operator's life as well as the life and well-being of the surrounding community is needlessly at risk.

It is essential that every chlorination facility be equipped with the following safety devices:

- Self-contained breathing equipment
- Emergency repair kit(s)
- Adequate ventilation equipment.

In addition, a chlorine detector should always be installed at unattended chlorination stations; the detectors are also good safety investments at any other chlorination location.

Self-contained breathing apparatus. Self-contained breathing equipment should be available wherever gas or liquid chlorine is in use. Air packs, as shown in Figure 10-43, have a positive-pressure mask with a full, wide-view face piece and a cylinder of air or oxygen carried on the operator's back. These units should be approved by the National Institute of Occupational Safety and Health (NIOSH). Before purchasing any such equipment, make sure it is approved.

Figure 10-43. Air Pack with Positive-Pressure Mask

Canister-type gas masks (those that filter chlorine from incoming air) are not recommended for personal protection under any circumstances.

Every operator should be familiar with the location and use of the breathing apparatus available at the treatment plant. Operating instructions are provided with each unit. Operators should study and periodically review these instructions and have regular formal training and practice sessions in the proper use of the equipment. General instructions on equipment use are given in the following paragraphs.

An air pack is very similar to the equipment used by scuba divers and fire departments. The tank contains a 15 to 30 min supply of air, depending on size. The actual length of air supply will vary with the individual, depending on the individual's pattern of breathing under stress. The mask must fit tightly around the face; operators with beards or eye glasses may find it difficult to fit the mask. Operators should practice repairing simulated leaks while using breathing equipment, so they will know how long they can work before the air-tank is empty. A low air pressure alarm on the 30-min air pack is activated when approximately 5 min of air is left. When the alarm sounds, the wearer of the pack should immediately leave the contaminated area to get a fresh cylinder of air.

The air pack should be located at a readily accessible point, away from the area likely to be contaminated with chlorine gas. The unit should not be located in the chlorine feed or storage rooms. Usually, a wall cabinet mounted outside these rooms is an excellent location for storage (Figure 10-44). In case of emergency, the operator should be able to get to the air pack without going through or into the room containing the leak.

The mask and breathing-air supply tanks should be routinely inspected and maintained in good condition. Spare air-supply cylinders should be on the site for use during prolonged emergencies.

The mask and breathing apparatus should be cleaned after each use and at regular intervals. When needed, air-supply tanks should be refilled at stations

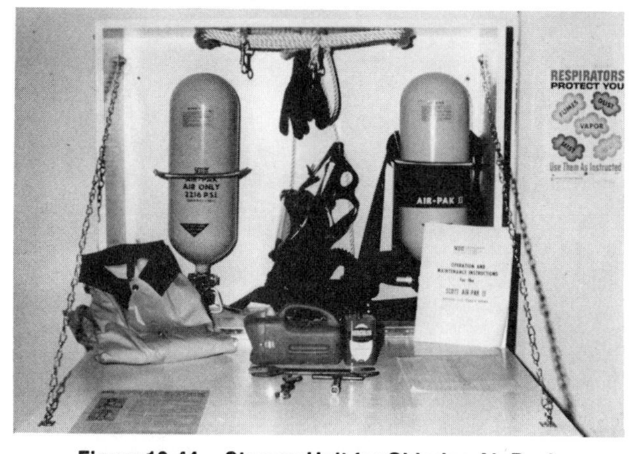

Figure 10-44. Storage Unit for Chlorine Air Pack

where proper air-compressor equipment is used. The local fire department usually has this equipment.

When putting away the equipment, the straps should be extended to their limits. This allows the equipment to be fitted quickly to the next user.

Emergency repair kits. Standardized emergency repair kits are available containing various devices and hardware for stopping leaks from chlorine-shipping containers meeting US and Canadian specifications. Currently, there are three standard Chlorine Institute emergency kits: Emergency Kit A for cylinders (Figure 10-45), Emergency Kit B for ton containers (Figure 10-46), Emergency Kit C for tank cars and tank trucks. Each kit is designed to be used in the repair of leaks that can occur in the shipping containers. The kits contain all the equipment needed to:

- Cap a leaking valve
- Seal a side-wall leak
- Cap a fusible plug.

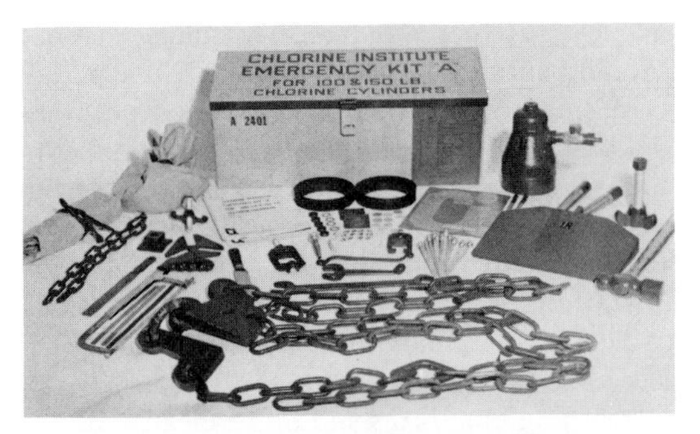

Courtesy of the Chlorine Institute, Inc.

Figure 10-45. Emergency Kit A for Use with Chlorine Cylinders

Courtesy of the Chlorine Institute, Inc.

Figure 10-46. Emergency Kit B for Use with Chlorine Ton Containers

The kits do not contain breathing equipment or other personal protective equipment for performing repairs. Instruction booklets and 35-mm slide presentations are available from the Chlorine Institute for training operators in the use of repair kits. It is important each operator be trained through an established training program.

Adequate ventilation. Since chlorine gas is 2.5 times as dense as air, it will settle and stay near the floor when leakage occurs. The rooms in which chlorine cylinders are stored and the enclosures that surround chlorinators should have sealed walls, and the doors to the rooms should open outward. The rooms should be fitted with chlorine-resistant power exhaust fans ducted at the floor level and with fresh air intake vents at the top of the room. In larger installations, it may be desirable to provide a fresh-air fan near the ceiling to force fresh air into the chlorinator enclosure. This fan will help in clearing the room of chlorine gas in case of leakage. The ventilating equipment should be capable of completely changing the volume of air in the room every 1 to 4 min; the required rate may be specified in the regulations of the state drinking water agency.

Exhaust-fan switches should be located outside the room and should be wired to room lights so that lights and fan go on at the same time. The fan and lights can be wired to a switch that activates both when the door is opened. Figure 10-47 shows a well designed chlorine room.

Chlorine detector. The chlorine detector is one of the most important pieces of chlorination equipment at the plant. It detects chlorine concentrations so small that they are impossible to smell. A chlorine detector can give early warning, so a small leak can be stopped before it becomes larger.

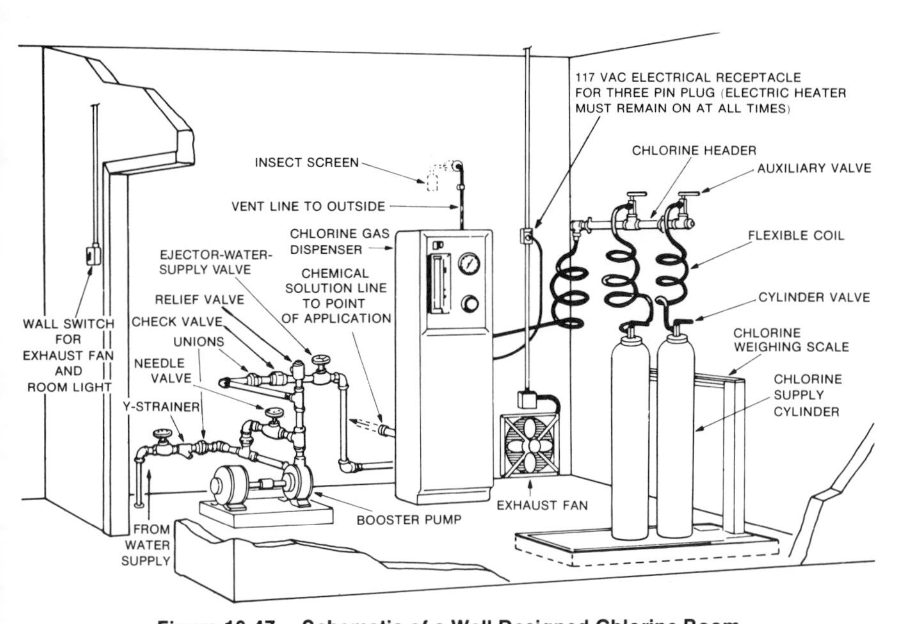

Figure 10-47. Schematic of a Well Designed Chlorine Room

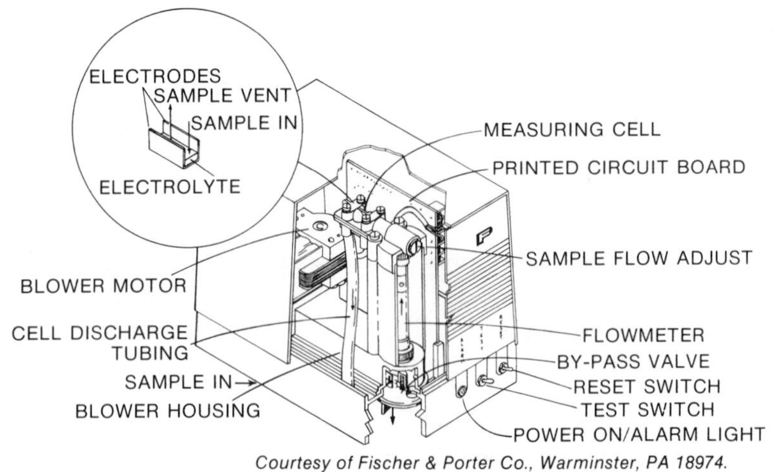

Courtesy of Fischer & Porter Co., Warminster, PA 18974.

Figure 10-48. Chlorine Detector

Various types of chlorine detectors are available. One type, shown in Figure 10-48, operates by measuring conductivity between two electrodes set in a liquid or solid electrolyte. Samples of air are drawn from the area being monitored and any chlorine in the sampled air will cause an increase in the current passing between the electrodes. A sensitive circuit detects this change and sets off a visual or audible alarm. The sensitivity of chlorine detectors is such that they will respond to less than 1 ppm of chlorine in air, by volume.

Proper Safety Procedures

Every facility using chlorine, regardless of size, should have established operating procedures covering the handling of chlorine and chlorine leaks. These procedures should cover:

- Safety precautions for storing and handling cylinders or containers
- Basic steps in connecting and disconnecting cylinders or containers
- Procedures to follow in case of a chlorine leak
- Emergency procedures to be taken if a chlorine leak threatens nearby residential areas
- First-aid procedures for persons exposed to chlorine.

Each operator should be made thoroughly familiar with all procedures through routine, in-plant training programs. These programs should particularly emphasize the use of respiratory protection equipment, leak repair kits, and emergency first aid.

It is helpful to post descriptions of important procedures near the chlorination facilities. Many of the chlorine and chlorine equipment manufacturers distribute wall charts on safety and handling procedures that can be used for training and display.

Selected Supplementary Readings

Babbitt, H.E.; Doland, J.J. & Cleasby, J.L. *Water Supply Engineering*. McGraw-Hill Book Company, New York. (6th ed., 1962). Chap. 25.

Baker, R.J. Maximum Withdrawal Rates From Chlorine, Sulfur Dioxide, and Ammonia Cylinders. *OpFlow*, 6:4:4 (Apr. 1980).

Basic Gas Chlorination Manual: I. *Jour. AWWA*, 64:5:319 (May 1972).

Basic Gas Chlorination Manual: II. *Jour. AWWA*, 64:6:395 (June 1972).

Basic Gas Chlorination Manual: III. *Jour. AWWA*, 64:8:522 (Aug. 1972).

Basic Gas Chlorination Manual: IV. *Jour. AWWA*, 64:10:683 (Oct. 1972).

CHEMTREC: 800-424-9300. *OpFlow*, 6:4:1 (Apr. 1980).

Chlorine—History and Characteristics. *OpFlow*, 9:4:3 (Apr. 1983).

Chlorine Institute Emergency Kit "A," "B," and "C." *OpFlow*, 6:9:3 (Sept. 1980).

Chlorine Leak Injures Two in Michigan. *OpFlow*, 2:6:7 (June 1976).

Chlorine Manual. The Chlorine Institute, Inc. New York. (4th ed., 1969).

Craun, G.F. Outbreaks of Waterborne Disease in the United States: 1971–1978. *Jour. AWWA*, 73:7:360 (May 1981).

Craun, G.F., et al. Waterborne Disease Outbreaks in the US—1971–1974. *Jour. AWWA*, 68:8:420 (Aug. 1976).

Craun, G.F. & Gunn, R.A. Outbreaks of Waterborne Disease in the United States: 1975–1976. *Jour. AWWA*, 71:8:422 (Aug. 1979).

Culp, G.L. & Culp, R.L. *New Concepts in Water Purification*. Van Nostrand Reinhold Company, New York. (1974). Chap. 6.

Introduction to Water Quality Analyses. AWWA, Denver, Colo. (1982). Modules 4-1, 4-2, and 5-3 and Appendix C.

Leidholdt, Ralph. Breakpoint Chlorination—The Chemistry of Chlorine. *OpFlow*, 8:10:1 (Oct. 1982).

Leidholdt, Ralph. Chlorine—'Special Agent' For Disinfecting Water. *OpFlow*, 8:9:1 (Sept. 1982).

Manual of Instruction for Water Treatment Plant Operators. New York State Dept. of Health, Albany, N.Y. (1975). Chap. 10.

Safety Practices for Water Utilities. AWWA Manual M3, AWWA, Denver, Colo. (4th ed., 1983). Chap. 10.

Safety Procedures for Working With Chlorine. *OpFlow*, 8:10:3 (Oct. 1982).

Simplified Procedures for Water Examination (Chlorine Demand & Residual; Bacteriological Examination). AWWA Manual M12, AWWA, Denver, Colo. (1977).

Testing the Microbiological Quality of Water. *OpFlow*, 9:5:4 (May 1983).

THMs—Cancer-Causing Agents in Drinking Water. *OpFlow*, 9:1:1 (Jan. 1983).

Treatment Techniques for Controlling Trihalomethanes in Drinking Water. AWWA, Denver, Colo. (1982).

Vogele, Lou. Safe Handling of Chlorine Emphasized in Toronto. *OpFlow*, 5:5:1 (May 1979).

Water Chlorination Principles and Practices. AWWA Manual M20, AWWA,

Denver, Colo. (1973). Chap. 1, 2, 3, 7, and 8.
Water Quality and Treatment. AWWA Handbook. McGraw-Hill Book Company, New York. (3rd ed., 1971). Chap. 5.
Water Works Operators Manual. Alabama Dept. of Public Health, Montgomery, Ala. (3rd ed., 1972). Chap. XVII.
Wesorick, Jim. Chlorination Systems—Feeding Chlorine Into Water. *OpFlow*, 8:11:1 (Nov. 1982).
White, G.C. *Handbook of Chlorination*. Van Nostrand Reinhold Company, New York. (1972). Chap. 2, 3, 4, 6, and 13.

Glossary Terms Introduced in Module 10

(Terms are defined in the Glossary at the back of the book.)

Auxiliary tank valve
Baume'
Breakpoint
Chlorination
Chlorinator
Combined chlorine residual
Container valve
Cylinder valve
Diffuser
Disinfection
Drip leg
Eductor
Ejector
Evaporator
Flow proportional control
Free chlorine residual
Hypochlorination
Injector
Muriatic acid
Organobromine compound
Pathogenic
Photochemically
Reducing agent
Residual flow control
Rotameter
Sterilization
THM
Tare weight
Trihalomethane
Trunnion
Turbidity
UV disinfection
Waterborne disease

Review Questions

(Answers to review questions are given at the back of the book.)

1. What is the difference between disinfection and sterilization?

2. What is a pathogen?

3. List at least five waterborne diseases.

4. Name the three broad classifications of disinfection by type of treatment.

5. List at least three chemicals that could be used as disinfectants.

6. Name the most common disinfectant in use today.

7. Explain what "Kill $\propto$ $C \times t$" means.

8. In addition to concentration and contact time, list other factors important to chlorination.

9. Describe the effect of temperature on chlorination.

10. List at least four substances in the water that might affect chlorination.

11. The measure of the amount of chlorine remaining after the disinfection reaction is complete is called ________.

12. List two types of chlorine residual.

13. What is the practical significance of chlorine residual?

14. The measure of the amount of chlorine used up during the chlorination reactions is called ________.

15. Plot a typical chlorination curve and describe the four significant parts.

16. Disinfection of potable water is the most important use of chlorination. List at least four others.

17. List the three common chemicals used in chlorination, with their chemical abbreviations.

18. List the types of liquid-chlorine (Cl_2) containers in common use.

19. How much does a full ton container of chlorine weigh?

20. Describe chlorine gas, including its health effects.

21. What are the most commonly used types of pressurized chlorine containers?

22. Describe sodium hypochlorite and calcium hypochlorite.

23. What devices are used to handle 150-lb cylinders and ton containers?

24. Where should cylinders and containers be stored?

25. Beginning with the chlorine supply, list in order the major components of equipment found in a typical chlorination system.

26. What is the purpose of a weigh scale?

27. What function does the injector serve?

28. List at least four typical pieces of auxiliary chlorination equipment.

29. What are the maximum gas withdrawal rates for a 150-lb cylinder and a ton container?

30. In what position are cylinders and containers used?

31. How many ton containers would have to be manifolded together in order to provide a continuous supply of 2700 lb of chlorine gas per day?

32. While a chlorine tank valve is open, where should the tank-valve wrench be kept?

33. What chemical is used to check for chlorine leaks?

34. Why should tank valves be closed before making a repair to the system?

35. Once a tank valve is closed, how is all chlorine removed from the feed system?

36. What is the best chlorine dosage to use in all cases?

37. List at least four common operating problems.

38. Consumers are complaining that their water tastes and smells like "chlorine bleach." Would you increase or decrease chlorine dosage?

39. Name the basic operational control test for chlorination.

40. What organism is measured to determine the effectiveness of disinfection? Why?

41. Explain the significance of water containing total coliform.

42. Three types of safety equipment are absolutely essential in every installation. What are they?

Study Problems and Exercises

1. You have just been employed at a water treatment plant that produces 6.0 mgd. The state health department has issued an order to the city to install chlorination equipment.
   a. What factors would you consider in selecting the type of chlorination equipment that would be the most cost-effective?
   b. Assuming a chlorine demand of 2.7 mg/L for this system, what factors would you take into consideration in selecting the type of chlorine container and chlorination equipment?

2. From operating data provided to you, determine the following:
   a. Chlorine contact time
   b. Chlorine demand
   c. Chlorine residual (type and amount)
   d. Amount of chlorine used daily.

3. Your system uses ground water as the source for drinking water and presently does not disinfect. As the operator, you recommend to the town council that disinfection equipment be installed to prevent any possible disease outbreaks. Recent bacteriological sampling results indicate contamination of the well has occurred. Discuss your recommendation in detail with respect to the chlorine source to be used, the type of equipment to be used, any necessary safety construction features, and the benefits of disinfecting the water and the distribution system.

Appendix A

AWWA Statement on Quality Goals for Potable Water

AWWA Statement on Quality Goals for Potable Water

(adopted in 1968)

At the 1965 AWWA Annual Conference, the Water Quality Division voted to set up quality goals for potable water, and guidelines for the goals to be established were derived. These actions subsequently were approved by the Executive Committee of the Association. An eleven-member task group (now the Committee on Water Quality Goals) was appointed, expanding the group of four that had previously been working in this area and had published a set of ideal criteria for water quality in the November 1962 *Journal AWWA*.[1]

Guidelines

The Committee on Water Quality Goals was directed to set up realistic quality goals for the water industry—goals that would tend to raise the quality of the water delivered to the consumer by being at once in accordance with advanced thinking and amenable to being grasped by non-technical personnel. They were not to be impractical objectives, even though they were to be substantially more exacting than the existing USPHS Drinking Water Standards with respect to aesthetic qualities. They were to be generally attainable by correct application of known treatment processes and methods.

It was recognized that the goals should not be static, but should be reviewed, and revised if warranted, perhaps every three years. The water industry is constantly being faced with the problem of meeting rising aesthetic demands. At the same time, because of increasing population and industrialization, the quality of water sources has deteriorated.

Responsibility for Application

Public water utilities should adopt quality criteria against which they can gage the effectiveness of their day-to-day operations. Managers of a given system must decide whether the cost of furnishing quality water is warranted. The expense of approaching or meeting the goals may be very high for some systems. For instance, removal of high dissolved-solids content or color may prove inordinately expensive. Management, then, has to weigh the value of the benefits to be derived against the cost of producing the finer quality water. It may conclude that consumers are satisfied with water meeting goals less rigorous than those recommended here.

The industry has accomplished much in the areas of water treatment, control, pumping, and delivery, but consumers judge quality at the tap, not at the source. Therefore, good consumer relations depend on maintaining high quality to the point of delivery. The goals stated herein are not merely for water entering the distribution system.

The definition of a functionally ideal water is as follows:

> Ideally, water delivered to the consumer should be clear, colorless, tasteless, and odorless. It should contain no pathogenic organisms and be free from biological forms that may be harmful to human health or aesthetically objectionable. It should not contain concentrations of chemicals that may be physiologically harmful, aesthetically objectionable, or economically damaging. The water should not be corrosive or incrusting to, or leave deposits on, water-conveying structures through which it passes, or in which it may be retained, including pipes, tanks, water heaters, and plumbing fixtures. The water should be adequately protected by satural processes, or by treatment processes, which ensure consistency in quality.

The Committee on Water Quality Goals believes it should not set goals for items that primarily and principally concern health, but should defer to USPHS and the medical profession. With respect to toxic substances, the USPHS Drinking Water Standards provide very broad safety factors, and the committee generally accepts these standards. This eliminated from consideration such items as the following:

Lead	Nitrates and Nitrites
Barium	Radium
Fluoride	Strontium
Arsenic	Phenolic compounds
Cyanide	Organic phosphorus
Silver	Chlorinated hydrocarbons
Selenium	Boron
Cadmium	Uranylion
Chromium	

Rationale of Goals

Turbidity. Today's consumer expects a sparkling, clear water. The goal of less than 0.1 unit of turbidity ensures satisfaction in this respect. There is evidence that freedom from disease-causing organisms is associated with freedom from turbidity, and that complete freedom from taste and odor requires no less than such clarity. Improved technology in the modern treatment processes make this a completely practical goal.

Nonfilterable residue. Water should be free of observable suspended particles of residue after settling. The goal indicates a virtually suspension-free state.

Macroscopic and nuisance organisms. It is obvious that macroscopic organisms such as larvae, crustacea, and numerous algae that may affect appearance should not be present. Nuisance organisms may affect appearance, taste, or odor, perhaps only after standing, heating, or freezing. They include, among others, the iron bacteria, sulfur bacteria, and slime growths.

Color. Because of difficulty in matching the colors of natural waters to the colors of standards, it is suggested that, when difficulty is encountered, a photometric transmittancy method should be employed. Color of less than 3 units will not be noticed, even in a filled bathtub, whereas color of 5 units may be noted by many. Fifty-five percent of 102 interstate waters in 1961 showed color of less than 3 units.[2] For the 100 largest cities in 1962, the median color was reported as 2 units.[3]

Odor. Odor is a nebulous characteristic, difficult to quantify; agreement is seldom obtainable as to the presence of odor or its character in a given potable water. The goal of water utilities should be elimination of all odor. The presence of odor is to be evaluated by difference before and after contact with carbon. Some materials may be removed by carbon only after treatment with a strong oxidant, which should be utilized when required to demonstrate the difference.

Taste. Taste is also a nebulous characteristic whose determination is complicated by the variability of perception of individuals from day to day. It is generally agreed that all potable waters do have some taste. If the taste is mild and not offensive in character, most individuals become accustomed to it. But water should be palatable to all; individuals first tasting any water should not be offended, and it should not be necessary that one become acclimated to the taste to regard it as characteristic of a good-quality water.

Aluminum. At levels exceeding 0.05 mg/L, precipitation may take place on standing, or in the distribution system. Turbidity and non-filterable residue will be affected.

Iron. With an iron content exceeding 0.05 mg/L, some color may develop, staining of fixtures may occur, and precipitates may form. The magnitude of such phenomena are directly proportional to the concentration of iron in the water.

Manganese. In concentrations of only a few hundredths milligrams per litre, manganese will cause buildup of coatings in distribution piping, which slough off. It causes staining of laundry items in brown blotches and forms black precipitates objectionable to consumers.

Copper. Copper content of 0.5 mg/L or less in some soft waters will cause staining of porcelain. In 1961, of 163 interstate supplies, 70 percent contained less than 0.2 mg/L;[2] of the 100 largest cities, 94 contained less than 0.1 mg/L.[4]

Zinc. In concentrations of 5 mg/L, a disagreeable taste may be noted. Zinc is undesirable in water passing through piping systems, as it may aid corrosion. The states of Ohio and North Dakota now limit zinc content to 1.0 mg/L. In 1961, of 163 interstate supplies, 45 percent contained less than 1.0 mg/L.[2]

Filterable residue. Low dissolved-solids content is desirable, if one is to avoid precipitations in boilers or other heating units, to reduce sludge in freezing processes, and to reduce sludge on utensils and precipitations on foods being cooked. The stated limit of 200.0 mg/L is in line with other goals stated herein.

Carbon-chloroform extract (CCE). This goal is based on equipment design and procedures utilized by USPHS throughout the early 1960s. Other designs need correlation with such units to determine relative adsorption efficiency to determine equivalent goal values. Materials adsorbed on activated carbon are organic. Toxic substances recovered may include chlorinated insecticides, nitrates, nitrobenzenes, aromatic ethers, and many others. Tastes and odors often may be correlated with the amounts of chloroform-soluble materials present, these materials having excessive odor thresholds. Most of the chloroform-soluble materials derive from man-made wastes. Waters from sources remote from concentrated industrial activities or human populations usually show CCE concentrations less than 0.04 mg/L. Where concentrations of CCE of 0.2 mg/L are found, the taste and odor of the water is always poor.[5] In 1961, USPHS found the average CCE in 139 cities to be 0.066 mg/L. In 1962–63, the average was 0.065 mg/L.[6] In 1961, of 172 supplies, 48 showed less than 0.05 mg/L.[1]

Carbon-alcohol extract (CAE). This requirement is supplementary to the preceding CCE. The proportion of materials most commonly found is roughly in the proportion indicated by the limits stated for the CCE and CAE. (See the comments on CCE.)

Methylene-blue-active substances. This classification replaces the designation of ABS previously in use. The change is required because of changes in composition of the new detergents. The analytic technique used determines not only ABS but also alkyl sulfates and related materials that react with methylene blue.

Hardness. It is not intended to imply that hardness of waters below 80 mg/L should be increased to that content when corrosion can be otherwise controlled. To the average water consumer, hardness of 80–100 mg/L is not objectionable. It is important that, whatever the goal chosen, the hardness should be maintained at a uniform level. Most important is the degree of stability attained. It has been fairly standard practice, for many years, to soften to roughly 5 grains or 85 mg/L hardness. The individual management, in choosing a standard, must consider that, the higher the hardness, the greater the cost to individual consumers; the less the hardness unless corrected, the greater the corrosion tendencies, and the greater the relative cost for treatment.

Alkalinity. This goal is a measure of alkalinity decrease or increase in the distribution system, and also after 12 hours at 130° F (54° C) in a closed plastic bottle, followed by filtration. This goal is a simple determination, indicating in a practical way that the alkalinity is stable. The maintenance of calcium carbonate stability is the most effective method of preventing corrosive action on iron water mains. Undersaturation will result in reactions causing iron pickup and development of red water. Oversaturation will result in carbonate deposition in utensils, water heaters, household piping, and even in water mains. The point of stability is quite variable in different waters, and even in water from a single source. Various methods have been utilized to determine the point of stability, including the Enslow Stability Indicator, the Langelier Index, the Ryznar Index, and Oxygen Depletion, the latter three covering only a few of the many chemical factors involved, none of which suffice in practice, being only hypothetical indicators.

Table A.1 Potable Water Quality Goals (not a standard)*

Characteristic	*Goal*
Physical Factors	
Turbidity	Less than 0.1 unit
Nonfilterable residue	Less than 0.1 mg/L
Macroscopic and nuisance organisms	No such organisms
Color	Less than 3 units
Odor	No odor
Taste	No taste objectionable
Chemical Factors mg/L	
Aluminum (Al)	Less than 0.05
Iron (Fe)	Less than 0.05
Manganese (Mn)	Less than 0.01
Copper (Cu)	Less than 0.2
Zinc (Zn)	Less than 1.0
Filterable residue	Less than 200.0
Carbon-chloroform extract (CCE)	Less than 0.04
Carbon-alcohol extract (CAE)	Less than 0.10
Methylene-blue-active substances (MBAS)	Less than 0.20
Corrosion and Scaling Factors	
Hardness (as $CaCO_3$)	80 mg/L; a balance between deposition and corrosion characteristics is necessary; a level of 80 mg/L seems best, generally, considering all the quality factors; however; for some supplies, a goal of 90 or 100 mg/L may be deemed desirable
Alkalinity (as $CaCO_3$)	Change of not more than 1 mg/L (decrease or increase in distribution system, or after 12 hours at 130° F (54° C) in a closed plastic bottle, followed by filtration)
Coupon tests (incrustation and loss by corrosion)	90-day tests (incrustation on stainless steel not to exceed 0.05 mg/sq cm; loss by corrosion of galvanized iron not to exceed 5.00 mg/sq cm)
Bacteriologic Factors	
Coliform organisms (by multiple-fermentation techniques)	No coliform organisms
Coliform organisms (by membrane filter techniques)	No coliform organisms
Radiologic Factors	
Gross beta activity	Less than 100 pCi/L

*For all health-related constituents not stated herein, these goals shall require complete compliance with all recommended and mandatory limits contained in current USPHS Drinking Water Standards. Unless other methods are indicated, analyses shall be made in conformance with the latest edition of *Standard Methods for the Examination of Water and Wastewater.*

Coupon tests. Coupon insertion in pipelines is now the recognized method for checking on corrosion properties of vapors and gases. It has been used to a very limited extent in the field of water supply where at least 15 factors affect corrosivity. Coupons measure the combined effects, both additive and neutralizing, of all the factors of corrosion, both known and unknown, including the physical factors of velocity and turbulence. Thus, coupons can provide valuable standards of comparison. Data on the use of coupons are still limited. Present data indicate levels should be those stated in Table A.1 as goals. One design of coupon test equipment for water mains has been published.[1]

Coliform organisms. Many water utilities have adopted high standards of operation and their water supplies have shown only a fraction of one coliform per litre over periods of many years. Municipalities that have so raised their bacteriologic quality far above the existing standards have established much improved health conditions with respect to certain significant illness, such as intestinal disturbances. Modern disinfection control procedures are such that a practical goal can be the destruction of all coliform organisms.

Gross beta activity. All evidence indicates the effects of radioactivity to be entirely harmful rather than beneficial. Therefore, it appears desirable to limit the intake of radioactivity as much as possible. A goal of 100 pCi/L is well below the existing standards, yet it is several degrees above present-day general levels. In 1961, of 136 potable water supplies tested, 132 showed no more than 16 pCi/L.[2] Of 100 largest cities, 92 showed less than 20 pCi/L in 1961.[4] In periods of bomb testing, natural waters have shown double this content; however, in periods free of bomb detonations, the natural background in most areas is only about 10 pCi/L, or one tenth this goal.

References

1. Bean, E.L. Progress Report on Water Quality Criteria. *Jour. AWWA*, 54:1313 (Nov. 1962).
2. Taylor, F.B. Effectiveness of Water Utility Control Practices. *Jour. AWWA*, 54:1257 (Oct. 1962).
3. Durfor, C.N. & Becker, E. Selected Data on Public Supplies of the 100 Largest Cities in the United States, 1962. *Jour. AWWA*, 56:236 (Mar. 1964).
4. *Public Water Supplies of the 100 Largest Cities in the United States.* USGS Water Supply Paper No. 1918. US Govt. Printing Office, Washington, D.C. (1962).
5. Tentative Methods for Carbon Chloroform Extract (CCE) in Water. *Jour. AWWA*, 54:223 (Feb. 1962).
6. Minutes of Advisory Committee on Use of USPHS Drinking Water Standards. Sept. 24–25, 1965.

Appendix B

Maintenance Guidelines for Gas Chlorination Equipment

Maintenance Guidelines for Gas Chlorination Equipment

These guidelines are suggested maintenance procedures. Procedures in manufacturers' manuals should be followed for specific equipment.

Maintenance

A system that is to operate safely and properly must be adequately maintained. By keeping the system in good repair, costly and lengthy breakdowns can be avoided. However, when, and if, a malfunction occurs, the operator should be able to recognize it, determine the most probable cause(s) of trouble, and know what to do about it. The comprehensive table included in these pages will help the operator root out malfunctions and take appropriate action.

A vigorous maintenance schedule should include:

- *Daily* checks of the automatic controls to see that they start and stop properly.
- *Weekly* cleaning of the screens and strainers on the water lines and checkout of alarm systems and shut-off and rate-control valves.
- *Annual* refurbishing of parts and packing as required by inspection, and cleaning and painting of paint-flaked areas on metal parts; also, disassembly of ejector and inspection of all parts (as described later).

Should problems develop, the operator should run a checkout of the affected piece(s) of equipment.

Vent fan. The motor should be inspected.

- Visually check to see if shaft turns when energized.
- If shaft does not turn, check to see if relay is "on" and fuses are not blown. Correct as necessary.
- Check heater.
- Bearing or fan blade may be seized; disconnect power and proceed to find and correct problem.
- If motor still does not run, have it repaired or changed.
- Disconnect fan from power supply and inspect the fan blades and ascertain that the blades are free to turn tight on shaft, symmetrical and balanced. (If balanced, blades will not always stop in the same position when rotated and released.)
- Inspect louvers to be sure they will open when required and that no debris is piled against the unit.

Injector (or ejector).

- Check nozzle (throat) for plugging by rust or dirt. Remove blockage as necessary.
- Check tailway (discharge section) of injector for abrasion (particularly when or where fine sand may be in suspension in water).
- Inspect ball check. It should be clean and free to move.
- Inspect diaphragm for breakage. If broken, replace.
- Visually inspect for leaks at joints. Replace gaskets when required, tighten bolts, and check springs for corrosion.

Rate controller (or feed rate adjustor or rate valve).

- Inspect for possible linkage disconnection, rack-and-pinion failure, stripped gears, or broken shaft. (Since operator must get behind machine to look, machine should not be installed too close to wall.)
- Manually operate controller to see if gas flow can be regulated (increased or decreased).
- Check all seal gaskets and replace if necessary.
- Check for foreign material on stem and seat.
- Check automatic control by using manufacturer's procedures manual. (If chlorinator works in manual mode, problem may be in automatic control.)

Residual analyzer. Since the time involved in the transfer of water and sample through the control loop system is critical, it is important to adhere very closely to the manufacturer's specifications for setting up and regulating the equipment, which will give the best operating conditions for a particular plant.

- The cell assembly should be cleaned as required by the suggested maintenance program or as dictated by operating conditions. Methods should follow manufacturer's specifications.
- Leads from the cell (analyzer) to the recorder unit should be inspected regularly for corrosion. This is particularly important for older models of equipment.
- Electric contacts should be cleaned whenever they appear dirty or perform erratically.
- The use of a voltmeter with a low-range d-c millivolt scale can indicate whether or not the cell and electrode assembly is producing an electric signal and whether the amplifier is converting this signal into the motor's requirements. The use of a voltmeter also will verify if the contacts are energizing the required alarm and control circuits. An examination of the manufacturer's electrical drawings will indicate checkpoints and what readings should be expected.
- If for any reason gears or mechanical linkages are taken apart, they should be marked with check lines to facilitate reassembly.

Rotameter.

- Visually inspect for buildup of foreign material inside the rotameter and on the float.
- Determine if the rotameter is properly seated at top and bottom. If the rotameter is off center, a leak may develop, leading to faulty operation.
- Inspect gaskets for such defects as cracking or flattening. Replace if required.

Presssure and vacuum gauge.

- Check for leaks.
- Open line at joint and check for plugging.
- Open gauge by removing glass and dial face, and check if gearing is meshing properly. However, do not disconnect gauge from its diaphragm assembly.
- Check bellows for possible rupture as indicated by oil at connection to the tube.

Evaporator.

- Check temperature. If temperature is high (approximately 160° F, 70° C), shut off power supply to heaters. If temperature is low (approximately 120° F, 50° C), check to see if heater is functioning and determine cause for malfunction.
- Check water level. It should be within the limits set by the manufacturer. If the level is low, check makeup valve. If valve is a manual one, open it. If valve operates automatically, it may need repair. Low level may also be caused by evaporation. Pour a light film of oil over surface to prevent evaporation. If the water level is high, drain water to level required and be sure makeup valve is not leaking.
- Check pressure. If pressure is high, check high-temperature control to see if it is functioning. (High pressure is caused by high temperature.) If pressure is low, check chlorine supply, which may be low. Increase chlorine supply, as necessary, changing cylinder if necessary. Check to see if thermostat turns on heaters at low temperature. Check incoming liquid pressure and ascertain that it is at a proper level.

Alarm systems. These systems are most important to safe operation.

Chlorinator malfunction. The chlorine and water systems should be checked.

- Turn off water to injector and wait for alarm to sound (approximately 15-20 s); when alarm rings, turn water on again; if alarm does not sound, check linkage to switch, then check diaphragm.
- Turn off chlorine gas supply and wait for alarm to sound; when alarm rings, turn chlorine supply on again; if alarm does not sound, check linkage to switch or positioning of switch.

For atmosphere in chlorination room. If alarm system uses sensitive paper, use finger to block light path; listen for alarm to sound; unblock light. (Be certain that sensitive paper is not too aged; it loses its sensitivity to light with time. Remember, its shelf life is four to six months if kept away from light). Systems using a cell assembly can be tested by holding a beaker of household chlorine bleach near the air intake. This will activate a working alarm. The system should slowly return to normal operation after removal of the beaker.

SOURCE: Basic Gas Chlorination Manual: III. Ontario Water Resources Com., *Jour. AWWA*, 64:8:522 (Aug. 1972).

Glossary

Glossary

Words defined in the glossary are set in SMALL CAPITAL LETTERS where they are first used in the text.

Activated alumina, 205 The chemical compound aluminum oxide, which is used to remove fluoride and arsenic from water by ADSORPTION.

Activated carbon, 205 A highly adsorptive material used to remove ORGANIC SUBSTANCES from water. (See ADSORPTION.)

Activated silica, 69 A COAGULANT AID used to form a denser, stronger FLOC.

Activation, 206 The process of producing a highly porous structure in carbon by exposing the carbon to high temperatures in the presence of steam.

Adhesion, 205 Sticking together.

Adsorbent, 205 Any material, such as ACTIVATED CARBON, used to adsorb substances.

Adsorption, xi, 169, 203 The water treatment process used primarily to remove ORGANIC contaminants from water. Adsorption involves the ADHESION of the contaminants to an ADSORBENT such as ACTIVATED CARBON.

Aeration, ix, 41 The process of bringing water and air into close contact to remove or modify constituents in the water.

After-precipitation, 142 The continued PRECIPITATION of a chemical compound (primarily $CaCO_3$ in the SOFTENING process) after leaving the SEDIMENTATION or SOLIDS CONTACT BASIN. This can cause scale formation on the filter media and in the distribution system.

Aggressive, 255 See CORROSIVE.

Air binding, 57 A condition that occurs in filters when air comes out of solution as a result of pressure decreases and temperature increases. The air clogs the voids between the media grains, which causes increased HEAD LOSS through the filter and shorter filter runs.

Air gap, 242 A method to prevent BACKFLOW by physically disconnecting the water supply and source of contamination.

Alum, 67 The most common chemical used for COAGULATION. It is also called ALUMINUM SULFATE.

Aluminum sulfate, 67 See ALUM.

Alizarin-visual test, 245 A laboratory procedure for determining the fluoride concentration in water.

Anaerobic, 25 The absence of air or free oxygen.

Anionic, 188 Having a negative IONIC charge.

Anionic polyelectrolyte, 70 A POLYELECTROLYTE that forms negatively charged IONS when dissolved in water.

Anode, 256 Positive end (pole) of an electrolytic system.

Anti-siphon device, 242 See VACUUM BREAKER.

Arching, 207 A condition that occurs when dry chemicals bridge over the opening from the hopper to the dry feeder, clogging the hopper.

Aspirator feeder, 72 A feeder using a hydraulic device that creates a negative gauge pressure by forcing water through a restriction (VENTURI TUBE). The negative pressure draws the material to be fed into the flow of water.

Auxiliary scour, 174 See FILTER AGITATION.

Auxiliary tank valve, 300 In a chlorination system, a union or yoke-type valve connected to the chlorine container or cylinder. It acts as a shut-off valve in case the CONTAINER VALVE is defective.

Backflow, 242 A hydraulic condition, caused by a difference in pressures, that causes nonpotable water or other fluid to flow into a potable water system.

Backwash, 170 The reversal of flow through a filter to remove the material trapped on and between the grains of filter media.

Baffle, 99 A metal, wooden, or plastic plate installed in a flow of water to slow the water velocity and provide a uniform distribution of flow.

Bar screen, 4 A series of straight steel bars welded at their ends to horizontal steel beams forming a grid. Bar screens are placed on intakes or in waterways to remove large debris.

Baume′, 314 The Baume′ scale is a means of expressing strength of a solution based on the solution's specific gravity.

Bed life, 209 The time it takes for a bed of ADSORBENT to lose its adsorptive capacity. When this occurs it must be replaced with fresh adsorbent.

Bivalent ion, 67 ION that has a valence charge of two. The charge can be positive or negative.

Body feed, 180 In DIATOMACEOUS EARTH FILTERS, the continuous addition of diatomaceous earth during the filtering cycle to provide a fresh filtering surface as the suspended material clogs the PRECOAT.

Breakpoint, 290 The point at which the chlorine dosage has satisfied the chlorine demand.

Breakthrough, 185 The point in a filtering cycle at which TURBIDITY causing material starts to pass through the filter.

Calcium carbonate, 117 The principal HARDNESS and scale-causing compound in water.

Calcium hardness, 117 The portion of total HARDNESS caused by calcium compounds such as $CaCO_3$ and $CaSO_4$.

Carbon dioxide, 43 A common gas in the atmosphere that is very soluble in water. High concentrations of CO_2 in water can cause it to be CORROSIVE. Carbon dioxide is added to water after the lime-softening process to lower the pH in order to reduce CALCIUM CARBONATE scale formation. This process is known as RECARBONATION.

Carbonate hardness, 117 HARDNESS caused primarily by compounds containing carbonate (CO_3), such as $CaCO_3$ and $MgCO_3$.

Carcinogen, 16, 204 Chemical compound that can cause cancer in animals or humans.

Cathode, 256 Negative end (pole) of an electrolytic system.

Cation, 116 A positive ION.

Cation exchange, 143 ION EXCHANGE involving IONS with positive charges, such as calcium and sodium.

Cationic, 188 Having a positive IONIC charge.

Cationic polyelectrolyte, 70 POLYELECTROLYTE that forms positively charged IONS when dissolved in water.

Centrate, 131 The water that is separated from SLUDGE and discharged from a centrifuge.

Centrifugation, 129 In water treatment, a method of DEWATERING SLUDGE using a mechanical device (centrifuge), which spins the sludge at a high speed.

Chlorination, 284 The process of adding chlorine to water to kill disease-causing organisms or to act as an oxidizing agent.

Chlorinator, 302 Any device that is used to add chlorine to water.

Clarification, 96 Any process or combination of processes that reduces the amount of suspended matter in water.

Clarifier, 97 See SEDIMENTATION BASIN.

Coagulant, 63 A chemical used in water treatment for COAGULATION. Common examples are ALUMINUM SULFATE and FERRIC SULFATE.

Coagulant aid, 69 A chemical added during COAGULATION to improve the process by stimulating FLOC formation or by strengthening the floc so it holds together better.

Coagulation, 63 The water treatment process that causes very small suspended particles to attract one another and form larger particles. This is accomplished by adding a chemical, called a COAGULANT, that neutralizes the electrostatic charges on the particles that cause them to repel each other.

Coagulation/flocculation, ix The water treatment process that converts small particles of suspended solids into larger more settleable clumps.

Colloidal solid, 64 Finely divided solid that will not settle out of water for very long periods of time unless the COAGULATION/FLOCCULATION process is used.

Combined chlorine residual, 288 The chlorine residual produced by the reaction of chlorine with substances in the water. Since the chlorine is "combined" it is not as effective a disinfectant as FREE CHLORINE RESIDUAL.

Concentration cell corrosion, 256 A form of localized CORROSION, which can form deep pits and TUBERCULES.

Contactor, 209 A vertical steel cylindrical pressure vessel used to hold the ACTIVATED CARBON bed.

Container valve, 300 The valve mounted on a chlorine container or cylinder.

Corrosion, 255 The gradual deterioration or destruction of a substance or material by chemical action. The action proceeds inward from the surface.

Corrosive, 255 To deteriorate material, such as pipe, through electrochemical processes.

Coupon test, 276 A method of determining the rate of CORROSION or scale formation by placing metal strips (coupons) of a known weight in the pipe.

Cross connection, 242 Any connection between a safe drinking water and a nonpotable water or fluid.

Cylinder valve, 300 See CONTAINER VALVE.

DE filter, 179 See DIATOMACEOUS EARTH FILTER.

Debris rack, 4 See BAR SCREEN.

Decant, 107, 129 To draw off the liquid from a basin or tank without stirring up the sediment in the bottom.

Decanted, 107 See DECANT.

Density current, 109 A flow of water that moves through a larger body of water, such as a reservoir or SEDIMENTATION BASIN, and does not become mixed with the other water due to a density difference. This difference is usually because the incoming water has a different temperature or SUSPENDED SOLIDS content than the water body.

Detention time, 105 The average length of time a drop of water or a suspended particle remains in a tank or chamber. Mathematically, it is the volume of water in the tank divided by the flow rate through the tank.

Dewater, 107 See DEWATERING (of sludge).

Dewatering (of reservoirs), 19 A physical method for controlling aquatic plants in which a water body is completely or partially drained and the plants allowed to die.

Dewatering (of sludge), 95 A process to remove a portion of water from SLUDGE.

Diaphragm pump, 234 See DIAPHRAGM-TYPE METERING PUMP.

Diaphragm-type metering pump, 73 A pump in which a flexible rubber, plastic, or metal diaphragm is fastened at the edges in a vertical cylinder. As the diaphragm is pulled back, suction is exerted and the liquid is drawn into the pump. When it is pushed forward, the liquid is discharged.

Diatom, 9, 180 A type of algae characterized by the presence of silica in its cell walls. Diatoms can cause filter-clogging.

Diatomaceous earth filter, 179 A pressure filter using a media made from DIATOMS. The water is forced through the diatomaceous earth by pumping.

Diffuser, 50, 304 Section of a perforated pipe or POROUS PLATES used to inject a gas, such as CO_2 or air, under pressure into water. (Also, a type of pump.)

Disinfection, xi, 283 The water treatment process that kills disease-causing organisms in water, usually by adding chlorine.

Dissolved solid, 66 Any material that is dissolved in water and can be recovered by evaporating the water after filtering the suspended material.

Divalent, 116 See BIVALENT ION.

Dredging, 19 A physical method for controlling aquatic plants in which a dragline or similar mechanical equipment is used to remove plants and the bottom mud in which they are rooted.

Drip leg, 313 A small piece of pipe installed on a chlorine cylinder or container that prevents collected moisture from draining back into the container.

Eductor, 207, 297 A device used to mix a chemical with water. The water is forced through a constricted section of pipe (venturi) to create a low pressure, which allows the chemical to be drawn into the stream of water.

Effluent launder, 99 A trough that collects the water flowing from a basin (effluent) and transports it to the effluent piping system.

Ejector, 303 The portion of a chlorination system that feeds the chlorine solution into a pipe under pressure.

Emergent weed, 17 An aquatic plant, such as cattails, that is rooted in the bottom mud of a water body but projects above the water surface.

Empty bed contact time (EBCT), 213 The volume of the tank holding an ACTIVATED CARBON bed divided by the flow rate of water. The EBCT is expressed in minutes and corresponds to DETENTION TIME in a SEDIMENTATION BASIN.

Erosion, 255 The wearing away of a material by physical means.

Evaporator, 307 A device used to increase the chlorine feed rate by heating the liquid chlorine.

Excess-lime treatment, 122 A modification of the LIME-SODA ASH METHOD that uses additional lime to remove magnesium compounds.

Ferric sulfate, 67 A common chemical used for COAGULATION.

Filter agitation, 174 A method used to achieve more effective cleaning of a filter bed. The system usually uses nozzles attached to a fixed or rotating pipe installed just above the filter media. Water or an air–water mixture is fed through the nozzles at high pressure to help agitate the media and break loose accumulated suspended matter. It can also be called AUXILIARY SCOUR or SURFACE WASHING.

Filter sand, 172 Sand that is prepared according to detailed specifications for use in filters.

Filter tank, 172 The concrete or steel basin that contains the filter media, gravel support bed, UNDERDRAIN, and WASH-WATER TROUGHS.

Filtration, xi The water treatment process involving the removal of suspended matter by passing the water through a porous medium such as sand.

Flash mixing, 74, 135 See RAPID MIXING.

Floating weed, 17 An aquatic plant, such as water lilies, that entirely or in part floats on the surface of the water.

Floc, 63 Collections of smaller particles that have come together (agglomerated) into larger, more settleable particles as a result of the COAGULATION/FLOCCULATION process.

Flocculation, 63 The water treatment process following COAGULATION, which uses gentle stirring to bring suspended particles together so they will form larger, more settleable clumps called FLOC.

Flow measurement, viii A measurement of the quantity of water flowing through a given point in a given amount of time such as gallons per day.

Flow proportional control, 306 A method of controlling chemical feed rates by having the feed rate increase or decrease as the flow increases or decreases.

Flow tube, 28 One type of primary element used in a PRESSURE DIFFERENTIAL METER. It measures flow velocity by the amount of pressure drop through the tube. It is similar to a VENTURI TUBE.

Fluoridation, xi The water treatment process in which a chemical is added to the water to increase the concentration of fluoride IONS to an optimum level. The purpose of fluoridation is to reduce the incidence of dental cavities in children.

Fluorosis, 227 Staining or pitting of the teeth due to excessive amounts of fluoride in the water.

Free chlorine residual, 287 The residual formed once all the chlorine demand has been satisfied. The chlorine is not combined with other constituents in the water and is free to kill microorganisms.

Galvanic corrosion, 258 A form of localized CORROSION caused by the connection of dissimilar metals in an electrolyte such as water.

Galvanic series, 258 A listing of metals and alloys according to their CORROSION potential.

Granular activated carbon (GAC), 206 ACTIVATED CARBON in a granular form, which is used in a bed, much like a conventional filter, to adsorb ORGANIC SUBSTANCES from water.

Gravel bed, 173 Layers of gravel of specific sizes that support the filter media and help distribute the BACKWASH water uniformly.

Gravimetric dry feeder, 233 See GRAVIMETRIC FEEDER.

Gravimetric feeder, 73 Chemical feeder that adds specific weights of dry chemical.

Hardness, 115 A characteristic of water, caused primarily by the salts of calcium and magnesium. Hardness causes deposition of scale in boilers, damage in some industrial processes, and sometimes objectionable taste.

Harvesting, 19 A physical method for controlling aquatic plants in which the plants are pulled or cut and raked from the water body.

Head loss, 28, 175 The amount of energy used by water in moving from one point to another.

Herbicide, 204 A compound, usually a synthetic ORGANIC SUBSTANCE, used to stop or retard plant growth.

Humic substance, 204 Material resulting from the decay of leaves and other plant matter.

Hydrofluosilicic acid, 230 A strongly acidic liquid used to FLUORIDATE drinking water.

Hydrogen sulfide (H_2S), 44, 57 A toxic gas produced by the ANAEROBIC decomposition of organic matter and by sulfate-reducing bacteria. Hydrogen sulfide has a very noticeable rotten-egg odor.

Hypochlorination, 309 Chlorination using solutions of calcium hypochlorite or sodium hypochlorite.

Injector, 303 See EJECTOR.

Inlet zone, 98 The initial zone in a SEDIMENTATION BASIN, which decreases the velocity of the incoming water and distributes it evenly across the basin.

Insecticide, 204 A compound, usually a synthetic ORGANIC SUBSTANCE, used to kill insects.

Ion, 227 An atom that is electrically unstable because it has more or less electrons than protons. A positive ion (one with more protons than electrons) is called a CATION. A negative ion (one with fewer protons than electrons) is called an ANION.

Ion-exchange process, 115 A process used to remove HARDNESS from water, which depends on special material known as ion-exchange RESINS. The resins trade non-hardness-causing IONS (usually sodium) for the hardness-causing ions of calcium and magnesium. The process will remove practically all the hardness from water.

Ion-exchange water softener, 239 A treatment unit used to remove calcium and magnesium from water using ion-exchange RESINS.

Ionize, 57 To change or be changed into IONS.

Iron, 44 An abundant element found naturally in the earth. As a result, dissolved iron is found in most water supplies. When the concentration of iron exceeds 0.3 mg/L, it causes red stains on plumbing fixtures and other items in contact with the water. Dissolved iron can also be present in water due to CORROSION of cast-iron or steel pipes. This is usually the cause of RED-WATER problems.

Iron bacteria, 254 Bacteria that use dissolved IRON as an energy source. They can create serious problems in a water system since they form large, slimy masses that clog well screens, pumps, and other equipment.

Jar tests, 108 A laboratory procedure for evaluating COAGULATION, FLOCCULATION, and SEDIMENTATION processes used to estimate the proper COAGULANT dose.

Lamella plates, 103 A series of thin, parallel plates installed at a 45-degree angle for SHALLOW-DEPTH SEDIMENTATION.

Langelier Index (LI), 268 A numerical index that indicates whether CALCIUM CARBONATE will be deposited or dissolved in a distribution system. The index is also used to indicate the CORROSIVITY of water.

Leopold filter bottom, 173 A patented filter UNDERDRAIN system using a series of perforated vitrified clay blocks with channels to carry the water.

Lime-soda ash method, 115 The process used to remove CARBONATE and NONCARBONATE HARDNESS from water.

Lining, 20 A physical method for controlling aquatic plants by placing a permanent lining, such as synthetic rubber, in the water body.

Loading rate, 151 The flow rate per unit area (gpm/ft^2) of a filter or ion-exchange unit at which the water is passed through them.

Localized corrosion, 260 A form of CORROSION that attacks a small area.

Magnesium hardness, 117 The portion of total HARDNESS caused by magnesium compounds such as $MgCO_3$ and $MgSO_4$.

Magnetic flow meter, 32 A flow measuring device in which the movement of water induces an electrical current proportional to the rate of flow.

Manganese, 44 An abundant element found naturally in the earth. Dissolved manganese is found in many water supplies. At concentrations above 0.05 mg/L, it causes black stains to plumbing fixtures, laundry, and other items in contact with the water.

Manifold, 50 A pipe with several branches or fittings to allow water or gas to be discharged at several points. In AERATION, manifolds are used to spray water through several nozzles.

Manual solution feed, 235 A method of feeding a chemical solution for small water systems. The chemical is dissolved in a small plastic tank, transferred to another tank and fed to the water system using a POSITIVE-DISPLACEMENT PUMP.

Metering pump, 73 A chemical solution feed pump that adds a measured volume of solution with each stroke or rotation of the pump.

Methane (CH_4), 44, 59 A colorless, odorless, flammable gas formed by the ANAEROBIC DECOMPOSITION of organic matter. When dissolved in water, methane causes a garlic-like taste. It is also called natural gas.

Microfloc, 66 The initial FLOC formed immediately after COAGULATION, composed of small clumps of solids.

Microstrainer, 25 A rotating drum lined with a finely woven material such as stainless steel. Microstrainers are used to remove algae and small debris before they enter the treatment plant.

Milk of lime, 126, 266 The lime SLURRY formed when water is mixed with calcium hydroxide.

Monovalent ion, 67 ION having a valence charge of one. The charge can be either positive or negative.

Mottling, 229 The staining of teeth due to excessive amounts of fluoride in the water.

Mudball, 190 Accumulation of media grains and suspended material that creates clogging problems in filters.

Muriatic acid, 318 Another name for hydrochloric acid (HCl).

NTU, 104, 168 See NEPHELOMETRIC TURBIDITY UNIT.

Negative head, 192 A condition that can develop in a filter bed when the HEAD LOSS gets too high. When this occurs, the pressure in the bed can drop to less than atmospheric.

Nephelometric turbidimeter, 194 An instrument that measures TURBIDITY by measuring the amount of light scattered by turbidity in a water sample. It is the only instrument approved by the US Environmental Protection Agency to measure turbidity in treated drinking water.

Nephelometric turbidity unit (NTU), 104, 168 The amount of TURBIDITY in a water sample as measured using a NEPHELOMETRIC TURBIDIMETER.

Noncarbonate hardness, 117 HARDNESS caused by the salts of calcium and magnesium.

Nonionic, 188 Not having an IONIC charge.

Nonionic polyelectrolyte, 70 POLYELECTROLYTE that forms both positively and negatively charged IONS when dissolved in water.

Nonsettleable solid, 64 Finely divided solid, such as bacteria and fine clay particles, that will stay suspended in water for long periods of time.

Nucleus (plural: nuclei), 211 The center of an atom, made up of positively charged particles called protons and uncharged particles called neutrons.

On-line turbidimeter, 176 A turbidimeter that continuously samples, monitors, and records TURBIDITY levels in water.

Organic substance (organic), 203 A chemical substance of animal or vegetable origin, having carbon in its molecular structure.

Organobromine compound, 286 The chemical compound formed when chlorine reacts with bromine.

Orifice plate, 29 A type of primary element used in a PRESSURE DIFFERENTIAL METER, consisting of a thin plate with a precise hole through the center. The plate responds to flow velocity by causing a pressure drop as the water passes through the hole.

Outlet zone, 98 The final zone in a SEDIMENTATION BASIN, which provides a smooth transition from the SETTLING ZONE to the effluent piping.

Overflow weir, 99 A steel or fiberglass plate designed to evenly distribute flow. In a SEDIMENTATION BASIN, the weir is attached to the EFFLUENT LAUNDER.

Oxidation, 42 (1) The chemical reaction in which the valence of an element increases due to the loss of electrons from that element. (2) The conversion of ORGANIC SUBSTANCES to simpler, more stable forms by either chemical or biological means.

Oxidize, 179 To chemically combine with oxygen.

Pathogenic, 283 See PATHOGEN.

Pathogen, 283 A disease-causing organism.

Percolation, 167 The movement or flow of water through the pores of soil, usually downward.

Permanent hardness, 118 Another term for NONCARBONATE HARDNESS derived from the fact that the hardness-causing noncarbonate compounds do not PRECIPITATE when the water is boiled.

Pesticide, 203 Any substance or chemical used to kill or control troublesome organisms including insects, weeds, and bacteria.

Photochemically, 286 Referring to chemical reactions that depend on light.

Photosynthesis, 8, 43 The process by which plants, using the chemical chlorophyll, convert the energy of the sun into food energy. Through photosynthesis, all plants, and ultimately all animals that feed on plants or on other plant-eating animals, obtain the energy of life from sunlight.

Piezometer, 176 An instrument for measuring pressure head in a conduit, tank, or soil, by determining the location of the free water surface.

Pipe-lateral system, 173 A filter UNDERDRAIN system using a main pipe (header) with several smaller perforated pipes (laterals) branching from it on both sides.

Piston pump, 234 A POSITIVE-DISPLACEMENT PUMP that uses a piston moving back and forth in a cylinder to deliver a specific volume of liquid being pumped.

Plain sedimentation, 96 The SEDIMENTATION of suspended matter without the use of chemicals or other special means.

Polyelectrolyte, 70, 188 High-molecular-weight synthetic ORGANIC compound that forms IONS when dissolved in water. It is also called a POLYMER.

Polymer, 70, 188 High-molecular-weight synthetic ORGANIC compound that can be either soluble or insoluble in water.

Polystyrene resin, 119 The most common RESIN used in the ION-EXCHANGE PROCESS.

Porous plate, 174 A patented filter UNDERDRAIN system using a ceramic plate supported by perforated clay saddles. This system is often used without a gravel layer so the plates are directly beneath the filter media.

Positive-displacement pump, 72, 233 A pump that delivers a specific volume of liquid for each stroke of the piston or rotation of the impeller.

Powdered activated carbon (PAC), 206 ACTIVATED CARBON in a fine powder form. It is added to water in a slurry form to remove those ORGANIC compounds primarily causing tastes and odors.

Precipitate, 118, 168 A substance separated from a solution or suspension by a chemical reaction. Precipitate can also be defined as the separation of a substance from a solution or suspension, which is caused by a chemical reaction.

Precoating, 180 The initial step in DE FILTRATION. A thin coat of diatomaceous earth is applied to a support surface called a septum. This provides an initial layer of media for the water to pass through.

Precursor compound, 219 Any of the ORGANIC SUBSTANCES that react with chlorine to form TRIHALOMETHANES.

Preliminary treatment, viii, 3 Any physical, chemical, or mechanical process used before the main water treatment processes. It can include SCREENING, PRE-SEDIMENTATION, and chemical addition. It is also called PRETREATMENT.

Presedimentation, 22, 96 A PRETREATMENT process used to remove gravel, sand, and other gritty material from the raw water before it enters the main treatment plant. This is usually done without the use of coagulating chemicals.

Presedimentation impoundment, 22 A large earthen or concrete basin used for PRESEDIMENTATION of raw water. It is also useful for storage and reducing the impact of raw water quality changes on water treatment processes.

Pressure differential meter, 28 Any flow measuring device that creates and measures a difference in pressure proportionate to the rate of flow. Examples include the venturi meter, orifice meter, and flow nozzle.

Pressure-sand filter, 179 A sand filter placed in a cylindrical steel pressure vessel. The water is forced through the media by pumping.

Pretreatment, 3 See PRELIMINARY TREATMENT.

Primary element, 28 The part of a PRESSURE DIFFERENTIAL METER that creates a signal proportional to the water velocity through the meter.

Progressive cavity pump, 74 A type of POSITIVE-DISPLACEMENT PUMP.

Propeller meter, 31 A meter for measuring flow rate by measuring the speed at which a propeller spins, and hence the velocity at which the water is moving through a conduit of known cross-sectional area.

Proportional meter, 32 Any flow meter that diverts a small portion of the main flow and measures the flow rate of that portion as an indication of the rate of the main flow. The rate of the diverted flow is proportional to the rate of the main flow.

Quicklime, 124 Another name for calcium oxide (CaO), which is used in water SOFTENING and STABILIZATION.

Radial flow, 98 Direction of flow across a basin from center to outside edge or vice versa.

Rapid mixing, 74 The process of quickly mixing a chemical solution uniformly through the water.

Rate-of-flow controller, 175 A control valve used to maintain a fairly constant flow through the filter.

Reactivate, 206 To remove the adsorbed materials from spent ACTIVATED CARBON and restore the carbon's porous structure so it can be used again. The reactivation process is similar to that used to activate carbon.

Recarbonation, 121 The reintroduction of CARBON DIOXIDE into the water, either during or after lime-soda ash softening.

Receiver, 28 The part of a PRESSURE DIFFERENTIAL METER that converts the signal from the receiver into a flow rate that can be read by the operator.

Rectilinear flow, 97 Uniform flow in a horizontal direction.

Red water, 254 A term used to describe rust-colored water due to the formation of ferric hydroxide from IRON naturally dissolved in the water or as a result of the action of IRON BACTERIA.

Reducing agent, 288 Any chemical that decreases the positive valence of an ION.

Regeneration, 143 The process of reversing the ion-exchange softening reaction of ion-exchange materials, removing the HARDNESS IONS from the used materials and replacing them with nontroublesome ions, thus rendering the materials fit for reuse in the SOFTENING process.

Regeneration rate, 151 The flow rate per unit area (gpm/ft^2) of an ion-exchange RESIN at which the regeneration solution is passed through the resin.

Residual flow control, 307 A method of controlling chlorine feed rate based on the residual chlorine after the feeder.

Resin, 143 In water treatment, the synthetic bead-like material used in the ION-EXCHANGE PROCESS.

Respiration, 43 The process by which a living organism takes in oxygen from the air or water, uses it in OXIDATION, and gives off the products of oxidation, especially CARBON DIOXIDE. Breathing is an example of respiration.

Rotameter, 303 A flow measurement device used for gases.

Sand boil, 191 The violent washing action in a filter caused by uneven distribution of BACKWASH water.

Sand trap, 22 An enlargement of a conduit carrying raw water that allows the water velocity to slow down so that sand and other grit can settle.

Saturation point, 261 The point at which a solution can no longer dissolve any more of a particular chemical. Precipitation of the chemical will occur beyond this point.

Saturator, 237 A piece of equipment that feeds a SODIUM FLUORIDE solution into water for FLUORIDATION. A layer of sodium fluoride is placed in a plastic tank and water is allowed to trickle through the layer forming a constant solution, which is fed to the water system.

Schmutzdecke, 170 The layer of solids and biological growth that forms on top of a slow sand filter, allowing the filter to remove TURBIDITY effectively without chemical COAGULATION.

Screening, 4 A PRETREATMENT method using coarse screens to remove large debris from the water to prevent clogging of pipes or channels to the treatment plant.

Sedimentation, ix The water treatment process that involves reducing the velocity of water in basins so the suspended material can settle out by gravity.

Sedimentation basin, 96, 97 A basin or tank in which water is retained to allow settleable matter, such as FLOC, to settle by gravity. It is also called a SETTLING BASIN, SETTLING TANK, or SEDIMENTATION TANK.

Sedimentation tank, 97 See SEDIMENTATION BASIN.

Sequestering agent, 118, 262 A chemical compound such as EDTA or certain POLYMERS that chemically tie up (sequester) other compounds or IONS so they cannot be involved in chemical reactions.

Settleability test, 25 A determination of the settleability of solids in a suspension by measuring the volume of solids settled out of a measured volume of sample in a specified interval of time, usually reported in millilitres per litre.

Settling basin, 96, 97 See SEDIMENTATION BASIN.

Settling tank, 97 See SEDIMENTATION BASIN.

Settling zone, 98 The zone in a SEDIMENTATION BASIN that provides a calm area so the suspended matter can settle.

Shading, 19 A physical method for controlling aquatic plants by limiting the amount of sunlight reaching the bottom of the water body.

Shallow-depth sedimentation, 97 A modification of the traditional SEDIMENTATION process using inclined tubes or plates to reduce the distance the settling particles have to travel to be removed.

Short-circuit, 105 See SHORT-CIRCUITING.

Short-circuiting, 22 A hydraulic condition in a basin in which the actual flow time of water through the basin is less than the design flow time (DETENTION TIME).

Slake, 124 The addition of water to QUICKLIME (calcium oxide) to form calcium hydroxide, which can then be used in the SOFTENING or STABILIZATION processes.

Slaker, 266 The part of a QUICKLIME feeder that mixes the quicklime with water to form hydrated lime (calcium hydroxide).

Sludge, 95, 119 The accumulated solids separated from water during treatment.

Sludge-blanket clarifier, 103 See SOLIDS-CONTACT BASIN.

Sludge blow down, 136 The controlled withdrawal of SLUDGE from a SOLIDS-CONTACT BASIN to maintain the proper level of settled solids in the basin.

Sludge zone, 98 The bottom zone of a SEDIMENTATION BASIN, which receives and stores the settled particles.

Slurry, 207 A thin mixture of water and any insoluble material, such as ACTIVATED CARBON.

Sodium fluoride, 230 A dry chemical used in the FLUORIDATION of drinking water. It is commonly used in SATURATORS.

Sodium silicofluoride, 231 A dry chemical used in the FLUORIDATION of drinking water. It is derived from HYDROFLUOSILICIC ACID.

Softening, xi The water treatment process that removes the hardness-causing constituents in water—calcium and magnesium.

Solids-contact basin, 103 A basin in which the COAGULATION, FLOCCULATION, and SEDIMENTATION processes are combined. The flow of water is upward through the basin. It is used primarily in the lime SOFTENING of water. It can also be called an UPFLOW CLARIFIER or SLUDGE-BLANKET CLARIFIER.

Solids-contact process, 97 A process combining COAGULATION, FLOCCULATION, and SEDIMENTATION in one treatment unit, in which the flow of water is vertical.

SPADNS method, 245 A colorimetric procedure used to determine the concentration of fluoride ION in water. SPADNS is the chemical reagent used in the test.

Spray tower, 50 A tower built around a spray aerator to keep the wind from blowing the spray and to prevent the water from freezing during cold temperatures.

Stabilization, xi, 116, 253 The water treatment process intended to reduce the CORROSIVE or scale-forming tendencies of water.

Sterilization, 283 The destruction of all organisms.

Submerged weed, 17 An aquatic plant, such as pondweed, that grows entirely beneath the surface of the water.

Supersaturation, 46, 56 A condition under which water contains very high concentrations of dissolved oxygen.

Surface-overflow rate, 105 A measurement of the amount of water leaving a SEDIMENTATION TANK per square foot of tank surface area. Mathematically, it is the gallons-per-day flow rate from the tank divided by the square feet of tank surface.

Surface washing, 174 See FILTER AGITATION.

Surface weed, 17 See FLOATING WEED.

Suspended solid, 64 Solid ORGANIC and inorganic particle that is held in suspension by the action of flowing water.

Synthetic resin, 205 See RESIN.

THM, 319 See TRIHALOMETHANE.

Tare weight, 312 The initial weight of an item.

Temporary hardness, 118 Another term for CARBONATE HARDNESS, derived from the fact that the hardness-causing carbonate compounds PRECIPITATE when water is heated.

Terminal head loss, 184 The HEAD LOSS in a filter at which water can no longer be filtered at the desired rate, due to the suspended matter filling the voids in the filter and greatly increasing the resistance to flow (head loss).

Total organic carbon (TOC), 218 The amount of carbon bound in ORGANIC compounds in a water sample as determined by a standard laboratory test.

Toxic, 10 Poisonous.

Tracer study, 109 Study using a substance that can be readily identified in water (such as a dye) to determine the distribution and rate of flow in a basin, pipe, or channel.

Transmitter, 28 The part of a PRESSURE DIFFERENTIAL METER that measures the signal from the primary element and sends another signal to the RECEIVER.

Trash rack, 4 See BAR SCREEN.

Trihalomethane (THM), 16, 143, 203, 290, 319 Compound formed when natural ORGANIC SUBSTANCES from decaying vegetation and soil (such as humic and fulvic acids) react with chlorine.

Trivalent ion, 67 ION having three valence charges. The charges can be positive or negative.

Trunnion, 297 A roller device, placed under ton containers of chlorine to hold them in place.

Tube settlers, 101 A series of plastic tubes about 2 in. square used for SHALLOW-DEPTH SEDIMENTATION.

Tube-settling, 97 A SHALLOW-DEPTH SEDIMENTATION process that uses a series of inclined tubes.

Tubercules, 254 Knobs of rust formed on the interior of cast-iron pipes due to CORROSION.

Turbidity, 124, 293 A physical characteristic of water making the water appear cloudy. The condition is caused by the presence of suspended matter.

Turbine meter, 31 A meter for measuring flow rates by measuring the speed at which a turbine spins in water, indicating the velocity at which the water is moving through a conduit of known cross-sectional area.

Turbulence, 95 A flow of water in which there are constant changes in flow velocity and direction resulting in agitation.

Turbulently, 51 See TURBULENCE.

UV disinfection, 286 Disinfection using an ultraviolet light.

Ultrasonic flow meter, 33 A water meter that measures flow rate by measuring the difference in the velocity of sound beams directed through the water.

Underdrain, 173 The bottom part of a filter that collects the filtered water and uniformly distributes the BACKWASH water.

Uniform corrosion, 260 A form of CORROSION that attacks a material at the same rate over the entire area of its surface.

Unstable, 253 To be CORROSIVE or scale-forming.

Upflow clarifier, 103 See SOLIDS-CONTACT BASIN.

Vacuum breaker, 242 A mechanical device that prevents BACKFLOW due to a siphoning action created by a partial vacuum that allows air into the piping system, breaking the vacuum.

Van der Waals force, 66 The attractive force existing between colloidal particles that allow the COAGULATION process to take place.

Velocity meter, 30 A water meter using a rotor with vanes (such as a propeller) and operating on the principle that the vanes move at about the same velocity as the flowing water.

Venturi tube, 28 A type of primary element used in a PRESSURE DIFFERENTIAL METER that measures flow velocity by the amount of pressure drop through the tube. Also used in a filter RATE-OF-FLOW CONTROLLER.

Viscosity, 82 The resistance of a fluid to flowing due to internal molecular forces.

Viscous, 107 Having a sticky quality.

Volatile, 45 Capable of turning to vapor (evaporating) easily.

Volumetric dry feeder, 231 See VOLUMETRIC FEEDER.

Volumetric feeder, 73 Chemical feeder that adds specific volumes of dry chemical.

Wash-water troughs, 174 Trough placed above the filter media to collect the BACKWASH water and carry it to the drainage system.

Waterborne disease, 283 A disease caused by a waterborne organism or TOXIC substance.

Weighting agent, 70 A material, such as bentonite, added to low-turbidity waters to provide additional particles for good FLOC formation.

Weir overflow rate, 105 A measurement of the number of gallons per day of water flowing over each foot of weir in a SEDIMENTATION TANK or circular clarifier. Mathematically, it is the gallons-per-day flow over the weir divided by the total length of the weir in feet.

Wheeler bottom, 173 A patented filter UNDERDRAIN system using small porcelain spheres of various sizes in conical depressions.

Wire-mesh screen, 6 Screen made of a wire fabric attached to a metal frame. The screen is usually equipped with a motor so it can move continuously through the water and be automatically cleaned with a water spray. It is used to remove finer debris from the water than the BAR SCREEN is able to.

Zeta potential, 66, 84 A measurement (in millivolts) of the particle charge strength surrounding colloidal solids. The more negative the number, the stronger the particle charge and the repelling force between particles.

Answers to Review Questions

Answers to Review Questions

Module 1 Preliminary Treatment

1. Screening, chemical pretreatment, presedimentation, microstraining, sometimes pre-aeration, and prechlorination.

2. By removing certain impurities or changing certain objectionable characteristics of the water before it enters the main treatment plant.

3. To remove logs, twigs, leaves, and other large debris in order to prevent it from entering the treatment plant and damaging equipment.

4. Bar screens and wire-mesh screens.

5. Clogging and corrosion.

6. No. Algae are a part of the total aquatic environment. To eliminate them entirely would upset the ecological balance, causing serious problems with other aquatic plants and animals.

7. Taste and odor; filter clogging; slime; color; corrosion; interference with other treatment processes; toxicity.

8. Application of copper sulfate, $CuSO_4$.

9. 5.4 lb of commercial copper sulfate per acre of lake surface area where methyl-orange alkalinity is greater than 50 mg/L; 0.9 lb/acre-ft of total lake volume where methyl-orange alkalinity is less than 50 mg/L.

10. Chemical application by dragging bags from a boat, spray application from boat, spray application from shore.

11. Emergent weeds, floating or surface weeds, submerged weeds.

12. Tastes and odors, color, filter clogging, breeding areas for disease-causing and nuisance insects.

13. a, c, and d.

14. Reason for pretreating (such as taste and odor problems, or filter clogging); type of algae or weed treated; algal count or estimated weed coverage; chemical used, concentration, and dosage; date of pretreatment; length of time since last treatment; weather conditions; other water conditions (such as temperature, pH, and alkalinity); method of application; personnel involved; results of pretreatment (such as taste and odor following treatment, filter conditions).

15. To remove gravel, sand, silt, and other gritty material from the raw water before it enters the plant and damages equipment.

16. b, d, and e.

17. a, c, and d.

18. Control the flow rate to each treatment process; adjust chemical feed rates; determine pump efficiency and power requirements; calculate detention times; monitor the amount of water treated; calculate the unit cost of treatment.

19. a, c, and e.

20. The primary element; the transmitter; the receiver.

21. Record instantaneous flow rate, totalize flow, indicate instantaneous flow rate.

22. Propeller and turbine meters.

23. Magnetic flow meters.

24. Ultrasonic flow meters.

25. 5.8 mgd. (See *Basic Science Concepts and Applications*, Hydraulics Section, Flow Rate Problems—Flow Measuring Devices; also Mathematics Section, Conversions—Box Method: Flow Conversions.)

26. 390 lb (Since methyl-orange alkalinity is greater than 50, dosage should be 1

mg/L calculated for the volume of water in the upper 2 ft of the lake, without regard to total volume. This is about the same as 5.4 lb of commercial copper sulfate per acre of surface area for the 72 acre lake. 72 acre × 5.4 lb/acre = 388.8 lb)

27. Detention time = 1.2 hr (See *Basic Science Concepts and Applications*, Mathematics Section, Detention Time.)

Module 2 Aeration

1. The two removal processes of aeration are: (1) The *scrubbing action* caused by the turbulence of air and water mixing together; and (2) the *introduction of oxygen* (DO) into the water, which results in the oxidizing of certain metals and gases in the water.

2. Carbon dioxide (CO_2), hydrogen sulfide (H_2S), methane (CH_4), iron (Fe), manganese (Mn), and taste- and odor-causing materials.

3. *Carbon dioxide* increases acidity of water resulting in more corrosive water, makes iron removal more difficult, and increases the cost of softening. *Hydrogen sulfide* increases the cost of chlorination, causes disagreeable taste, and attacks cement and concrete. *Methane* is highly flammable and explosive, and causes a garlic-like taste in water. *Iron and manganese* cause a metallic, medicinal, or astringent taste in water; produce stains in sinks, laundry, etc.—yellowish to reddish brown for iron, black for manganese; and can deposit on surfaces such as pipe walls and well screens.

4. (a) Air diffuser, (b) Slat and coke tray, (c) Draft aerator, (d) Cascade, (e) Mechanical aerator.

5. *Corrosion:* This problem is the result of higher levels of DO in the water. *Floating floc in clarifiers:* Excess DO may result in small bubbles attaching to floc particles in the clarifier. This causes floc particles to float rather than settle. *False clogging of filters:* Excess DO may result in small bubbles that can attach to sand grains in a sand filter, blocking the void spaces and resulting in apparent clogging. *Problems with H_2S removal:* At pH values of 8 or above, H_2S can oxidize, resulting in the precipitation of sulfur. The presence of sulfur can result in higher chlorination costs and the appearance of a milky-blue water. *Algae problems:* Exposure to direct sunlight of slat trays, coke trays, or draft aerators may result in the growth of algae. *Clogged diffusers:* Diffusers are easily clogged and, if not properly maintained and cleaned, can cause poor aeration and excessive energy costs.

Wasted energy: When water is over-aerated or when aeration is of questionable benefit to the water, the energy required for pumps or blowers is wasted energy, which can result in considerable unnecessary expense.

6. *Corrosion:* do not over-aerate; avoid supersaturating the water with oxygen. *Floating floc in clarifiers:* do not over-aerate. *False clogging of filters:* do not over-aerate; evaluate the effectiveness of aeration in removing or modifying desired constituents; know the costs associated with this removal. *Problems with H_2S removal:* lower the pH during aeration to favor the removal of H_2S gas, or shorten the aeration time to prevent the oxidation of sulfur to sulfate. *Algae problems:* provide a roof or canopy to shade the aerator from direct sunlight. *Clogged diffusers:* maintain clean air filters, do not over-lubricate blowers, prevent backflow of water into diffusers, and inspect and clean diffusers routinely. *Wasted energy:* do not over-aerate.

7. Temperature, DO, pH (also iron, manganese, CO_2, taste and odor).

8. Hydrogen sulfide and methane are extremely dangerous gases. Since these gases may be scrubbed out of the water and may accumulate in confined spaces, it is of utmost importance that the aeration process is well ventilated.

9. H_2S is heavier than air and will accumulate in low areas. An unsuspecting operator entering such an area can be killed within minutes.

Module 3 Coagulation/Flocculation

1. Suspended, colloidal, dissolved.

2. The electrical force that keeps particles apart.

3. The natural physical force that pulls two particles together.

4. The neutralization of the electrical charge on the particle followed by the gathering together (agglomeration) of the neutral particles to form floc particles.

5. The gentle mixing of the microflocs formed during coagulation to form visible floc particles, which can easily be settled or filtered.

6. Visible particles made up of the original nonsettleable particles and the chemicals added. Floc is easily settled or filtered.

7. Alum and ferric sulfate.

8. Mixing conditions during coagulation and the water pH, alkalinity, and temperature.

9. Jar test.

10. A chemical used to improve the performance of a coagulant.

11. (1) Activated silica, (2) weighting agents and adsorbents, (3) polyelectrolytes.

12. Cationic, anionic, nonionic.

13. Dry and liquid.

14. (1) Chemical storage, (2) solution preparation equipment, (3) dosing equipment, (4) flash mixing equipment, and (5) the flocculation basin.

15. This gives the operator more precise control over the dosage.

16. Volumetric and gravimetric. Volumetric is less expensive and less accurate.

17. The chemical metering pump.

18. Selecting the chemicals, applying the chemicals, and monitoring process effectiveness.

19. Low water temperature, weak floc, slow floc formation. In all cases, mixing, detention time, and choice of coagulant and coagulant aid should be altered as necessary to achieve good coagulation/flocculation.

20. Jar test, pH, turbidity (continuous monitoring), filtrability, and zeta potential.

21. Some of the hazards associated with handling coagulant chemicals include: (a) The handling of dry chemicals can create a dust, which is very dangerous if inhaled or allowed on the skin or in the eyes (especially a problem with lime, soda ash, and alum). (b) Accidental mixing of some chemicals can create tremendous heat and could result in an explosion. (c) Some chemicals, such as alum, can irritate skin, eyes, and mucous membranes. (d) Lump alum can cause cuts. (e) Polyelectrolytes are slick and spills create dangerously slippery surfaces.

22. 57.6 min. The detention time is adequate; it is at the high end of the 20 to 60 min range. (Refer to *Basic Science Concepts and Applications,* Math-

ematics Section, Detention Time, for additional examples of detention time calculations.)

23. 22.4 lb/hr. 4.2 gal/hr. (Refer to *Basic Science Concepts and Applications,* Chemistry Section, Chemical Dosage Problems—Milligrams-per-Litre to Pounds-per-Day Problems, for additional examples of concentration to dosage calculations.)

Module 4 Sedimentation

1. The purpose of sedimentation is to remove settleable solids from the water and reduce loading on the filters and subsequent treatment processes.

2. Plain sedimentation removes solids without use of chemicals and is used primarily as a pretreatment process. Sedimentation following chemical addition depends on the use of chemicals to make the solids more settleable. This type is typically used after coagulation/flocculation and before filtration.

3. Rectangular: rectiliniar flow; circular: radial flow.

4. (a) Inlet zone—decreases the velocity of the incoming water and evenly distributes the flow. (b) Settling zone—provides a calm area necessary for the solids to settle. (c) Outlet zone—provides a smooth transition from the settling zone to the effluent flow area. (d) Sludge zone—receives the settled solids and keeps them separate from particles in the settling zone.

5. Detention time is the time it theoretically takes for water to move through the basin. It also can be defined as the time it takes to fill the basin at a given flow rate.

6. Alum sludge is very sticky and difficult to dewater so it cannot be disposed of in landfills. Usually a great deal of land is required for lagoons or drying beds and ultimate disposal. Alum sludge can be reduced by making sure over-dosing does not occur and by using coagulant aids so the alum dose can be further reduced.

7. (a) Detention time = 1.9 hours (See *Basic Science Concepts and Applications,* Mathematics Section, Detention Time.)
   (b) Surface overflow rate = 943 gpd/sq ft (See *Basic Science Concepts and Applications,* Mathematics Section, Surface Overflow Rate.)

(c) Weir overflow rate = 10.610 gpd/ft (See *Basic Science Concepts and Applications,* Mathematics Section, Weir Overflow Rate.)

8. (a) Detention time = 1.36 hours (See *Basic Science Concepts and Applications,* Mathematics Section, Detention Time.)
   (b) Surface overflow rate = 1320 gpd/sq ft (See *Basic Science Concepts and Applications,* Mathematics Section, Surface Overflow Rate.)
   (c) Weir overflow rate = 16,500 gpd/ft (See *Basic Science Concepts and Applications*, Mathematics Section, Weir Overflow Rate.)

9. Flow-through velocity = 0.66 ft/min (See *Basic Science Concepts and Applications,* Hydraulics Section, Flow Rate Problems [Instantaneous Flow Rate Calculations].)

10. Short circuiting occurs when the water flows through the basin in less than the calculated detention time. It can be caused by poor inlet conditions, density currents, and wind.

11. Turbidity is the key operational test. It can be used to determine the effluent quality and the efficiency of the basin by comparing raw-water and effluent quality.

Module 5 Softening

1. Soluble, divalent, metallic cations—primarily calcium and magnesium and to a much lesser extent, strontium, aluminum, barium, iron, manganese, and zinc.

2. Calcium is dissolved as water passes over and through limestone deposits. Magnesium is dissolved as water passes over and through dolomite and other magnesium-bearing minerals.

3. The degree of hardness consumers consider objectionable varies, depending both on the water and on the degree of hardness to which consumers have become accustomed.

4. Calcium and magnesium hardness and carbonate and noncarbonate hardness.

5. Total Hardness = Carbonate Hardness + Noncarbonate Hardness.

6. Hard water has the following disadvantages: (1) forms water spots on

glassware, plumbing fixtures, etc.; (2) forms scale in pipelines and residential hot-water heaters; (3) deposits residue (curd) in laundered fabrics; (4) may force consumers to use softer but unsafe alternative water supplies. Soft water has the following disadvantages: (1) it can be very corrosive; (2) toxic metals can be dissolved from piping materials; (3) the useful life of plumbing can be shortened considerably.

7. The lime-soda ash process and the ion-exchange process.

8. Calcium carbonate—$CaCO_3$
   Magnesium hydroxide—$Mg(OH)_2$.

9. Although carbon dioxide does not cause hardness, it reacts with and consumes the lime added to remove hardness. It is also used in recarbonation to stabilize the water.

10. Chemical storage facilities; chemical feed facilities; rapid-mix basin; flocculation basin; sedimentation basin and related equipment; solids-contact basin (can be used in place of the rapid-mix, flocculation, and sedimentation basins); sludge recirculation, dewatering, and disposal equipment; recarbonation basin.

11. Calcium hydroxide [$Ca(OH)_2$], called hydrated lime or slaked lime; and calcium oxide (CaO), called quicklime or unslaked lime.

12. Calcium oxide—called quicklime or unslaked lime.

13. 30.

14. Flocculation—40 to 60 min; sedimentation—2 to 4 hours.

15. Lime = 319 mg/L as 89-percent pure CaO, soda ash = 376 mg/L as 100-percent pure Na_2CO_3. (See *Basic Science Concepts and Applications*, Chemistry Section, Chemical Dosage Problems [Lime-Soda Ash Softening Calculations].)

16. Lime—2660 lb/mil gal, 4860 lb/day; soda ash—3136 lb/mil gal, 5720 lb/day. (See *Basic Science Concepts and Applications,* Chemistry Section, Chemical Dosage Problems [Feed Rate Conversions].)

17. Lime = 978 lb/mil gal, soda ash = 840 lb/mil gal. (See *Basic Science Concepts and Applications,* Chemistry Section, Chemical Dosage Problems [Lime-Soda Ash Softening Calculations].)

18. Lime = 720 lb/day, soda ash = 618 lb/day. (See *Basic Science Concepts and*

Applications, Chemistry Section, Chemical Dosage Problems [Feed Rate Conversions].)

19. (1) Excess calcium carbonate, (2) magnesium hydroxide scale, (3) after-precipitation, (4) carryover of sludge solids, (5) unstable water, (6) interference with other treatment processes.

20. (1) Eliminate incomplete coagulation/flocculation by improving the mixing process, increasing detention times, or pretreating the water to remove organic contaminants; (2) if fine particules of calcium carbonate cannot be eliminated at the source, they must be removed (dissolved) by recarbonation or controlled by applying a recommended dose of 0.25 to 1.0 mg/L of sodium hexametaphosphate (a sequestering agent).

21. Magnesium hydroxide scale can form inside boilers and household water heaters operated at 140° F (60° C) or higher. This scale reduces the efficiency of heat transfer, thereby increasing heating costs.

22. Calcium—338 mg/L as $CaCO_3$
    Total hardness—360 mg/L as $CaCO_3$
    Bicarbonate alkalinity—246 mg/L as $CaCO_3$
    Carbon dioxide—57 mg/L as $CaCO_3$.
    (See *Basic Science Concepts and Applications,* Chemistry Section, Chemical Dosage Problems [Lime-Soda Ash Softening Calculations].)

23. Magnesium ions—22 mg/L as $CaCO_3$. No. (See *Basic Science Concepts and Applications,* Chemistry Section, Chemical Dosage Problems [Lime-Soda Ash Softening Calculations].)

24. Magnesium hydroxide—pH of 10.6, calcium carbonate—pH of 9.4.

25. In the ion-exchange process, water to be softened is passed through a filter-like bed of ion-exchange material called resin. The ion-exchange resins exchange sodium ions (Na^+) that they hold for hardness-causing calcium and magnesium ions that are carried in the water passing through the resins. Water that has passed through ion-exchange resins has zero or near-zero hardness.

26. (1)Ion-exchange materials, (2) ion-exchange unit, (3) salt storage tanks, (4) brine-feeding equipment, (5) devices for blending hard and soft water.

27. Polystyrene resin.

28. (1) Softening, (2) backwashing, (3) regeneration, and (4) rinse.

29. 6356 min or 106 hours. (See *Basic Science Concepts and Applications,* Chemistry Section, Chemical Dosage Problems [Ion Exchange Softening Calculations].)

30. 330,000 grains removed per cycle. 21,384,000 mg removed per cycle. (See *Basic Science Concepts and Applications,* Chemistry Section, Chemical Dosage Problems [Ion Exchange Softening Calculations].)

31. 23,838 gal per cycle. (See *Basic Science Concepts and Applications,* Chemistry Section, Chemical Dosage Problems [Ion Exchange Softening Calculations].)

32. 102 lb salt. (See *Basic Science Concepts and Applications,* Chemistry Section, Chemical Dosage Problems [Ion Exchange Softening Calculations].)

33. 204 gal of brine solution. (See *Basic Science Concepts and Applications,* Chemistry Section, Chemical Dosage Problems [Ion Exchange Softening Calculations].)

34. (1) Resin breakdown; (2) iron fouling; (3) turbidity, organic color, and bacterial slime fouling; (4) stabilization.

35. Total hardness and alkalinity test. If total hardness is greater than total alkalinity, then total alkalinity equals carbonate hardness and total hardness minus total alkalinity equals noncarbonate hardness.

Module 6 Filtration

1. To remove suspended material from water.

2. A measurement of suspended material in water, which can include floc, microorganisms, silt, and chemical precipitates.

3. Turbidity (1) interferes with the disinfection process by shielding the microorganisms from the disinfectant; (2) combines chemically with the disinfectant (uses up the disinfectant) leaving less disinfectant to kill the microorganisms; (3) causes deposits in the distribution system that create tastes, odors, and bacterial growths.

4. Adsorption. As the water passes through the filter bed, the suspended

materials contact and adsorb onto the surface of the grains of filter media or onto previously deposited material.

5. (1) Water must be wasted until the schmutzdecke forms; (2) chemical coagulation cannot be used because the floc would quickly clog the filter; (3) flow rates are very low; therefore, large areas of land are required for these filters.

6. (a) Evenly distribute the backwash water.
(b) Uniformly collect the filtered water.

7. They assist in breaking up and washing away the accumulated material in the upper part of the filter bed.

8. Filters that use a combination of filter media (graded coarse-to-fine) in the filter bed. High-rate filters operate at rates three to four times those of rapid-sand filters.

9. The filter bed cannot be observed during operation.

10. The skeletal remains of microscopic aquatic plants called diatoms.

11. Conventional filtration is effective for practically any level of turbidity in raw water. Flocculation, coagulation, and sedimentation occur ahead of the filtration process.

    Direct filtration does not involve the use of a sedimentation process. Consequently, it should only be used when raw-water turbidity is below 25 NTU and color below 25 color units (CUs). Dual- or multi-media filters should always be used and should be monitored closely to prevent turbidity breakthrough since filter rates are higher.

12. (1) Filtering, (2) backwashing, (3) filtering to waste.

13. The need for frequent maintenance. A malfunctioning controller can damage the filter bed and cause poor treated-water quality.

14. When any one of the following occurs: (a) terminal head loss is reached, (b) breakthrough starts to occur, (c) a filter run reaches 36 hours.

15. To prevent any filtered material that remains in the filter bed after backwashing from entering the treated water.

16. To strengthen the bonds between the floc particles and coat the media grains

to improve adsorption. This allows the floc to resist the shearing forces of water as it flows through the media.

17. (1) Chemical treatment before filtration, (2) control of filter flow rate, (3) backwashing.

18. (a) Mudball formation, (b) filter bed shrinkage and cracking, (c) gravel displacement.

19. Mudballs are clumps of filter media grains stuck together with floc material. They can be prevented by using adequate backwash flow rates and filter agitation.

20. Head loss and filtered water turbidity.

Module 7 Adsorption

1. To remove organic compounds.

2. Organics cause color, taste, and odor. Some organics are toxic to humans. Others are suspect carcinogens.

3. Some natural organic materials (humic compounds) react with chlorine to form trihalomethanes. Trihalomethanes are suspect carcinogens.

4. Many man-made organics, such as pesticides and herbicides, are known to be toxic to humans.

5. Adsorption, aeration, coagulation, flocculation, sedimentation, filtration, and oxidation.

6. A dull black, dusty, adsorbent material available in powder or granular form. It is made by carbonizing wood, nutshells, coal, peat, sawdust, or petroleum products and then activating the carbon by exposing it to a steam-air mixture that oxidizes all particles on the carbon's surface, leaving the surface free to attract and hold organics.

7. Adsorption is the adhesion or "sticking" of a dissolved organic compound to the surface of a material by a combination of complex physcial and chemical forces.

8. Activated carbon, activated alumina, synthetic resins.

9. Granular activated carbon and powdered activated carbon.

10. PAC is a finely divided powder with a particle size less then 0.1 mm. GAC on the other hand has much larger particle size (1.2–1.6 mm) and more total surface area.

11. To minimize dust problems throughout the plant.

12. Slurry is a mixture of activated carbon and water (1 lb [0.45 kg] of carbon per 1 gal [3.8 L] of water). It is used to transport carbon from storage to the point of use to make handling easier and reduce dust problems.

13. Corrosion- and erosion-resistant material such as rubber, plastic, or stainless steel.

14. As a replacement for filter media in a conventional filter or in a separate closed pressure vessel called a contactor.

15. When carbon is needed to continuously remove organics.

16. 24 in. (610 mm).

17. The process of heating spent activated carbon in a furnace with a controlled atmosphere, which oxidizes the adsorbed impurities.

18. (1) Contact time, (2) surfaces of the PAC particles can lose adsorbing capacity if coated with coagulants, (3) PAC will adsorb chlorine.

19. EBCT is equal to the volume of the empty bed into which the GAC will be placed divided by the flow rate.

20. (1) Entrapped air within the carbon bed, (2) excessive backwash rate.

21. The type and concentration of organic compounds being removed.

22. Contactors can be designed and manufactured to provide the desired EBCT.

23. Dust problems during routine handling of the chemical.

24. High bacterial counts in the effluent from the bed.

25. Modified jar-test procedure.

26. If mixed with either chemical, spontaneous combustion can occur.

27. With a fine water spray or a chemical foam so as not to spread burning carbon particles.

Module 8 Fluoridation

1. To increase the fluoride ion concentration in water to the level needed to help prevent tooth decay in children.

2. Air temperature.

3. Basically, the warmer the climate, the more water consumers drink. The more water consumed, the lower the fluoride concentration needs to be in order for the optimum daily amount of fluoride to be consumed.

4. There is no difference. The ion is the same regardless of how it occurs in water.

5. Dental fluorosis is a condition that develops during childhood and is caused by consuming too much fluoride. It appears as a discoloration of the teeth, beginning with opaque white spots, turning to gray (black in more advanced cases), and in the most serious cases, leading to pitting and an overall softening of the teeth causing them to wear down to the gum line.

6. The MCLs are twice the optimum levels for each temperature range.

7. Sodium fluoride, NaF; Sodium silicofluoride, Na_2SiF_6; hydrofluosilicic acid, H_2SiF_6.

8. Dry feeders and solution feeders.

9. Gravimetric dry feeder—extremely accurate, more expensive, regulates volume of chemical added. Volumetric feeder—less expensive, lower capacity, regulates volume of chemical added.

10. Sodium fluoride has a nearly constant solubility of about 4 g/100 mL within the temperature range of 32 to 77° F (0 to 25° C). Constant solubility permits the use of a saturator. Without any sophisticated controls or tests, the saturator allows the operator to always be certain that the solution contains 4 g of sodium fluoride in every 100 mL of solution.

11. 21.3 lb F in every 50 lb NaF, 95 percent pure. (See *Basic Science Concepts*

and Applications, Chemistry Section, Dosage Calculations—Fluoridation Calculations.)

12. 60 percent F. (See *Basic Science Concepts and Applications*, Chemistry Section, Dosage Calculations—Fluoridation Calculations.)

13. 2.9 gph. (See *Basic Science Concepts and Applications*, Chemistry Section, Dosage Calculations—Fluoridation Calculations.)

14. The fluoride injection point should be located after filtration since considerable fluoride loss can result in the filter. It should be at a point where all water flows at nearly constant rates or in a clear well for more accurate feeding; not near a point where chemicals containing calcium are located.

15. Daily or continuous fluoride test results from the plant, service (if soft water or variable ground water), and the distribution systems. Daily records should also be kept on the volume of water treated and the fluoride fed. Theoretical concentration can be calculated and compared with actual concentration. This can help identify problems.

16. Do not eat, drink, or smoke in any area where fluoride is stored, handled, or applied; do not put hands to face unless hands are washed thoroughly; and use only designated areas for eating, drinking, or smoking.

17. Chemical goggles, respirator or mask, rubber gloves, rubber apron, rubber boots, protective clothing.

Module 9 Stabilization

1. To control the corrosive or scale-forming tendencies of drinking water.

2. Public health, aesthetics, and economics.

3. The wearing away or deterioration of a material due to chemical reactions with the environment.

4. DO, TDS, pH, alkalinity, temperature, type of metal in pipe, electrical insulation and current, and bacteria.

5. As the TDS concentration increases water becomes a better conductor of electricity, thereby increasing the corrosion rate.

6. When dissimilar metals are connected in a common flow of water, the metal highest in the galvanic series will corrode while the other metal will be protected.

7. Certain types of bacteria accelerate the corrosion rate since they produce CO_2 and H_2S. The slimes they produce can also prevent a protective coating from forming.

8. Localized corrosion attacks the metal surfaces unevenly and can produce rapid failure of pipe. Uniform corrosion takes place at an equal rate over the entire surface.

9. Divalent metallic cations associated with hardness combine with other minerals dissolved in water and precipitate.

10. Calcium carbonate, calcium sulfate, magnesium carbonate, and magnesium chloride.

11. As temperature increases, the solubility of $CaCO_3$ and other scale-forming compounds decrease, which increases formation.

12. pH and alkalinity adjustment, use of protective coatings, and use of corrosion inhibitors and sequestering agents.

13. Lime.

14. To reduce the pH so $CaCO_3$ will not form deposits in the distribution system.

15. They chemically tie up the scale-forming ions and prevent them from reacting to form scale.

16. A slaker.

17. The lime slurry will form deposits on all feeding equipment.

18. $CaCO_3$ will be dissolved and the water has corrosive tendencies.

19. Water quality data, distribution system conditions, and condition of home plumbing systems.

20. 6.8–7.3.

21. Poor control of the stabilization process.

22. pH and alkalinity.

23. Pipe and coupon testing.

24. Proper storage and handling of chemicals.

Module 10 Disinfection

1. Disinfection is the destruction of disease-causing microorganisms; sterilization is the destruction of all microorganisms.

2. A disease-causing microorganism.

3. Gastroenteritis, typhoid fever, dysentery, amebic dysentery, cholera, giardiasis, infectious hepatitis.

4. Heat treatment; radiation treatment; chemical treatment.

5. Bromine; iodine; ozone; chlorine.

6. Chlorine.

7. The effectiveness of the kill is related to chlorine concentration C and contact time t. As either C or t or both increase, the effective kill increases. From an operations standpoint, the important point is that chlorine concentration must be increased as contact time is decreased and vice versa.

8. Temperature; pH; substances in the water.

9. Generally, the higher the water temperature, the faster the kill.

10. Organic matter; iron; manganese; nitrite; hydrogen sulfide; any material that causes turbidity.

11. Chlorine residual.

12. Free available chlorine residual; combined chlorine residual.

13. It is a quick way to evaluate chlorination effectiveness. A residual is evidence that enough chemical was added. In chlorination, practical experience indicates that if a certain chlorine residual is present after a certain period of time, then enough chlorine was added to accomplish disinfection.

14. Chlorine demand.

15. As chlorine is added

- Between point 1 and point 2—No chlorine residual results since chlorine is used up in reactions with reducing agents such as iron and manganese.
- Between point 2 and point 3—As more chlorine is added, it reacts with ammonia and organic matter forming monochloramines and chloroorganic compounds. These are called combined chlorine residuals.
- Between point 3 and point 4—As more chlorine is added, the combined residual decreases as some of the chloroorganic compounds are oxidized and the monochloramines are converted to dichloramines and trichloramines.
- Point 4—This is the breakpoint—the point at which the amount of chloramines reaches a minimum concentration. Beyond the breakpoint free chlorine residual will form even though small amounts of combined residual remain.

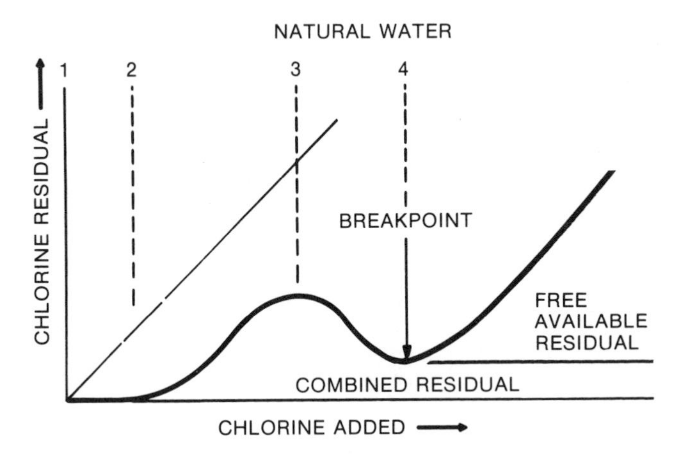

16. Disinfection of pipelines and tanks; iron and manganese oxidation; hydrogen sulfide oxidation; taste and odor destruction; algae control; slime control; coagulation aid; filtration aid.

17. Pure chlorine (liquid or gas), Cl_2; calcium hypochlorite, $Ca(OCl)_2$; sodium hypochlorite, NaOCl.

18. 100-lb cylinder; 150-lb cylinder; ton container; tank truck; railroad tank car; tank barge.

19. About 3700 lb (1700 kg).

20. A greenish-yellow gas about 2.5 times as dense as air; nonexplosive; nonflammable; highly irritating to smell; can kill at concentrations as low as 0.1 percent; very hazardous; turns to a liquid under pressure; noncorrosive if dry; extremely corrosive mixed with some moisture.

21. 150-lb cylinders and ton containers.

22. $Ca(OCl)_2$, calcium hypochlorite, is a dry, white to yellow-white granular material containing about 65 percent available chlorine by weight. It is also available in 0.01-lb (5-g) tablets. Calcium hypochlorite will react with organic materials to generate heat and possibly start a fire; it should be stored accordingly.

    NaOCl, sodium hypochlorite, is a clear, greenish-yellow liquid solution. Common liquid household bleach is an example of a 5-percent available chlorine solution. Solutions up to 15 percent are available commercially. There is no storage fire hazard, but the chemical is quite corrosive.

23. Cylinders are handled with hand trucks; ton containers are handled with lifting beams mounted on hoists suspended from overhead monorails.

24. Indoors or outdoors, but protected from direct sunlight and from objects that might strike and puncture the pressure vessels. This applies to either full or empty cylinders.

25. Chlorine cylinder or container; weigh scale; auxiliary valve; feed lines; manifold; feed line; chlorinator; injector; solution line; diffuser.

26. The weigh scale is the only accurate means the operator has of knowing exactly how much chlorine is left in a container.

27. The injector is a venturi-like device that creates a vacuum, which pulls the gas through the chlorinator and into the solution water.

28. Booster pumps; automatic controls; evaporators; flow recorders; chlorinator alarms; safety equipment.

29. 42 lb/day (19 kg/d) for 150-lb cylinders and 400 lb/day (180 kg/d) for ton containers.

30. Cylinders in a vertical position; ton containers in a horizontal position.

31. Seven: [(2700 lb/day) ÷ (400 lb/day/container)] = (6.75 cylinders, rounds to 7).

32. On the tank valve, so valves may be closed quickly in an emergency.

33. Ammonia.

34. Repairs usually require disassembly and reassembly of the piping. If the feed pipeline were opened while the tank valve was open, a major and possibly catastrophic leak would occur.

35. With injector running, disconnect yoke, leaving auxiliary tank valve open. The vacuum created by the injector will pull in fresh air and thereby force all remaining chlorine gas into the injector water.

36. There is no best chlorine dosage for all cases. Dosage varies depending on water quality and the reason for chlorination. The dose needed for disinfection will probably not be the best to use for algae control.

37. Chlorine leaks; stiff container valves; hypochlorinator problems; taste and odor; sudden change in residual; trihalomethane formation.

38. Generally, increase the chlorine dosage until it just passes breakpoint.

39. The chlorine residual test.

40. Total coliform bacteria, which is excreted with fecal matter. Total coliform bacteria are present in far greater numbers than the pathogens that might be present and are much easier to measure than the pathogens themselves. In addition, they are more resistant to environmental changes than pathogens; so if coliforms are not present, it is assumed pathogens are not present.

41. Total coliform in water indicates that pathogenic organisms might be present.

42. Self-contained breathing apparatus; container emergency repair kit; adequate ventilation.